Internet of Things

Technology, Communications and Computing

The series Internet of Things - Technologies, Communications and Computing publishes new developments and advances in the various areas of the different facets of the Internet of Things. The intent is to cover technology (smart devices, wireless sensors, systems), communications (networks and protocols) and computing (theory, middleware and applications) of the Internet of Things, as embedded in the fields of engineering, computer science, life sciences, as well as the methodologies behind them. The series contains monographs, lecture notes and edited volumes in the Internet of Things research and development area, spanning the areas of wireless sensor networks, autonomic networking, network protocol, agent-based computing, artificial intelligence, self organizing systems, multi-sensor data fusion, smart objects, and hybrid intelligent systems.

Indexing: *Internet of Things* is covered by Scopus and Ei-Compendex **

Giancarlo Fortino • Massimo Mecella
Editors

Internet of Things Meets Business Process Management

A Synergistic Integration

Springer

Editors
Giancarlo Fortino
University of Calabria
Arcavacata, Rende CS, Italy

Massimo Mecella
DIAG
Sapienza University of Rome
Roma, Italy

ISSN 2199-1073 ISSN 2199-1081 (electronic)
Internet of Things
ISBN 978-3-031-90745-6 ISBN 978-3-031-90746-3 (eBook)
https://doi.org/10.1007/978-3-031-90746-3

This Springer imprint is published by the registered company Springer Nature Switzerland AG
The registered company address is: Gewerbestrasse 11, 6330 Cham, Switzerland

Acknowledgments The book editors would like to thank all the expert authors of the 12 chapters for their valuable contribution and all the anonymous reviewers for providing useful feedback to improve the quality of the chapters.

Introduction

The Internet of Things (IoT) refers to a network of connected devices that collects and exchanges data through the Internet for implementing specific activities and services. These things can be artificial or natural and interact as autonomous agents that form a complex system. In turn, business process management (BPM) was established to analyze, discover, design, implement, execute, monitor, and evolve collaborative business processes within and across organizations. While the IoT and BPM have been regarded as separate topics in research and in practice, we strongly believe that, on the one hand, the management of IoT applications will greatly benefit from BPM concepts, methods, and technologies. On the other hand, the IoT poses challenges that will require enhancements and extensions of the current state of the art in the BPM field.

In this book, we discuss the extent to which these two paradigms can be combined, and we present emerging challenges and intersections from a research and practitioner's perspective in terms of complex software systems development and process-aware methods, including process mining. This book includes 12 chapters authored by well-known researchers pioneering works on the integration of IoT and BPM/Process Mining. Chapters "On the Interplay Between Business Process Management and Internet-of-Things–Behaviour-Centred Design of IoT Systems" are related to IoT and BPM, chapters "LACII: Enhancing Decision-Making in Business Processes with the IIoT: A Cognitive Load Aware Design Approach–NICE: The Native IoT-Centric Event Log Format for Process Mining" include IoT and Process Mining integration issues, and chapters "The Internet of Production: Accelerating the Optimization of Production Processes via Models-in-the-Middle–A Digital Twin Framework for Bioprocess Development Using IoT Sensor Data" concern new approaches based on the concepts of digital twin and Internet-of-Production. In the following, we provide a brief overview of the chapters.

The chapter entitled "On the Interplay Between Business Process Management and Internet-of-Things," authored by F. De Luzi, F. Leotta, A. Marrella, M. Mecella, and F. Monti, focuses on a Systematic Literature Review (SLR) aimed at illustrating the current state and maturity of IoT-aware BPM approaches. Unlike previous surveys that focus on specific facets of this integration, this study aims to deliver a broad, holistic perspective. The SLR addressed several key research questions: (a) identification of the main topics explored in the literature; (b) categorization of how IoT integrates within various BPM life cycle phases; (c) pinpointing the application domains where IoT and BPM integration is most impactful; (d) highlighting the research challenges outlined in the *IoT-Meets-BPM Manifesto*[1] that have been tackled to advance the management of IoT-aware BPs.

In the chapter "Human-Centric Processes: Where IoT and People Meet," A. Burattin et al. focus on the integration of "Human-centric processes" (HCPs) and IoT. In fact, HCP bring together observations and interactions from both humans and IoT. Specifically, the chapter deals with the challenges and opportunities that the IoT would offer in the context of human-centric processes, that is, business processes heavily involving and centered around interaction with humans. To this aim, a conceptual model that captures the interactions between the IoT, BPM, and humans is proposed. Notably, two different use cases of IoT-enabled human-centric processes discussing how they can be framed within the proposed model are presented: (a) the first use case relied on the IoT to monitor the behavior of humans involved in a process, and their impact on the correct execution of the process; (b) the second use case relied on IoT to observe the daily routines that humans when a process representation is leveraged.

The chapter "Behaviour-Centred Design of IoT Systems," written by U. Kannengiesser, C. Stary, R. Heininger, and T. E. Jost, focuses on one important challenge associated with integrating BPM and IoT systems, i.e., the IoT and BPM design expertise involved is partitioned across different communities of practice and research. Specifically, authors exploited the well-established ontology of designing, named the Function-Behavior-Structure (FBS) framework, to bring BPM and IoT perspectives together in a single model. The framework represents designing as involving eight fundamental transformations of design issues targeting the problem and its solution in separate spaces. BPM supports informed problem elaboration in the respective space, while IoT supports the development of working solutions in the respective space. Consequently, behavior-driven process modeling meets both design issues: understanding and refining the problem, and linking possible IoT solutions. Both become increasingly important when propagating the Internet of Behaviors and Web3, due to the increase of complexity and variability of system designs.

M. Nebel, S. A. Ciftci, and C. Janiesch, authors of the chapter entitled "LACII: Enhancing Decision Making in Business Processes with the IIoT-A Cognitive Load Aware Design Approach," aim to develop, in the context of Industrial Internet-of-Things (IIoT), design principles and an architecture that is aware of the cognitive load of decision-makers, with the goal of reducing the cognitive load of decision-makers. The objective of this study is to propose a design approach of a "load-aware

[1] C. Janiesch *et al.*, "The Internet of Things Meets Business Process Management: A Manifesto," in *IEEE Systems, Man, and Cybernetics Magazine*, vol. 6, no. 4, pp. 34–44, Oct. 2020, doi: 10.1109/MSMC.2020.3003135.

co-pilot for IIoT" (LACII) that acts as a computer-in-the-loop by observing the human-in-the-loop, learning, and intervening if the human decision-maker exhibits signs of cognitive load or task load. The goal of LACII is to enable decision-makers to engage in productive work while maintaining their well-being. To demonstrate and evaluate the potential application of LACII in a realistic context, a hypothetical production company and the associated business process are presented using the scenario technique.

The authors, J. Mangler et al., of the chapter "From Internet of Things Data to Business Processes: Challenges and a Framework" address the fundamental questions about extracting and abstracting process-related data from streams of IoT sensor values, e.g., which sensor values can be clustered together as part of process events? Which sensor values signify the start and end of such events? Which sensor values are related but not essential? Specifically, they propose a framework to semi-automatically perform a set of structured steps to convert low-level IoT sensor data into higher-level process events that are suitable for process mining. In order to demonstrate the potential of the framework, the authors also introduced a sample scenario and exemplary IoT dataset to discuss how a semi-automatically generated process model based on the steps in the framework can be (a) of similar quality as a process model provided by domain experts, (b) used to point out problems in existing process models, and (c) utilized to refine existing processes. The authors point out that this framework can serve as means to focus research on the overall goal of allowing domain experts to make IoT datasets fully process aware without the involvement of data engineers and data scientists.

The authors, D. Thakur, A. Guzzo, and G. Fortino, in the chapter "Exploring Process Mining in Human Activity Recognition: Challenges and Future Directions" explore the challenges and solutions of using Process Mining in Human Activity Recognition (HAR). The integration of process mining into HAR enhances the ability to analyze and interpret complex human behaviors, leading to better insights and applications across various fields, including healthcare, security, and smart environments. This integration deals with different issues such as how to effectively organize event logs to represent complex human activities in HAR systems based on the diversity of data sources, types of activities, and sensor modalities. The chapter shows that the combined utilization of process mining and HAR represents a potent alliance capable of revolutionizing various industries by offering actionable insights into human behavior and activity patterns, ultimately enhancing efficiency, safety, and compliance.

The chapter entitled "Event-Log Granularity for IoT Process Discovery," written by M. de Leoni, F. A. Khan, B. Depaire, G. van Houdt, and N. Martin, specifically builds on experiences on real-life IoT data and reports on the application of event log abstraction, aiming to discover IoT process models that are readable and accurate. These models are a prerequisite to provide actionable insights for a subsequent process' monitoring and optimization. In fact, the discovery, analysis, and improvement of processes executed via devices of a IoT system requires setting the right event granularity. The proposed techniques are applied in the second part of the chapter; specifically, experiences are provided on applying a clustering-based technique to two case studies, related to smart manufacturing and home automation, respectively.

D. Janssen, A. Koschmider, and F. Mannhardt, in the chapter "Process Mining on Sensor Location Event Data," propose an approach for process discovery on sensor location event data in single occupation settings by leveraging unsupervised learning in the form of clustering to raise the abstraction level of events. The main contribution is a feedback loop with a sensitivity analysis that measures how sensitive the discovered process model is to changes when processing and aggregating IoT sensor location event data in a (semi)automated manner. The evaluation results are promising to extract valuable insights from sensor location event data. The chapter showed that user involvement is only necessary for the labeling of the activities at a later stage. The rest of the process of the framework can be automated. The evaluation phase showed that combining clustering with process discovery algorithms is promising, and their isolated use would lead to fewer insights.

The chapter entitled "Process Mining Meets Robotic Systems: Challenges and Opportunities," written by F. Corradini, S. Pettinari, B. Re, L. Rossi, and F. Tiezzi, focuses on automation of robotic systems that involve missions defined as sequences of tasks enabling both robot actions and interactions. In this context, process mining techniques can be applied to leverage data generated during robot activities to discover and analyze a robotic mission. Specifically, this chapter examines the characteristics of robotic systems and discusses their interplay with process mining to foster the effective application of process mining. The presented challenges range from data preparation to process enhancement and arise from use cases and the current state of the art concerning approaches for analyzing robotic systems. By addressing the challenges and exploring new research directions, robotic systems can be empowered by techniques belonging to the process mining discipline.

The chapter entitled "NICE: The Native IoT-Centric Event Log Format for Process Mining," written by Y. Bertrand, S. Veneruso, F. Leotta, M. Mecella, and E. Serral, focuses on IoT data and process data integration issues. First, it discusses some recurrent challenges in the integration of IoT and process event data, from which ten concrete requirements for a suitable IoT-enhanced event log format are derived. Then, it proposes the Native IoT-Centric Event (NICE) log, a new event log format designed to incorporate IoT data into a process event log ensuring traceability, flexibility, and limiting data loss. In particular, this format combines different event types available in the literature with the notion of object from OCEL to form a very flexible format, capable of integrating IoT data with process data with minimal information loss. A first practical application demonstrated how generating a log following the proposed format enables process analysis integrating IoT data. Authors plan to publish NICE logs from real-life processes and further develop the format by creating more tool support to generate and manipulate logs.

In the chapter "The Internet of Production: Accelerating the Optimization of Production Processes via Models-in-the-Middle," L. Tacke Genannt Unterberg, I. Koren, and W. M. P. van der Aalst focus on the internet of production domain, where effective data management as well as utilization are necessary to efficiently and effectively work on digitalizing and optimizing (production) processes, and introduce "Models-in-the-Middle" as an approach that calls for a limited number of intermediary minimal data models between data source and data science processing. Such work was inspired by the need to bridge the gap between data source and data consumer in the context of digitalized production using Digital Shadows. Moreover, they introduced two example MitM that are especially relevant for the Internet of Production research project and the BPM discipline: OCED and MAED. Their applicability using initial tool support and by collaborating with data owners in the IoP is demonstrated.

The authors, Z. D. Bozorgi, A. Polyvyanyy, M. La Rosa, E. Otte, and A. Armas-Cervantes, in the chapter "A Digital Twin Framework for Bioprocess Development Using IoT Sensor Data" propose a comprehensive digital twin framework that harnesses the power of IoT technologies, specifically sensor data, to enhance efficiency and performance in process industries, specifically bioprocesses. This envisioned digital twin not only mirrors the current state of the system but also enables predictive analysis and proactive decision-making. Through a case study conducted in collaboration with a prominent pharmaceutical company, the authors have demonstrated the practical applicability and tangible benefits of the proposed framework. The results of the case study showed that through an approach based on the proposed framework, high prediction performance in the process monitoring task compared to a time series forecasting baseline can be achieved.

Contents

On the Interplay Between Business Process Management and Internet-of-Things

Francesca De Luzi, Francesco Leotta, Andrea Marrella, Massimo Mecella, and Flavia Monti

1 Introduction

Over the last decade, advancements in manufacturing have enabled the production of microcontrollers and CPUs with compact dimensions and at lower costs, making it feasible to integrate these components into a wide range of everyday physical objects, such as home appliances and industrial machinery, on a large scale. By combining this enhanced computing power with sensors, actuators, and communication modules, these objects can interact with their environment and operate independently, effectively transforming into *smart objects*. In this context, the Internet of Things (IoT) refers to the network of interconnected smart objects capable of collecting and exchanging data over the Internet, enabling advanced functionalities such as real-time monitoring, data sharing, and remote control [57].

Industries such as manufacturing, logistics, and healthcare are often early adopters of smart objects and IoT technologies. In these sectors, Business Process Management (BPM) tools and practices play a significant role in optimizing and automating processes [61]. BPM is a research domain focused on the identification, modeling, analysis, redesign, implementation, monitoring, and mining of business processes (BPs) [24]. A business process consists of a series of interrelated, structured activities, events, and decisions that achieve a particular organizational objective. Within BPM, processes are formalized through explicit business process models, typically represented using a graphical notation such as BPMN, the

F. De Luzi · F. Leotta · A. Marrella · M. Mecella (✉) · F. Monti
Sapienza Università di Roma, Rome, Italy
e-mail: deluzi@diag.uniroma1.it; leotta@diag.uniroma1.it; marrella@diag.uniroma1.it; mecella@diag.uniroma1.it; monti@diag.uniroma1.it

G. Fortino, M. Mecella (eds.), *Internet of Things Meets Business Process Management*, Internet of Things, https://doi.org/10.1007/978-3-031-90746-3_1

standard Business Process Model Notation[1]. To support the execution, monitoring, and analysis of business processes, a specialized type of information system known as a Process Management System (PMS) is employed, facilitating both real-time operations and the evaluation of historical process data [110].

The integration of IoT into BPM activities has been steadily increasing, as IoT data enables more precise automated monitoring, analysis, and control during business process execution. To leverage these capabilities, organizations need to align their business processes with IoT functionalities. This trend has sparked numerous studies on IoT-aware BPM, emphasizing the interplay between these fields [10, 11, 36]. A major catalyst for the convergence of IoT and BPM has been the Industry 4.0 movement [77]. In addition to promoting the use of smart objects, Industry 4.0 emphasizes high levels of customization in manufactured product—an aspect well-supported by IoT-aware BPM [35]. In this context, Janiesch et al., in their manifesto "*The Internet of Things Meets Business Process Management*" (in the following referred to as the *IoT-Meets-BPM Manifesto*), highlighted a range of open challenges that must be addressed to achieve a more effective and integrated application of IoT and BPM principles [46].

This chapter builds on the results of a Systematic Literature Review (SLR) published in *Business & Information Systems Engineering* in 2024 [17], which seeks to offer detailed insights into the development and maturity of existing approaches to IoT-aware BPM. The study presented in [17] includes results from an SLR[2] conducted to systematically assess literature on IoT-aware BPM. This assessment focused on the phases of the business process life cycle addressed by various approaches, the specific topics covered, the relevant application domains, and the degree to which these approaches address the research challenges identified in the *IoT-Meets-BPM Manifesto* [46].

Several reviews and surveys on this topic exist in the literature. For instance, [8, 18, 97] analyze the interaction between IoT and BPM. Furthermore, most available works focus on the IoT-aware BP modeling phase [1, 14, 15, 101]. However, while these studies examine specific aspects of IoT and BPM integration, they lack a holistic view of the integration impact across different BP life cycle phases and BPM-related application domains, which is instead presented in this chapter.

Several reviews and surveys have examined the intersection of IoT and BPM. For example, studies such as [8, 18, 97] explore the interaction between IoT and BPM and [1, 14, 15, 101] focusing, however, on the IoT-aware business process modeling phase. While these studies offer insights into specific aspects of IoT-BPM integration, they tend to overlook a comprehensive analysis of how IoT impacts various phases of the business process life cycle and its applications across different BPM domains, a broader perspective that this chapter aims to provide.

[1] BPMN is the ISO standard (ISO/IEC 19510:2013) for business process modeling.

[2] The SLR followed established guidelines from [50] and reviewed 93 publications from 1999 to June 2023. A comprehensive description of the methodology can be found in [17].

This chapter presents insights addressing the following research questions:

RQ1: *Which phases of the BP life cycle were mainly impacted by the integration of IoT?*
RQ2: *Which topics relating to IoT integration in BPM are discussed in the literature?*
RQ3: *Which application domains are used to study IoT and BPM integration?*
RQ4: *What are the research challenges addressed so far to support the management of IoT-aware BP?*

The next sections present a detailed discussion of the SLR findings corresponding to each research question. Specifically, Sect. 2 examines the BP life cycle phases most affected by IoT integration, Sect. 3 delves into the main research topics on BPM and IoT, Sect. 4 explores the application domains utilizing BPM and IoT solutions, and Sect. 5 outlines the research challenges addressed to date. To conclude, Sect. 6 offers a comprehensive discussion of the findings, and Sect. 7 provides concluding insights.

2 Which Phases of the BP Life Cycle Were Mainly Impacted by the Integration of IoT?

The selected studies refer to, or in some cases explicitly mention, one or more phases of the BPM's life cycle to contextualize their work. In detail, phases are considered by the identified studies with regard to the following aspects (cf. Fig. 1): identification (3.8%), modeling (35.4%), analysis (7%), redesign (5.7%), implementation (31.6%), and monitoring and mining (16.5%).

- *Identification:* [40] propose a dynamic integration mechanism of BPs to adapt to the dynamic characteristics of IoT, redirect and restructure the process logic using rules that can be edited at runtime, and design a dynamic integration algorithm to implement this mechanism. Schief et al. [85] propose a centralized framework that extends the process design and execution phases of BPM considering events generated by smart objects.
- *Modeling:* Based on the analysis, the studies are categorized according to several key areas. *Integration of IoT concepts in BPMN 2.0:* [31] propose a BPMN 2.0 extension with sensor and smart device business functions to enhance the integration between the physical world and BPs. Meyer et al. [73] extend the BPMN 2.0 to explicitly consider IoT basic concepts, enabling the modeling and development of IoT services in terms of BPs. Meyer et al. [74] integrate IoT devices as BP resources in the form of dedicated lanes and extend the BPMN 2.0 specification to include such IoT devices. Meyer et al. [75] explore the representation of the IoT domain component "Thing" in a process model, introducing potential BPMN extension for IoT concepts. *Verification and validation of IoT-BPs:* [33] extend BPMN to sustain the various cyber-

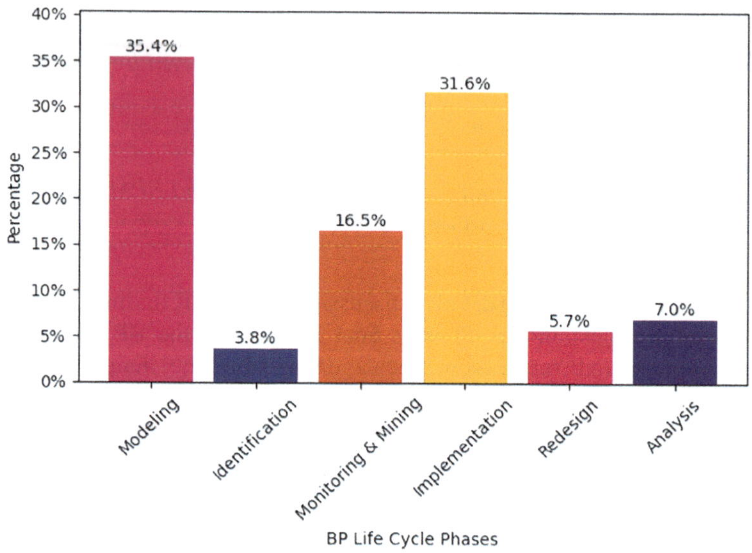

Fig. 1 Distribution of selected studies according to the aspect of the integration of IoT into the BP life cycle phases

physical system concepts and a verification approach at design time to detect the mistakes in specifying the CPS process. Senderovich et al. [93] derive events from sensor data, aligning them with process activities. Hornsteiner and Schönig [38] extend BPMN to map security aspects and thus obtain security- and IIoT-aware BPs. *Context management and IoT adaptation:* [86] propose a set of concepts for an IoT-enhanced process model reengineering, addressing model changes and adaptation tasks. Suri et al. [99] propose configuration concepts for handling IoT resource variability in the Configurable Process Model. *IoT-aware BP decentralization and distribution:* [23] present an automatic approach to decentralize IoT-aware BPs defined using the BPMN 2.0. Jain and Tata [45] have designed a distributed IoT application with annotated components and location information. *IoT event integration and management:* [7] define the role of the IoT concept of smart business processes modeling SBPM, using an ad hoc approach with active help to detect data flow anomalies. Li et al. [62] propose a business user-oriented BP modeling method supporting IoT event stream integration, with a framework that separates CEP and BP execution. Wang et al. [108] monitor and model IoT sensing data in real time, transforming it into IoT services for BP systems. Kirikkayis et al. [55] model IoT-driven events using BPMN 2.0 extension and DMN concepts. Kirikkayis et al. [53] introduce a BPMN 2.0 extension with IoT artifacts, enabling data acquisition and actuator control. *Other aspects related to IoT and BPM:* [114] present a GSM[3]-IoT framework which aims to represent the physical workflows in logistics, extending them by adding physical workflow concepts and considering particularities of IoT smart entities and their mutual interactions. Mottola et al. [78] propose a BPMN extension for wireless sensor network context modeling. Bocciarelli et al. [5] model and manage resource management during runtime. Hasić et al. [37] compare standard BPMN and BPMN + DMN for IoT process modeling. Grefen et al. [36] introduce a time and space specification for IoT-aware collaborative BPs. Hou et al. [39] suggest fragmenting IoT-aware BP for performance improvement. Grambow et al. [34] propose context-awareness integration in IIoT BPs, visual AR support, and enhanced BPMN modeling.

- *Analysis:* [2] model and simulate the cardiac remote patient monitoring process, identifying weaknesses needing improvement. Vitali and Pernici [107] present an approach based on the temporal events analysis for building a graph of dependencies between the whole set of events, unveiling interconnections between processes of cooperating organizations. Seiger et al. [92] develop an interactive method for analyzing low-level IoT data to detect higher-level process activity executions based on sensor-actuator-activity patterns. Diamantini et al. [20] discuss the role of semantic models when integrating IIoT data streams into an Industry 4.0 context. They propose a process-aware knowledge graph that enriches sensor data using ontological descriptions of IIoT sensors, processes, and KPIs. Gómez-Valiente et al. [32] present the DOMINIoT architecture to address scheduling, resource allocation, and state management challenges.
- *Redesign:* The redesign phase is often applied across other BPM life cycle stages. In [83], a redesign of a given clinical process is presented, while in [44], interdisciplinary healthcare processes are redesigned to ensure direct and real-time interaction between the patient and the medical staff. Song et al. [95] mainly addresses this phase by proposing an IoT-enabled context-aware BPM (IoT-CaBPM) framework, which facilitates the evolution of BPs in combination with IoT advances.
- *Implementation:* [9] propose to extend WS-BPEL[4] with IoT characteristics for a new process definition language. Friedow et al. [29] implement and write adapters using the Bosch IoT Things service to connect IoT devices and BPs via event influence. Kikuchi et al. [51] develop a cloud-IoT orchestration framework with IoT devices. Kunz et al. [60] provide novel components and architectural concepts for improved IoT communication. These include a

[3] The GSM (Guard-Stage-Milestone) refers to an approach to structuring Event-Condition-Action rules in the BPM context [41].

[4] The WS-BPEL (Web Services Business Process Execution Language) is an XML-based language that allows web services, APIs, and human processes to interconnect and share data in a business workflow.

scalable discovery service and a decentralized control structure, cloud EPCIS,[5] which facilitates access to RFID[6] data, and middleware that enables flexible adaption of object-specific event processing rules. Marrella and Mecella [67] automate process adaptation in cyber-physical domains during the occurrence of exceptions and exogenous events. Mass et al. [71] propose a system architecture enabling continuous, delay-tolerant BP execution on mobile nodes. Muhsin et al. [79] presents a mobile workflow management system for seamless and accurate data exchange between mobile devices and remote hubs. Seiger et al. [88] propose an IoT workflow management system with dynamic service selection, complex event processing, human interactions, and self-adaptation. Varga et al. [106] propose a supporting system for service chaining and workflow execution using a Petri net-based method, similar to a recent article that provides further insights into this area [59]. Wieland et al. [111] define and implement two smart workflows and provide an architecture for transforming sensor data into business-level information. Zhu et al. [115] design and implement a dynamic adaptation framework using an adaptive algorithm, allowing the system to infer the process when the environment changes. Di Martino et al. [19] provide a methodology for the smart management of irrigation systems using ontologies, BPMN semantic annotation, and logical inference. Park et al. [80] propose modeling and implementation methods for IoT execution design. Pastor et al. [81] integrate IoT sensors for improved user comfort in shared spaces, emphasizing semantic representation and process mining (PM) techniques. Valderas et al. [104] apply the SoC principle for the interdisciplinary development of IoT-enhanced BPs using BPMN, ontology-based technologies, and a microservice architecture.
- *Monitoring and mining:* [90] propose a framework correlating IoT sensor streams with process events and activities using complex event processing. Shamsuzzoha et al. [94] discusses collaborative BP monitoring across organizations using IoT and cloud-based data repositories. Wombacher [112] monitors workflow-state correlation with sensor data for online conformance checking. de Leoni and Pellattiero [16] propose a technique to discover readable human-habit models for optimizing human experience with IoT systems. Janssen et al. [47] presents a technique that combines clustering and PM to discover activities and process models from motion sensors. Koschmider et al. [58] presents a framework to discover activities and process models from event location sensor data. van Eck et al. [105] map sensor measurements to human activities, group them into process instances, and convert them into an event log as input for any PM technique. Sora et al. [96] transform raw movement measurements into actions and adapt sensor logs to apply BPM techniques. Bertrand et al. [3]

[5] EPC Information Services is a standard enables disparate applications to create and share visibility event data, both within and across enterprises.

[6] RFID (radio-frequency identification) refers to a technology whereby digital data are captured by a reader via radio waves.

propose a model bridging the gap between IoT and PM using IoT ontologies and BP context models. Elali et al. [25] enhance process models derived from PM with contextual information from sensor data. Elkodssi et al. [27] present a PM approach for discovering and analyzing IoT-aware BPs and transforming sensor logs into structured event logs enriched with IoT concepts using a BPMN ontology.

- *More phases:* A BPMN 2.0 IoT-enabled extension for modeling, executing, and monitoring IoT-aware BPs is proposed by Gallik et al. [30], while [54] also presents a framework for IoT-driven business rules. Kirikkayis et al. [56] present a BPMN 2.0 extension for IoT, enabling real-time interaction with devices that aids decisions through collected data but lacks high-level aggregation consideration. Kirikkayis et al. [52] present a web-based framework integrating IoT data into BPs, enhancing real-time decision-making and monitoring.

3 Which Topics Relating to IoT Integration in BPM Are Discussed in the Literature?

One of the most effective techniques for latent topic distribution within a corpus is the latent Dirichlet allocation (LDA) technique [4]. LDA is a generative probabilistic model of a corpus where the documents are represented as a random mixture of latent topics, each characterized by a mix of words. To answer RQ2, LDA was applied to the final set of 93 papers.[7]

The analysis produced four primary topics, summarized in Table 1, where bolded words indicate the terms with the highest estimated frequencies for each topic. After training the model and identifying these four topics, an interactive web-based visualization was used to examine the keywords associated with each topic to infer their semantic meanings. Following this, each topic was contextualized within its broader semantic scope and discussed with reference to relevant studies.

Table 1 The four topics mostly discussed in the 93 selected studies and terms related to them

Topic	Topic label	Relevant terms
1	Event mining from sensors	**Sensor** + **event** + **mining** + task + system + mobile + discover + activity + concept + service
2	Physical resource management	**Resource** + **management** + **physical** + time + system + event + object + cyber + physical + activity + property
3	Context-aware execution	**Context** + sensor + **execution** + application + system + environment + service + information + technology + user
4	IoT-aware process modeling	**bpmn** + context + information + **extension** + system + sensor + concept + execution + language + event

[7] For the complete description of the conducted study refer to [17].

- *Topic 1—event mining from sensors:* Knowledge extraction from sensor data has gained significant attention in the BPM community. Process mining techniques can greatly enhance BPM when applied to analyze and transform raw data into usable information. The information provided by the sensors can be expressed through a functional business view, as demonstrated in [31], representing the classification, storage, and distribution procedure of floral products within a flower company. This model in BPMN is connected to a functional model to import an ontology of the sensors and related instance data and then integrate the information modeled using the Semantic Sensor Network ontology with BPs. Similarly, various frameworks address the correlation between workflow states, sensor data, and events generated by IoT devices that initiate instances of BPs [7, 20, 44, 81, 83, 84, 90, 112]. Koschmider et al. [58] detect high-level events from raw event data and discover BPs from derived instances. Zhu et al. [115] present a multi-agent framework that chooses the best step to move in the BP when an event caused by agents occurs. In addition, [96] presents an approach to discover process models from activities in the field of smart spaces where sensors must turn the sensor log into an event log consisting of human actions. Another interesting challenge when applying PM techniques in the context of sensor data is mapping sensor measurements to human activities and grouping activities into process instances [16, 105]. The problem of mapping sensor data to event logs based on process knowledge is solved by Senderovich et al. [93], which maps location-based events to activities by recognizing interactions between various agents and transforming historical sensor data registers into standardized event logs. Another contribution to translating sensor data into higher-level ones comes from [47]. They discretize sensor data into activities using unsupervised learning through clustering, while [92] analyzes sensor data to detect higher-level process activities.
- *Topic 2—physical resource management:* This topic tackles the specification and management of the resources associated with the BPs supporting cyber-physical systems (CPSs). Graja et al. [33] propose a verification framework to support the various CPS concepts and properties, which enables the designer to handle CPS process features. To integrate and utilize smart devices as BP resources to support the modeling, a service-oriented BPM system architecture was developed by Wehlitz et al. [109]. The resource management during runtime of a system to model a CPS-aware resource associated with a BP activity is proposed by Bocciarelli et al. [5], while the DOMINIoT architecture is designed by Gómez-Valiente et al. [32]. In scenarios where physical resources are exchanged, knowing how a resource owned by a party is managed on another party's premises is impossible. Thus, possible misalignments can be detected too late. In [76], the authors investigated an approach for compliance checking that mixes commitments and smart devices. Suri et al. [98] developed a framework that describes IoT resources (e.g., extending the BP models with energy cost parameters to enable the energy-aware management of IoT resources in BPs).

- *Topic 3—context-aware execution:* A context-aware execution environment involves processes continuously observed and adapted according to the model specified when required by context changes. Contextual information enhances process execution, incorporating situation-aware insights for improved effectiveness. Such information is often gathered from sensors within intelligent environments, enabling pervasive execution known as intelligent workflows [111]. Frameworks that enhance process awareness and self-adaptation are proposed [28, 29, 66, 82, 89]. To support context variables and sensors as well as communication paradigms for IoT, [22] extend the WS-BPEL workflow language, dynamically selecting IoT services based on availability, functionality, and context [25, 88]. Other studies focused on the adaptability of BPs during the execution phase. A dynamic integration mechanism for coping with changes in BPs is proposed by Hu et al. [40], Domingos et al. [21], Tôn and Lê [100]. An application to the predictive process monitoring field using IoT events as a process context and developing a predictive model to predict the next activity is provided by Elhami et al. [26]. A new trend in running the IoT-aware BP is running on *fog computing*.[8] In this context, [10] dealt with IoT-aware BPs at the execution level, where they introduced a new intermediate layer consisting of a set of distributed fog nodes to perform certain parts of the process.
- *Topic 4—IoT-aware process modeling:* To create IoT-aware BPs and exploit the full potential of IoT and BPM, this must be integrated into process models. Modeling strategies for IoT-aware BPs fall into two categories [6]: those utilizing known modeling languages and those proposing new domain-specific languages. The former includes integrating BPMN with ontologies [3, 19, 27, 103, 104] and extending BPMN with IoT devices [13, 38, 53], as well as using Petri nets [106], or DMN[9] [37, 55]. In the latter, [9] provides a new process definition language for IoT-enabled BPs, encapsulating physical devices as SOA services. IoT device interaction is critical. To this end, the application of mixed reality as a new interaction paradigm to facilitate the modeling and configuration of processes among IoT devices was elaborated on [91]. From an access IoT data perspective, [12] presents a gateway for transferring IoT events to BPs, managing device heterogeneity. Meyer et al. [74] map IoT concepts to process models, and [69] use BPMN to model IoT device behaviors. Other studies focus on collaborative structures to improve physical and digital collaboration between multiple actors to achieve a common goal [62, 65, 94, 108]. For example, [36] discusses synchronizing physical objects for successful IoT-aware digital processes.

[8] Fog computing provides data, compute, storage, and application services to end users where the difference is where data processing occurs.

[9] DMN (decision model and notation) is a standard approach for describing and modeling repeatable decisions within organizations to ensure that decision models are interchangeable across organizations.

4 Which Application Domains Are Used to Study IoT and BPM Integration?

RQ3 was answered based on data from research conducted by HSPI.[10] Thus, from the HSPI process mining application database, the studies are collected and analyzed to map the applications provided through different industries and markets and understand the adoption of BPM and IoT. The application domains which the found studies refer to are (cf. Fig. 2): healthcare (17.6%), industrial sector (41.2%), public (7.1%), consumer services (5.9%), consumer goods (3.5%), and services (24.7%). Only eight studies (11%) do not refer to any application domain.

- *Healthcare:* Most healthcare applications are related to remote patient monitoring systems, within hospitals or at home, and constantly enriched with IoT devices [2, 10, 32, 45, 53, 82–84, 88, 100, 107]. Maamar et al. [64] use the example of a hospital that is on high alert being close to a severe car accident, where departments equipped with environmental sensors or smart wrists enable the real-time transmission of patients' vital signs to appropriate recipients, demonstrating how IoT can facilitate operations and improve efficiency. Senderovich et al. [93] presents an example of a process of treatment in a day hospital for cancer patients through Real-Time Locating System

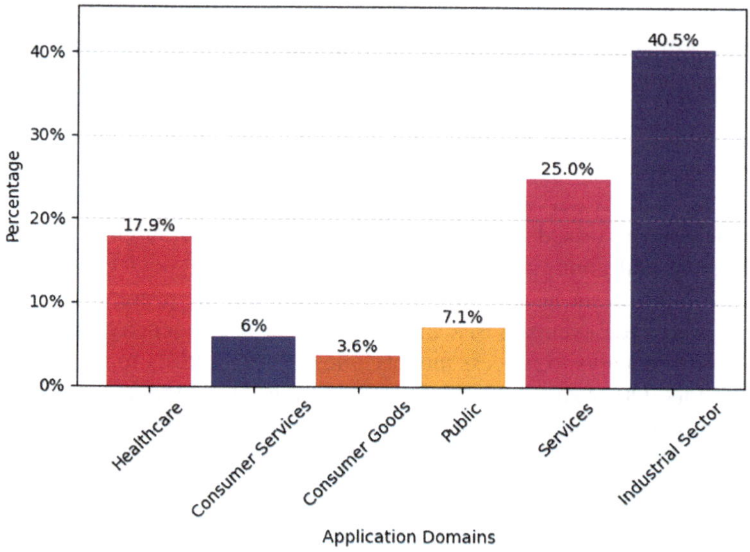

Fig. 2 Distribution of selected studies by application domains

---

[10] Process mining: a database of applications, HSPI SpA—Management Consulting. https://www.hspi.it/wp-content/uploads/2020/01/HSPI_Process_Mining_Database2020.pdf.

(RTLS) receivers that monitor all corporate entities involved in the process (e.g., patients, doctors, nurses) as well as some of the medical devices that record the data emitted and use it for real-time monitoring of process entities and equipment. Another frequent scenario is elderly care. Integrating the real-time event streams from bed pressure sensors into the BP field can reflect the health status of the elderly in real time and thus monitor them through immediate warnings [62].

- *Industrial sector:* In the domain of transportation, IoT might be used to realize monitoring of goods, such as dangerous or perishable goods [20–23, 30, 92, 98, 103, 104]. These systems react to events that occur during transport using sensors and location technologies to improve the monitoring of transport conditions, such as temperature and pressure [71]. Sensors and smart devices have great potential to provide further automation in various domains including logistics [31, 36, 72, 85, 95, 112, 114]. Take, for example, the maritime transport scenario of heavy fog where disasters can be prevented in sea-land transport [108]. Montali and Plebani [76] uses an example of a seafood company that organizes fish delivery and aims to verify if all actors operate correctly because of the complexity of the delivery process (deviations to the plan may occur). Several studies focus on smart factories as representatives of IoT environments [5, 34, 52, 54–56, 66, 90, 111]. In the manufacturing industry, BPM and IoT support could improve both management by closely linking digital production and machine data [49, 86, 87, 94] and cybersecurity by integrating security requirements into modeling using a "security by design" paradigm [38].
- *Public:* [11] develop an actual sensor-aware BP application and validate it by using the designed system to protect a large area of forest in North China. Graja et al. [33] evaluate their study through an example of disaster recovery systems. Maamar et al. [65] present a city that runs many systems like transportation for traffic control and environment for air pollution monitoring; [67] rely on a case study previously conducted by the same authors in [42, 43, 68] to evaluate their study in an emergency management domain, in which teams of first responders act in disaster locations with the primary purpose of assisting potential victims and stabilizing the situation. Chadli et al. [7] propose a classroom case study involving the deployment of hardware device and management software to measure the environmental parameters and control ventilation, lighting, and air-conditioning. Pastor et al. [81] employ a smart camera with active sensors for vehicle identification in road traffic.
- *Consumer services:* [48] test their study for common BPs in the retail domain within the innovative retail lab of a large European supermarket chain; [73] evaluate their study using a use case from the domain of retail and show how sensors monitor perishable goods in a store; [74] test their study using a dynamic pricing process in the retail domain and show how the IoT device temperature sensor monitors the perishable good orchid in a store; [99] evaluate their study using a process family from the Retail/Supply Chain Management domain; [102] test their study using a Smart City Tourism Organization (SCTO) scenario to

automate its BPs related to visitor mobility by communicating with them using social networks and collecting their feedback.

- *Consumer goods:* [29] use a simple coffee machine billing system to automate the process of counting the coffee amount for each user; [60] use the life cycle of a fish fillet within the different domains of manufacturer, logistics service provider, wholesaler distribution hub, wholesaler store, and the customer; [105] use a case study performed at Philips where a smart baby bottle has been developed.
- *Services:* Scenarios that use smart device detection and implementation capabilities in smart home environments have been extensively addressed [13, 25, 40, 47, 63, 89, 109, 113]. Examples of instances where service applications play a crucial role include the intruder detection system in a security scenario [51], office supply processes [12], temperature control processes [91], warehouse management system monitoring processes [39], automatic irrigation control processes [19, 69, 70], processes for adaptive ventilation in a student dormitory [78] or on student thesis defense [115], and processes for repair and maintenance of computers and peripherals [79], while [80] implement smart toilets for companion animals by simply attaching sensors and introducing BPM technology.

5 What Are the Research Challenges Addressed so far to Support the Management of IoT-Aware BP?

Enriching BPs with IoT technologies brings several advantages and many challenges [46]. Figure 3 and Table 2 report the distribution of the studies according to the *IoT-Meets-BPM Manifesto*. Not all studies face one of these challenges; actually, only 38% of them do.

- *C1—placing sensors in a process-aware way:* The research challenge C1 is considered by 25% (i.e., 12) of the studies. The link between the position of the sensors and that of the corresponding actuator is discussed in [78]. Using a physical simulation model, [66] conducted research in BPM and Industry 4.0, aiming to program a smart factory for BPs by selecting hardware components to detect process-relevant events. To place IoT sensors in a process-aware way and link them to the running process, a BPMS must be aware of the current values of IoT objects, and based on an established mapping from IoT variables to process models, IoT data is sent to a BPMS [87]. Process model discovery from sensors or wearable devices depends on their location and amount [58, 93, 105]. Other interesting studies on how to integrate sensors with other services and into existing BPs have been conducted [19, 30, 84, 95]. Regarding sensors, the presented ontological model covers domain configuration, including IoT resources' placement and gathered observations over time [20, 104].

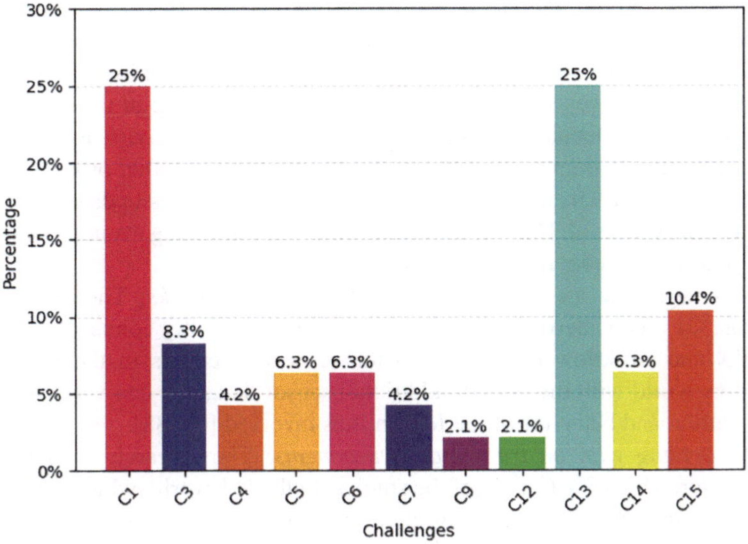

Fig. 3 Distribution of the studies by research challenges addressed in the *IoT-Meets-BPM Manifesto*

Table 2 Definitions of the research challenges of the *IoT-Meets-BPM Manifesto*

ID	Definition	Literature
C1	Placing sensors in a process-aware way	[19, 20, 30, 58, 66, 78, 84, 87, 93, 95, 104, 105]
C2	Support for managing manually executed physical processes	
C3	Connection of analytical processes with the IoT	[26, 55, 62]
C4	Integrating the IoT with process correctness checks	[66, 88]
C5	Dealing with unstructured environments	[21, 66]
C6	Managing the link between microprocesses	[31, 56, 66]
C7	Breaking down end-to-end processes	[48, 66]
C8	Detecting new process from data	
C9	Specifying the autonomy level of things	[66]
C10	Specifying the roles of things	
C11	Concretizing abstract process models	
C12	Dealing with new situations	[66].
C13	Bridging the gap between event-based and process-based systems	[3, 16, 26, 47, 49, 58, 66, 86, 93, 96, 105, 107]
C14	Improving online conformance checking	[66, 90, 112]
C15	Improving resource utilization optimization	[5, 44, 74, 98, 99]
C16	Improving resource monitoring and the quality of task execution	

- *C3—connection of analytical processes with IoT:* The research challenge C3 is considered by 8.3% (i.e., 4) of the studies. Elhami et al. [26] uses contextual events in the decision point to control the process execution and incorporate contextual changes with process rules and exaction logic at the runtime. With the proposed extension [55] can specify event-driven behavior at various points in a BP but also demarcate IoT-based data sources from other process inputs for monitoring and error handling. BP execution necessitates updated information availability. Li et al. [62] proposed a framework separating event processing and BP execution, reducing BPM engine load.
- *C4—integrating the IoT with process correctness checks:* The research challenge C4 is considered by 4.2% (i.e., 2) of the studies. By connecting individual tasks and workflow executions to the respective effects in the physical and virtual world with the help of additional sensor data, you can verify the correct execution and behavior of the IoT entities involved [66, 88].
- *C5—dealing with unstructured environments:* The research challenge C5 is considered by 6.3% (i.e., 3) of the studies. In the IoT world, which is much more ad hoc and situative, [21] introduce mechanisms to perform ad hoc changes in IoT-aware BPs by identifying change primitives that support the change operations needed to modify parts of the BP.
- *C6—managing the link between microprocesses:* The research challenge C6 is considered by 6.3% (i.e., 3) of the studies. An example is to achieve an efficient and flexible production line by investigating the microprocesses at the individual machines and stations and their interconnections to achieve a more flexible composition of smaller processes [66]. By extending BPMN 2.0, [31] links process models with business functions, connecting them to external open data, including sensor and instance ontology data; another extension suggests using low-level IoT data for decision-making, not including high-level data aggregation [56].
- *C7—breaking down end-to-end processes:* The research challenge C7 is considered by 4.2% (i.e., 2) of the studies. Malburg et al. [66] suggests relaxing and detailing the static and coarse-grained "hardwired" processes to achieve a more flexible composition of smaller processes. Kahl et al. [48] propose a system that provides an agent-assisted realization and adaptation of BPs with semantic service selection and facilitates an event-driven selection and controlled execution of relevant BPs in intelligent environments.
- *C9—specifying the autonomy level of things:* The research challenge C9 is considered by 2.1% (i.e., 1) of the studies. Ensuring appropriate autonomy for resource constrained IoT device is investigated by Malburg et al. [66].
- *C12—dealing with new situations:* The research challenge C12 is considered by 2.1% (i.e., 1) of the studies. Malburg et al. [66] use past successful process executions to automatically learn possible adaptations of process instances to deal with new or similar situations.
- *C13—bridging the gap between event-based and process-based systems:* The research challenge C13 is considered by 25% (i.e., 12) of the studies. The complex IoT system is event-driven due to the large number of sensors and process-based [66]. While in the past, process models have been detected by

documents or interviews with domain experts, today the challenge of automatically discovering the process model from the network of sensors or wearable devices also depends on their location and amount [58, 93, 105]. Therefore, considering the physical context and revealing the correlation between IoT data and process events is crucial [3].
- *C14—improving online conformance checking:* The research challenge C14 is considered by 6.3% (i.e., 3) of the studies. Seiger et al. [90] mainly focuses on generating event logs from streams of IoT sensor data with smart factories as an application domain.
- *C15—improving resource utilization optimization:* The research challenge C15 is considered by 10.4% (i.e., 5) of the studies. Suri et al. [98] presents a framework that formalizes IoT properties and rules to optimize resource management in BPs, defining resource constraints to be mapped to a task during the design phase.

6 Discussion

This section discusses the relationships among the topics extracted from the analyzed studies, their application domains, BPM life cycle phases, and the research challenges highlighted by Janiesch et al. [46]. Lastly, it discusses key issues that appear to require further investigation.

- *BPM life cycle phases.*
  The findings indicate that many studies address multiple BPM life cycle phases, with the modeling and implementation phases frequently combined (43%, 40 studies). Significant research attention has focused on the impact of IoT technology on the modeling phase, leading to the development of IoT-aware BP modeling. Incorporating IoT entities into the modeling phase has posed challenges, particularly due to the heterogeneity of these entities. The studies reveal two main approaches for modeling IoT entities: some utilize standard elements of process modeling languages (though this is less common), while the more prevalent approach extends these languages with specific IoT elements. However, existing BPM solutions have yet to precisely define IoT entity behavior, a limitation that complicates the automation of BPs on IoT devices using traditional BPM systems. The analysis further highlights that an integrated workflow engine can quickly adapt the model and its execution, avoiding the need for re-converting and distributing executable source code when the process model changes. Another frequent approach is the implementation of middleware to decouple BP from device-specific technologies. Nonetheless, a comprehensive approach that addresses most phases of IoT-aware BP remains a pressing challenge. Based on these insights, a set of research directions has been identified for the life cycle phases of IoT-aware BPs (see Table 3).
- *Topic modeling:* One of the main challenges identified in the topic analysis, especially in extracting knowledge from sensor data, is closing the gap between

Table 3 Trends for research studies to support IoT-aware BP life cycle phases

Trends	BPM life cycle phase	Literature
The modeling of IoT-driven BPs must be integrated into a standard modeling language such as BPMN	Modeling	[33, 34, 37, 73–75, 78]
It should be possible to have a built-in workflow engine to make quick changes to the running model or to implement middleware that makes the process independent of device technology, i.e., aware of the involvement of IoT devices	Execution	[29, 49, 60]
It should be possible to monitor IoT devices in real time	Monitoring	[94, 112]

raw sensor data and structured event logs. This involves identifying specific events within raw data, recognizing their associated activities, and linking these events to the relevant process instances. Another interesting area that emerged from the review is the modeling of IoT-aware business processes. Most of the examined approaches extend BPMN to integrate IoT devices within the process model, allowing for a seamless representation of IoT data within traditional BP workflows. An alternative approach uses linked data mechanisms to bridge the BP model with external data sources, like sensor ontologies and real-time instance data, enhancing the integration between the digital and physical worlds. For many researchers, IoT serves as an essential data source in BPM, especially for providing process context. Lastly, managing physical resources was a central focus in IoT-enhanced BPM. Here, it is critical to specify the active IoT resources used for task execution, though a major challenge remains in formalizing these relationships to optimize resource management effectively.
- *Application domains and case studies:* Findings reveal that certain application domains gain more from integrating IoT concepts into BPM scenarios than others. Industrial applications (41.2%, 35 studies) and services (24.7%, 21 studies) stand out as the most frequently explored contexts for BPM and IoT integration. Within the industrial domain, studies span various sectors, with a notable focus on IoT-based logistics, where BPM aids in streamlining processes. While filtering and processing IoT events play a crucial role, they are only initial steps toward creating transparent BPs. Given that data in logistics is often incomplete or inconsistent, ensuring data quality becomes essential for effective BPM-IoT integration. Addressing data quality challenges is critical, as current solutions rely on persistent, relatively static data sets, which do not align well with the dynamic nature of physical-world data. Interestingly, only a few of the studies apply IoT-BPM integration to real-world scenarios, such as temperature control processes [13], small-scale smart factory models [52], smart hospital systems [45], and clinical hypertension management processes [83]. Additionally, two studies highlight plans to undertake comprehensive evaluations of real-world applications in future work [20, 64].

- *Research challenges:* An essential insight from a portion of the analyzed studies (38%, 35 studies) highlights the challenges noted in the BPM-IoT Manifesto [46]. Complex IoT environments typically involve numerous sensors, actuators, and control units, alongside an additional software stack that raises programming and research to a BP-oriented level of abstraction. This integration allows BPM to fully leverage IoT potential. Such systems are often event-driven, due to the volume of sensors, or process-driven, depending on operational needs (C13—Sect. 5). One of the research tasks associated with this challenge is analyzing sensor and actuator functionality and developing methods to group and abstract them within a BPM context. Identifying and refining IoT-generated process events and converting them into relevant events can be challenging, as ambiguities and uncertainties inherent in IoT environments often persist in event logs, which must then be accounted for in subsequent analyses (e.g., during compliance checks). The review of research challenges further emphasizes the importance of situating IoT sensors in a way that aligns with ongoing processes and connecting them to active workflows (C1—Sect. 5). Current efforts focus primarily on BPMS configurations that monitor real-time IoT data. Another crucial area for investigation is developing methods for obtaining real-time values from diverse data variables and mapping these IoT variables onto process models, facilitating data flow to BPMS. This step is critical as by using real-time variable values, tasks can be dynamically activated or canceled, decisions can be informed, and IoT data variables become more configurable and traceable within BPM.

7 Concluding Remarks

This chapter has offered a comprehensive overview of the dynamic relationship between BPM and IoT. Unlike previous surveys that focus on specific facets of this integration, this study aims to deliver a broad, holistic perspective. Through a systematic analysis of the reviewed studies, several key research questions were addressed: *(i)* identifying the main topics explored in the literature (such as event sensor mining, physical resource management, context-aware execution, and IoT-aware process modeling), *(ii)* categorizing how IoT integrates within various BP life cycle phases (including identification, modeling, analysis, redesign, implementation, and monitoring and mining), *(iii)* pinpointing the application domains where IoT and BPM integration is most impactful (notably in consumer goods, consumer services, healthcare, industrials, services, and public sectors), and *(iv)* highlighting the research challenges outlined in the *IoT-Meets-BPM Manifesto* that have been tackled to advance the management of IoT-aware BPs.

References

1. Abouzid, I., Bekali, Y.K., Saidi, R.: Modelling IoT behaviour in supply chain business processes with BPMN: a systematic literature review. J. ICT Stand. **10**, 439–468 (2022)
2. Antonius, N., Dachyar, M.: The internet of things (IoT) design for cardiac remote patient monitoring using business process re-eng. In: 2020 3rd International Conference on Applied Engineering (ICAE), pp. 1–7 (2020)
3. Bertrand, Y., De Weerdt, J., Serral, E.: A bridging model for process mining and IoT. In: International Conference on Process Mining (ICPM), pp. 98–110 (2021)
4. Blei, D.M., Ng, A.Y., Jordan, M.I.: Latent dirichlet allocation. J. Mach. Learn. Res. **3**, 993–1022 (2003)
5. Bocciarelli, P., D'Ambrogio, A., Giglio, A., Paglia, E.: A BPMN extension for modeling cyber-physical-production-systems in the context of industry 4.0. In: 2017 IEEE 14th International Conference on Networking, Sensing and Control (ICNSC), pp. 599–604 (2017)
6. Brouns, N., Tata, S., Ludwig, H., Asensio, E.S., Grefen, P.: Modeling IoT-aware business processes-a state of the art report. CoRR (2018)
7. Chadli, N., Elhoseny, M., Kabbaj, M.I., Bakkoury, Z.: Smart business process modeling: Toward an IoT to detect the data flow anomalies in ad hoc mesh network. In: Distributed Sensing and Intelligent Systems, pp. 13–27 (2022)
8. Chang, C., Srirama, S.N., Buyya, R.: Mobile cloud business process management system for the internet of things: a survey. ACM Comput. Surv. **49**(4), 1–42 (2016)
9. Chen, F., Ren, C., Wang, Q., Shao, B.: A process definition language for internet of things. In: 2012 IEEE International Conference on on Service Operations and Logistics, and Informatics (SOLI), pp. 107–110 (2012)
10. Cheng, Y., Zhao, S., Cheng, B., Chen, J.: A service-based fog execution environment for the IoT-aware business process applications. In: 2018 IEEE International Conference on Web Services (ICWS), pp. 323–326 (2018)
11. Cheng, Y., Zhao, S., Cheng, B., Chen, X., Chen, J.: Modeling and deploying IoT-aware business process applications in sensor networks. Sensors **19**(1), 11 (2019)
12. Cherrier, S., Deshpande, V.: From BPM to IoT. In: International Conference on Business Process Management (BPM), pp. 310–318 (2017)
13. Chiu, H.H., Wang, M.S.: Extending event elements of business process model for internet of things. In: 2015 IEEE International Conference on Computer and Information Technology; Ubiquitous Computing and Communications; Dependable, Autonomic and Secure Computing; Pervasive Intelligence and Computing, pp. 783–788 (2015)
14. Compagnucci, I., Corradini, F., Fornari, F., Polini, A., Re, B., Tiezzi, F.: Modelling notations for IoT-aware business processes: A systematic literature review. In: Business Process Management Workshops—BPM 2020 International Workshops. Lecture Notes in Business Information Processing, vol. 397, pp. 108–121 (2020)
15. Compagnucci, I., Corradini, F., Fornari, F., Polini, A., Re, B., Tiezzi, F.: A systematic literature review on IoT-aware business process modeling views, requirements and notations. Softw. Syst. Model. **22**(3), 969–1004 (2023)
16. de Leoni, M., Pellattiero, L.: The benefits of sensor-measurement aggregation in discovering IoT process models: A smart-house case study. In: Business Process Management Workshops, BPM 2021 International Workshops, vol. 436, pp. 403–415 (2021)
17. De Luzi, F., Leotta, F., Marrella, A., Mecella, M.: On the interplay between business process management and internet-of-things. Bus. Inf. Syst. Eng. **67**(2), 265–288 (2025)
18. Del Giudice, M.: Discovering the internet of things (IoT) within the business process management. Bus. Process. Manag. J. **22**(2), 263–270 (2016)

19. Di Martino, B., Cante, L.C., Esposito, A., Graziano, M.: Towards a methodology for the semantic representation of Iot sensors and BPMNs to discover business process patterns: A smart irrigation case study. In: International Conference on Broadband and Wireless Computing, Communication and Applications, pp. 248–258 (2022)
20. Diamantini, C., Mircoli, A., Potena, D., Storti, E.: Process-aware IIoT knowledge graph: a semantic model for industrial IoT integration and analytics. Futur. Gener. Comput. Syst. **139**, 224–238 (2023)
21. Domingos, D., Martins, F., Martinho, R., Silva, M.: Ad-hoc changes in IoT-aware business processes. In: 2010 Internet of Things (IoT), pp. 1–7 (2010)
22. Domingos, D., Martins, F., Cândido, C., Martinho, R.: Internet of things aware WS-BPEL business processes context variables and expected exceptions. J. Univers. Comput. Sci. **20**(8), 1109–1129 (2014)
23. Domingos, D., Martins, F., Caiola, L.: Decentralising internet of things aware bpmn business processes. In: Sensor Systems and Software, pp. 110–119 (2015)
24. Dumas, M., La Rosa, M., Mendling, J., Reijers, H.A.: Fundamentals of Business Process Management, vol. 2. Springer, Berlin (2018)
25. Elali, R., Kornyshova, E., Deneckère, R., Salinesi, C.: Mining contextual process models using sensors data: A case of daily activities in smart home. In: European, Mediterranean, and Middle Eastern Conference on Information Systems, pp. 409–425 (2022)
26. Elhami, E., Ansari, A., Farahani, B., Aliee, F.S.: Towards IoT-driven predictive business process analytics. In: 2020 International Conference on Omni-layer Intelligent Systems (COINS), pp. 1–7 (2020)
27. Elkodssi, I., Laanaoui, M.D., Sbai, H.: Toward a new framework for process-aware IoT discovery. In: International Conference on Smart City Applications, pp. 793–803 (2022)
28. Engels, G., Strothmann, T., Teetz, A.: Adapt cases 4 BPM—a modeling framework for process flexibility in IIoT. In: 2018 IEEE 22nd International Enterprise Distributed Object Computing Workshop (EDOCW), pp. 59–68 (2018)
29. Friedow, C., Völker, M., Hewelt, M.: Integrating IoT devices into business processes. In: International Conference on Advanced Information Systems Engineering, pp. 265–277 (2018)
30. Gallik, F., Kirikkayis, Y., Reichert, M.: Modeling, executing and monitoring IoT-aware processes with bpm technology. In: 2022 International Conference on Service Science (ICSS), pp. 96–103 (2022)
31. Gao, F., Zaremba, M., Bhiri, S., Derguerch, W.: Extending BPMN 2.0 with sensor and smart device business functions. In: 2011 IEEE 20th International Workshops on Enabling Technologies: Infrastructure for Collaborative Enterprises, pp. 297–302 (2011)
32. Gómez-Valiente, P., Pérez, J., Lillo-Castellano, J., Marina-Breysse, M.: smart-IoT business process management: a case study on remote digital early cardiac arrhythmia detection and diagnosis. IEEE Int. Things J. **10**(19), 16744–16757 (2023)
33. Graja, I., Kallel, S., Guermouche, N., Cheikhrouhou, S., Kacem, A.H.: Modelling and verifying time-aware processes for cyber-physical environments. IET Softw. **13**(1), 36–48 (2019)
34. Grambow, G., Hieber, D., Oberhauser, R., Pogolski, C.: A context and augmented reality BPMN and bpms extension for industrial internet of things processes. In: Business Process Management Workshops—BPM 2021 International Workshops. Lecture Notes in Business Information Processing, vol. 436, pp. 379–390 (2021)
35. Grefen, P., Ludwig, H., Tata, S., Dijkman, R., Baracaldo, N., Wilbik, A., D'hondt, T.: Complex collaborative physical process management: A position on the trinity of BPM, IoT and DA. In: Working Conference on Virtual Enterprises, pp. 244–253 (2018)
36. Grefen, P., Brouns, N., Ludwig, H., Serral, E.: Co-location specification for IoT-aware collaborative business processes. In: Information Systems Engineering in Responsible Information Systems, pp. 120–132 (2019)

37. Hasić, F., Serral, E., Snoeck, M.: Comparing BPMN to BPMN + DMN for IoT process modelling: A case-based inquiry. In: 35th Annual ACM Symposium on Applied Computing, SAC '20, pp. 53–60 (2020)
38. Hornsteiner, M., Schönig, S.: SIREN: Designing business processes for comprehensive industrial IoT security management. In: 18th International Conference on Design Science Research in Information Systems and Technology, (DESRIST). Lecture Notes in Computer Science, vol. 13873, pp. 379–393 (2023)
39. Hou, S.L., Zhao, S., Cheng, B., Cheng, Y.Y., Chen, J.L.: Fragmentation and optimal deployment for IoT-aware business process. In: 2016 IEEE International Conference on Services Computing (SCC), pp. 657–664 (2016)
40. Hu, G., Wu, B., Zou, H., Chen, J.: Dynamic integration mechanism of business process for internet of things. In: 2014 IEEE International Conference on Services Computing, pp. 852–853 (2014)
41. Hull, R., Damaggio, E., Fournier, F., Gupta, M., III, Heath, F., Hobson, S., Linehan, M.H., Maradugu, S., Nigam, A., Sukaviriya, P., Vaculín, R.: Introducing the guard-stage-milestone approach for specifying business entity lifecycles. In: 7th International Workshop, Web Services and Formal Methods WS-FM 2010. Lecture Notes in Computer Science, vol. 6551, pp. 1–24 (2010)
42. Humayoun, S.R., Catarci, T., de Leoni, M., Marrella, A., Mecella, M., Bortenschlager, M., Steinmann, R.: Designing mobile systems in highly dynamic scenarios: the WORKPAD methodology. Knowl. Technol. Policy **22**, 25–43 (2009)
43. Humayoun, S.R., Catarci, T., de Leoni, M., Marrella, A., Mecella, M., Bortenschlager, M., Steinmann, R.: The WORKPAD user interface and methodology: Developing smart and effective mobile applications for emergency operators. In: Universal Access in Human-Computer Interaction. Applications and Services, pp. 343–352 (2009)
44. Ismaili-Alaoui, A., Baïna, K., Benali, K., Baïna, J.: Towards smart incident management under human resource constraints for an IoT-BPM hybrid architecture. In: International Conference on Web Services, pp. 457–471 (2018)
45. Jain, R., Tata, S.: Cloud to edge: Distributed deployment of process-aware IoT applications. In: 2017 IEEE International Conference on Edge Computing (EDGE), pp. 182–189 (2017)
46. Janiesch, C., Koschmider, A., Mecella, M., Weber, B., Burattin, A., Di Ciccio, C., Fortino, G., Gal, A., Kannengiesser, U., Leotta, F.: The internet of things meets business process management: a manifesto. IEEE Syst. Man Cybern. Mag. **6**(4), 34–44 (2020)
47. Janssen, D., Mannhardt, F., Koschmider, A., Zelst, S.V.: Process model discovery from sensor event data. In: ICPM Workshops (2020)
48. Kahl, G., Klusch, M., Zinnikus, I., Schimmelpfennig, J., Zapp, M.: ADIGE: Semantic business process management for smart retail environments. In: 17th International Conference on Information Integration and Web-Based Applications & Services (iiWAS), pp. 53:1–53:5 (2015)
49. Keates, O.: Integrating IoT with BPM to provide value to cattle farmers in australia. In: Business Process Management Workshops—BPM 2019 International Workshops. Lecture Notes in Business Information Processing, vol. 362, pp. 119–129 (2019)
50. Keele, S.: Guidelines for performing systematic literature reviews in software engineering. Technical Report, Technical Report, Ver. 2.3 EBSE Technical Report. EBSE (2007)
51. Kikuchi, S., Thomas, I., Jallouli, O., Doerr, J., Morgenstern, A., Baccelli, E., Schleiser, K.: Orchestration of IoT device and business workflow engine on cloud. In: 2018 3rd Cloudification of the Internet of Things (CIoT), pp. 1–2 (2018). https://doi.org/10.1109/CIOT.2018.8627118
52. Kirikkayis, Y., Gallik, F., Reichert, M.: IoTDM4BPMN: An IoT-enhanced decision making framework for BPMN 2.0. In: International Conference on Service Science, (ICSS) 2022, pp. 88–95 (2022)
53. Kirikkayis, Y., Gallik, F., Reichert, M.: Towards a comprehensive BPMN extension for modeling IoT-aware processes in business process models. In: International Conference on Research Challenges in Information Science, pp. 711–718 (2022)

54. Kirikkayis, Y., Gallik, F., Reichert, M.: Modeling, executing and monitoring IoT-driven business rules. In: BPMDS/EMMSAD 2023. Lecture Notes in Business Information Processing, vol. 479, pp. 88–102 (2023)
55. Kirikkayis, Y., Gallik, F., Seiger, R., Reichert, M.: Integrating IoT-driven events into business processes. In: CAiSE Forum 2023, vol. 477, pp. 86–94 (2023)
56. Kirikkayis, Y., Gallik, F., Winter, M., Reichert, M.: BPMNE4IoT: a framework for modeling, executing and monitoring IoT-driven processes. Futur. Int. **15**(3), 90 (2023)
57. Kopetz, H., Steiner, W.: Internet of things. In: Real-Time Systems, pp. 325–341. Springer, Cham (2022)
58. Koschmider, A., Janssen, D., Mannhardt, F.: Framework for process discovery from sensor data. In: 10th International Workshop on Enterprise Modeling and Information Systems Architectures. CEUR Workshop Proceedings, vol. 2628, pp. 32–38 (2020)
59. Kozma, D., Varga, P., Larrinaga, F.: Data-driven workflow management by utilising BPMN and CPN in IIoT systems with the arrowhead framework. In: 24th IEEE International Conference on Emerging Technologies and Factory Automation, ETFA 2019, pp. 385–392 (2019)
60. Kunz, S., Fabian, B., Ziekow, H., Bade, D.: From smart objects to smarter workflows—an architectural approach. In: 2011 IEEE 15th International Enterprise Distributed Object Computing Conference Workshops, pp. 194–203 (2011)
61. Leotta, F., Marrella, A., Mecella, M.: IoT for bpmers. challenges, case studies and successful applications. In: 17th International Conference on Business Process Management (BPM), pp. 16–22 (2019)
62. Li, Q., Wang, J., Chen, G., Yuan, Y., Ye, S.: Business-oriented IoT-aware process modeling and execution. In: 18th International Conference on Web Information Systems and Applications (WISA). Lecture Notes in Computer Science, vol. 12999, pp. 747–755 (2021)
63. Loke, S., Ling, S., Indrawan, M., Kurniati, S.: Adaptive workflows for smart devices - a concrete approach towards device failures. In: Ninth International Conference on Enterprise Information Systems, (ICEIS), pp. 191–197 (2007)
64. Maamar, Z., Sellami, M., Faci, N., Ugljanin, E., Sheng, Q.Z.: Storytelling integration of the internet of things into business processes. In: International Conference on Business Process Management (BPM)—BPM Forum, pp. 127–142 (2018)
65. Maamar, Z., Kajan, E., Guidara, I., Moctar-M'Baba, L., Sellami, M.: Bridging the gap between business processes and IoT. In: 24th Symposium on International Database Engineering & Applications, IDEAS '20 (2020)
66. Malburg, L., Seiger, R., Bergmann, R., Weber, B.: Using physical factory simulation models for business process management research. In: Business Process Management Workshops—(BPM) 2020 International Workshops. Lecture Notes in Business Information Processing, vol. 397, pp. 95–107 (2020)
67. Marrella, A., Mecella, M.: Adaptive process management in cyber-physical domains. Adv. Intell. Process.-Aware Inf. Syst. Concepts Methods Technol. **123**, 15 (2017)
68. Marrella, A., Mecella, M., Russo, A.: Collaboration on-the-field: Suggestions and beyond. In: 8th International Conference on Information Systems for Crisis Response and Management, ISCRAM 2011 (2011)
69. Martins, F., Domingos, D.: Modelling IoT behaviour within BPMN business processes. Procedia Comput. Sci. **121**, 1014–1022 (2017)
70. Martins, F., Domingos, D., Vitoriano, D.: A graph pattern based approach for automatic decomposition of IoT aware business processes. In: 21st International Conference on Enterprise Information Systems, (ICEIS). Lecture Notes in Business Information Processing, vol. 378, pp. 498–513 (2020)
71. Mass, J., Chang, C., Srirama, S.N.: Wiseware: A device-to-device-based business process management system for industrial internet of things. In: 2016 IEEE International Conference on Internet of Things (iThings) and IEEE Green Computing and Communications (GreenCom) and IEEE Cyber, Physical and Social Computing (CPSCom) and IEEE Smart Data (SmartData), pp. 269–275 (2016)

72. Meroni, G., Baresi, L., Montali, M., Plebani, P.: Multi-party business process compliance monitoring through IoT-enabled artifacts. Inf. Syst. **73**, 61–78 (2018)
73. Meyer, S., Sperner, K., Magerkurth, C., Pasquier, J.: Towards modeling real-world aware business processes. In: Second International Workshop on Web of Things, pp. 1–6 (2011)
74. Meyer, S., Ruppen, A., Magerkurth, C.: Internet of things-aware process modeling: Integrating IoT devices as business process resources. In: 25th International Conference on Advanced Information Systems Engineering (CAiSE), pp. 84–98 (2013)
75. Meyer, S., Ruppen, A., Hilty, L.: The things of the internet of things in BPMN. In: Advanced Information Systems Engineering Workshops—CAiSE 2015 International Workshops. Lecture Notes in Business Information Processing, vol. 215, pp. 285–297 (2015)
76. Montali, M., Plebani, P.: IoT-based compliance checking of multi-party business processes modeled with commitments. In: Service-Oriented and Cloud Computing, pp. 179–195 (2017)
77. Monti, F., Mathew, J.G., Leotta, F., Koschmider, A., Mecella, M.: On the application of process management and process mining to industry 4.0. Softw. Syst. Model. **23**, 1–13 (2024)
78. Mottola, L., Picco, G.P., Oppermann, F.J., Eriksson, J., Finne, N., Fuchs, H., Gaglione, A., Karnouskos, S., Montero, P.M., Oertel, N., Römer, K., Spiess, P., Tranquillini, S., Voigt, T.: makeSense: simplifying the integration of wireless sensor networks into business processes. IEEE Trans. Softw. Eng. **45**(6), 576–596 (2019)
79. Muhsin, Z.J., Al-Taee, A., Al-Taee, M.A., Al-Nuaimy, W., Al-Ataby, A.: Mobile workflow management system based on the internet of things. In: 2016 13th International Multi-Conference on Systems, Signals Devices (SSD), pp. 436–441 (2016)
80. Park, M., Kim, H., Ahn, H., Kim, K.P.: A process-aware IoT application execution environment design. In: 2018 20th International Conference on Advanced Communication Technology (ICACT), pp. 724–727 (2018)
81. Pastor, F.J.F., Horcajadas, M.P., Davo, J.A.L., Iglesias, V.G.: Developing environmental adaptative comfort using internet of things and business process management: Application in a university building. In: International Conference on Ubiquitous Computing and Ambient Intelligence (UCAmI), pp. 535–546 (2022)
82. Pryss, R., Reichert, M., Bachmeier, A., Albach, J.: BPM to go: Supporting business processes in a mobile and sensing world. In: BPM Everywhere: Internet of Things, Process of Everything, pp. 167–182 (2015)
83. Ruiz-Fernández, D., Marcos-Jorquera, D., Gilart-Iglesias, V., Vives-Boix, V., Ramírez-Navarro, J.: Empowerment of patients with hypertension through BPM, IoT and remote sensing. Sensors **17**(10), 2273 (2017)
84. Ruppen, A., Meyer, S.: An approach for a mutual integration of the web of things with business processes. In: Workshop on Enterprise and Organizational Modeling and Simulation, pp. 42–56 (2013)
85. Schief, M., Kuhn, C., Rösch, P., Stoitsev, T.: Enabling business process integration of IoT-events to the benefit of sustainable logistics. In: International Conference on Logistics, Informatics and Service Science (LISS), pp. 40–49 (2011)
86. Schönig, S., Ackermann, L., Jablonski, S.: Internet of things meets BPM: A conceptual integration framework. In: SIMULTECH, pp. 307–314 (2018)
87. Schönig, S., Ackermann, L., Jablonski, S., Ermer, A.: IoT meets bpm: a bidirectional communication architecture for IoT-aware process execution. Softw. Syst. Model **19**(6), 1443–1459 (2020)
88. Seiger, R., Assmann, U., Huber, S.: A case study for workflow-based automation in the internet of things. In: 2018 IEEE International Conference on Software Architecture Companion (ICSA-C), pp. 11–18 (2018)
89. Seiger, R., Huber, S., Heisig, P., Assmann, U.: Toward a framework for self-adaptive workflows in cyber-physical systems. Softw. Syst. Model. **18**(2), 1117–1134 (2019)
90. Seiger, R., Zerbato, F., Burattin, A., García-Bañuelos, L., Weber, B.: Towards IoT-driven process event log generation for conformance checking in smart factories. In: IEEE 24th International Enterprise Distributed Object Computing Workshop (EDOCW), pp. 20–26 (2020)

91. Seiger, R., Kühn, R., Korzetz, M., Aßmann, U.: HoloFlows: modelling of processes for the internet of things in mixed reality. Softw. Syst. Model. **20**(5), 1465–1489 (2021)
92. Seiger, R., Franceschetti, M., Weber, B.: An interactive method for detection of process activity executions from iot data. Futur. Int. **15**(2), 77 (2023)
93. Senderovich, A., Rogge-Solti, A., Gal, A., Mendling, J., Mandelbaum, A.: The ROAD from sensor data to process instances via interaction mining. In: 28th International Conference on Advanced Information Systems Engineering (CAiSE), pp. 257–273 (2016)
94. Shamsuzzoha, A., Abels, S., Kuspert, S., Helo, P.: Smart collaborative processes monitoring in real-time business environment. In: 16th International Conference on Enterprise Information Systems (ICEIS) – Volume 2, ICEIS 2014, pp. 556–563 (2014)
95. Song, R., Cui, W., Vanthienen, J., Huang, L., Wang, Y.: Business process redesign towards IoT-enabled context-awareness: the case of a chinese bulk port. Bus. Process. Manag. J. **28**(3), 656–683 (2022)
96. Sora, D., Leotta, F., Mecella, M.: An habit is a process: A BPM-based approach for smart spaces. In: Business Process Management Workshops - BPM 2017 International Workshops. Lecture Notes in Business Information Processing, vol. 308, pp. 298–309 (2017)
97. Stoiber, C., Schönig, S.: Event-driven business process management enhancing IoT—a systematic literature review and development of research agendas. In: International Conference on Wirtschaftsinformatik, Innovation Through Information Systems, pp. 645–661 (2021)
98. Suri, K., Gaaloul, W., Cuccuru, A., Gerard, S.: Semantic framework for internet of things-aware business process development. In: 2017 IEEE 26th International Conference on Enabling Technologies: Infrastructure for Collaborative Enterprises (WETICE), pp. 214–219 (2017)
99. Suri, K., Gaaloul, W., Cuccuru, A.: Configurable IoT-aware allocation in business processes. In: International Conference on Services Computing, pp. 119–136 (2018)
100. Tôn, L.P., Lê, L.S.: Enacting a rule-based alert business process in smart healthcare using IoT wearables. In: 2019 IEEE 23rd International Enterprise Distributed Object Computing Workshop (EDOCW), pp. 104–107 (2019)
101. Torres, V., Serral, E., Valderas, P., Pelechano, V., Grefen, P.: Modeling of IoT devices in business processes: A systematic mapping study. In: 2020 IEEE 22nd Conference on Business Informatics (CBI), vol. 1, pp. 221–230 (2020)
102. Ugljanin, E., Stojanović, D., Kajan, E., Maamar, Z.: Re-engineering of smart city's business processes based on social networks and internet of things. Facta Universitatis Ser. Autom. Control Rob. **16**, 275 (2018)
103. Valderas, P., Torres, V., Serral, E.: Modelling and executing IoT-enhanced business processes through BPMN and microservices. J. Syst. Softw. **184**, 111139 (2022)
104. Valderas, P., Torres, V., Serral, E.: Towards an interdisciplinary development of IoT-enhanced business processes. Bus. Inf. Syst. Eng. **65**(1), 25–48 (2023)
105. van Eck, M.L., Sidorova, N., van der Aalst, W.M.P.: Enabling process mining on sensor data from smart products. In: 2016 IEEE Tenth International Conference on Research Challenges in Information Science (RCIS), pp. 1–12 (2016)
106. Varga, P., Kozma, D., Hegedús, C.: Data-driven workflow execution in service oriented IoT architectures. In: 2018 IEEE 23rd International Conference on Emerging Technologies and Factory Automation (ETFA), vol. 1, pp. 203–210 (2018)
107. Vitali, M., Pernici, B.: Interconnecting processes through IoT in a health-care scenario. In: 2016 IEEE International Smart Cities Conference (ISC2), pp. 1–6 (2016)
108. Wang, G., Fang, J., Wang, J., Yu, J., Zhang, L., Han, Y.: Service-based event penetration from IoT sensors to businesses: A case study. In: 2022 International Conference on Service Science (ICSS), pp. 72–79 (2022)
109. Wehlitz, R., Rößner, I., Franczyk, B.: Integrating smart devices as business process resources—concept and software prototype. In: International Conference on Service-Oriented Computing (ICSOC), pp. 252–257 (2017)
110. Weske, M.: Business Process Management. Concepts, Languages, Architectures. Springer, Berlin (2019)

111. Wieland, M., Kaczmarczyk, P., Nicklas, D.: Context integration for smart workflows. In: 2008 Sixth Annual IEEE International Conference on Pervasive Computing and Communications (PerCom), pp. 239–242 (2008)
112. Wombacher, A.: How physical objects and business workflows can be correlated. In: 2011 IEEE International Conference on Services Computing (SCC), pp. 226–233 (2011)
113. Xing, Z., Hong, Z., Yulong, L.: A petri-net based context-aware workflow system for smart home. In: 2012 IEEE 26th International Parallel and Distributed Processing Symposium Workshops PhD Forum, pp. 2336–2342 (2012)
114. Zanfack, D.G., Nait-Sidi-Moh, A., Durand, D., Fortin, J.: Modeling business entities and physical workflow in the context of internet of things: The GSM-IoT framework. In: 2015 Third World Conference on Complex Systems (WCCS), pp. 1–7 (2015)
115. Zhu, L., Cai, H., Jiang, L.: Minson: A business process self-adaptive framework for smart office based on multi-agent. In: 2014 IEEE 11th International Conference on e-Business Engineering (ICEBE), pp. 31–37 (2014)

From Internet of Things Data to Business Processes: Challenges and a Framework

Jürgen Mangler, Ronny Seiger, Janik-Vasily Benzin, Joscha Grüger, Yusuf Kirikkayis, Florian Gallik, Lukas Malburg, Matthias Ehrendorfer, Yannis Bertrand, Marco Franceschetti, Barbara Weber, Stefanie Rinderle-Ma, Ralph Bergmann, Estefanía Serral Asensio, and Manfred Reichert

1 Introduction

In IoT environments, large amounts of procedural data are generated from IoT devices, information systems, and other software applications. The use of this data can foster the development of innovative applications in process control [35, 43, 52, 54, 56, 63, 68, 75], process conformance checking [23, 28, 81, 83], and process enhancement [59, 67], among others. Particularly, the use of process mining techniques to analyze not only process data but also IoT-collected data could provide

J. Mangler (✉) · J.-V. Benzin · M. Ehrendorfer · S. Rinderle-Ma
Department of Informatics, Technical University of Munich, Garching, Germany
e-mail: juergen.mangler@tum.de; janik.benzin@tum.de; matthias.ehrendorfer@tum.de; stefanie.rinderle-ma@tum.de

R. Seiger (✉) · M. Franceschetti · B. Weber
Institute of Computer Science, University of St. Gallen, St. Gallen, Switzerland
e-mail: ronny.seiger@unisg.ch; marco.franceschetti@unisg.ch; barbara.weber@unisg.ch

J. Grüger · L. Malburg · R. Bergmann
Business Information Systems II, University of Trier, Trier, Germany

German Research Center for Artificial Intelligence (DFKI), Branch University of Trier, Trier, Germany
e-mail: grueger@uni-trier.de; joscha.grueger@dfki.de; malburgl@uni-trier.de; lukas.malburg@dfki.de; bergmann@uni-trier.de; ralph.bergmann@dfki.de

Y. Kirikkayis · F. Gallik · M. Reichert
Institute of Databases and Information Systems, Ulm University, Ulm, Germany
e-mail: yusuf.kirikkayis@uni-ulm.de; florian-1.gallik@uni-ulm.de; manfred.reichert@uni-ulm.de

Y. Bertrand · E. S. Asensio
Research Centre for Information Systems Engineering (LIRIS), KU Leuven, Leuven, Belgium
e-mail: yannis.bertrand@kuleuven.be; estefania.serralasensio@kuleuven.be

G. Fortino, M. Mecella (eds.), *Internet of Things Meets Business Process Management*, Internet of Things, https://doi.org/10.1007/978-3-031-90746-3_2

important insights into processes and interactions as shown in different applications in the manufacturing domain, such as [56, 58, 59, 67, 75]. In these applications, IoT actuators are used to realize and execute process activities, while IoT sensors and smart tags are used to closely monitor the execution environment and involved resources [26, 37, 54, 75, 79]. IoT technology can therefore capture the context in which certain process tasks are performed, allowing process mining techniques to better understand and analyze the processes [7, 12, 76].

As such, besides the procedural data generated from the process execution systems, the data captured by IoT should also be considered an integral part of the process execution in the form of *IoT-enriched event logs* [53, 57]. Both the procedural nature of sensor logs and the tight integration of these with the process executions and the executing resources [24] makes sensor data an integral part of process-based application scenarios in IoT [7, 75, 76].

However, the integration of IoT data and process data to be used for process mining is still often done ex post in a manual fashion during a separate preprocessing phase [53, 73, 95]. In these cases, the data from the IoT environment is still collected and stored separately, and only later it is explicitly connected to the notion of a process, which is nontrivial as pointed out in the challenge "Bridging the Gap Between Event-based and Process-based Systems" in the BPM-IoT manifesto [37]. In case no explicit process orchestration or monitoring happens via a BPM system, and thus no explicit process data is available [30], the IoT data is the sole provider of process-related data, which has to be separately identified, extracted, abstracted, and correlated [19, 73, 76].

We define the following two extreme states of a spectrum that IoT data sets may be in regarding their inclusion of process-related data (called **process awareness**), with certain levels of process awareness in between:

- An IoT data set (log) exhibits **full process awareness**, when all relations between sensor events and process events are known.
- An IoT data set (log) exhibits **no process awareness**, when it only contains raw sensor events and no process events at all.

In between *full process awareness* and *no process awareness*, IoT data sets may exhibit some known correlations with the process executions, e.g., the type of a process activity that a specific sensor event belongs to might be known, but not the specific instance, or the correlation of a process-level event identified from the sensor events with a specific process instance might be unknown [95].

In this work, we propose a framework to transition IoT data sets from **no process awareness** to **full process awareness**, in a structured, non-exploratory fashion. The contribution of the chapter is threefold:

- We highlight relevant challenges regarding the automatic derivation and connection of process data with IoT data to allow for a better subsumption of future work in this area of research.
- We propose a generic framework to address the identified challenges and to derive process data from IoT data step-by-step to facilitate process mining.

- We showcase the framework using a real-life example from smart manufacturing.

This work is structured as follows: Sect. 2 introduces the notion of *process awareness* in detail. Section 3 discusses related work. Section 4 presents challenges regarding the connection of process and IoT data. Section 5 presents our framework to (semi)automatically increase the level of process awareness in IoT data sets with a real-world example from smart manufacturing. Section 6 concludes the chapter and shows opportunities and starting points for future work.

2 IoT Data and Process Awareness

This work aims at bringing the world of BPM together with the world of IoT and cyber-physical systems (CPS). The conceptual model linking the worlds of BPM (highlighted in grey) with IoT/CPS that we base our work on is depicted in Fig. 1. A *business process* contains *events*, *decision points*, and *activities* [20]. It involves *actors* playing an active part in the process and *objects*. In the context of our work, components of a CPS (e.g., a production machine) can also be actors that are able to execute process activities, which in turn can be observed through *sensor events*.

IoT data is typically generated by a set of sensors that measure a set of values emitted as *sensor events* [4]. These sensors can be standalone or part of a larger CPS component (e.g., a production machine or production cell) resulting in a particular *sensor topology* for a CPS. Actuators interact with the physical world and manipulate its state. As actuators may also produce data regarding their current state and actions, we consider data produced from actuators as relevant *sensor events* for our analysis, too. Figure 2 shows a Fischertechnik model of a smart factory (i.e., an implementation for simulation purposes) that we use for research.[1] It represents

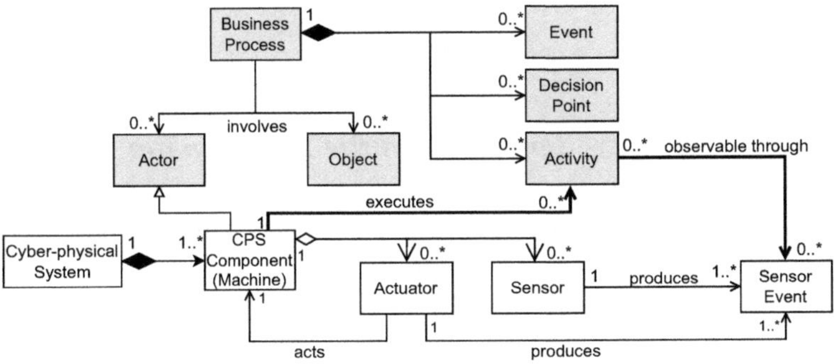

Fig. 1 Relation between concepts from BPM (in grey) and CPS/IoT. (Adopted from [73])

[1] https://www.fischertechnik.de/en/products/industry-and-universities/training-models, last access: 2024-05-13.

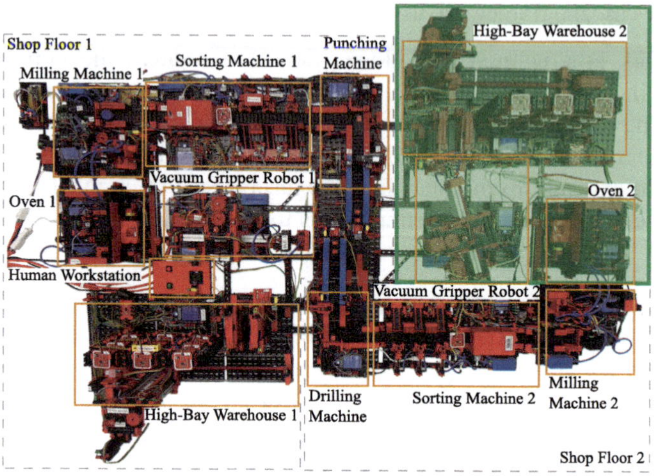

Fig. 2 Factory model representing a CPS with highlighted CPS components [56]

a CPS consisting of several (highlighted) CPS components that again consist of a multitude of sensors and actuators. Sensor events may also be directly related with actors and objects. However, we put process-related concepts in the center of our investigations, which is why the key relationship for our work to bridge the BPM and IoT worlds is the *observable through* association between *activity* and *sensor event* in Fig. 1.

The sensor events are typically either directly pushed to some IoT data log by the sensor or collected and stored in an IoT data log by an external mechanism. Thus, a set of IoT data logs, given some data retention policies, holds historical information regarding all timestamped observations from a set of sensors. For each sensor, the stream of events (called *raw sensor stream* in this work) is available as a time series [16].

2.1 Process Activity–Sensor–Machine Relationship

When associating data collected from IoT with process data to achieve full process awareness, one has to consider that sensors can be related to process activities without explicitly knowing the corresponding actions, i.e., which machine or human triggered a certain chain of events. This is because one activity might trigger, e.g., the start of a machine, while other activities might fine-tune the parametrization of the machine at runtime. A third activity might then wait for an event signifying the end of the operation. This also means that one sensor at any given time might produce data relevant to many activities, while one activity might be enacted in the context of many sensors (cf. Fig. 1). The same holds true for machines: a sensor might at any given time produce data relevant to the operation of many machines (e.g., a temperature sensor might deliver data relevant to the operation of multiple machines), while one machine might operate in the context of multiple sensors [42, 45, 59, 73].

The reason for collecting IoT data typically is to monitor the effects of a series of actuations, e.g., as part of a production process, the treatment of a patient, or as part of a building automation scenario. The collected data can be used to [18, 37, 44, 54, 76, 84]:

- Monitor, visualize, and track progress
- Find deviations from rules
- Find drifts in relation to goal functions

While these analysis targets are very generic and universally applicable, they can only be reached with detailed knowledge of the scenario, its rules, goal functions, and the underlying **business process** [20].

All IoT scenarios have in common a series of actuations, potentially influenced by humans and the environment, that are based on a blueprint that defines when something should happen and which order relations might exist between actuations. This blueprint, the **process** [20, 37], might either be:

- Implicit, i.e., the result of the interactions of some hard-coded software components with potential human involvement, or
- Explicit, i.e., a known artifact (e.g., the **process model**), which is the basis for enacting a certain scenario, e.g., through a process engine

The relationship between IoT data and processes can be understood as IoT data tracking changes in the observed environment while processes and activities are executed and thus allow to observe their effects (cf. Fig. 1 [4]).

Albeit, some caveats to complicate the situation exist:

- Sensors, e.g., temperature sensors, might collect data while no scenario is enacted. Thus, data might not be relevant for the process [73].
- Not all sensor data can be directly assigned to a specific scenario; only the combination of a set of individual observations from different IoT data logs (e.g., temperature sensors combined with the status of an oven) might allow identifying significance for a particular scenario [76].
- As IoT data from different sources is available in different granularity, substantial effort for aggregation of the data has to be made [76, 80].

Consequently, as briefly introduced in Sect. 1, we define **no process awareness** as a situation where only raw sensor streams from a set of data sources are available (cf. Fig. 3) [73]. **full process awareness**, on the other hand, **is the goal** as it represents the full contextualization of all data regarding a certain scenario, which is represented explicitly by a set of process models [59]. Each event in each available sensor stream time series has been either discarded due to irrelevance or assigned to an observation of an actuation. Each actuation can be related to many process-level events, and these events might be related to multiple process activities [31, 73].

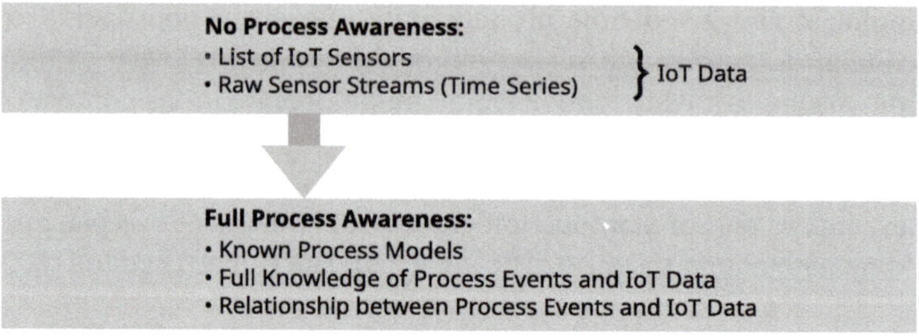

Fig. 3 Relation between IoT data and process awareness

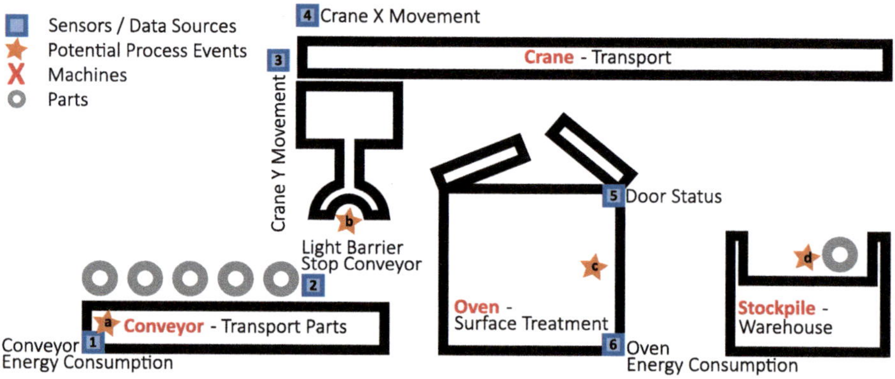

Fig. 4 Running example, part of factory simulation model shown in Fig. 2

For the remainder of this section, we further explore the two extreme states of the process awareness spectrum with different levels in between.

2.2 *Example: Smart Manufacturing Process*

In order to better exemplify the concepts discussed in this section, and following sections, for the remainder of this work we will focus on a specific slice of the scenario displayed in Fig. 2 (top right area, green box). This **running example** is displayed in simplified form in Fig. 4 and describes the interaction between a **Conveyor** which transports raw **parts** to a **Crane** ("Vacuum Gripper Robot 2"), which can move the parts to an **Oven** ("Oven 2") for surface treatment, and subsequently to a **Stockpile** ("High-Bay Warehouse 2").

Our running example utilizes a total of six different sensors, respectively data sources, which deliver the following values:

[1] is a continuous sensor that emits the energy consumption levels of the **Conveyor** in watts every second. When the **Conveyor** is not running, the sensor emits the value (close to) 0 every second.

[2] is a binary sensor that switches from 0 to 1 whenever a **part** is in its field. It is also used by the internal programming of the factory to start/stop the **Conveyor** to ensure that whenever a **part** is taken by the **Crane**, a new part is moved into a position where it can be picked up.

[3] is a continuous sensor that monitors the y position of the **Crane**, i.e., how far the gripper is extended downwards. It delivers values every second whenever the position changes, and it delivers no values, when the position is unchanged. Fully extended the y value for the **Crane** is 100, and fully retracted the value is 0.

[4] is a continuous sensor that monitors the x position of the **Crane**, i.e., in which position above the **Conveyor**, **Oven**, or **Stockpile** the crane is. It delivers values every second whenever the position changes, it delivers no values, when the position is unchanged. When the **Crane** is above the **Conveyor** the value is 0, when it is above the **Oven** it is 30, and when it is above the **Stockpile** it is 60.

[5] is a binary sensor that monitors if the door of the **Oven** is opened (1) or closed (0).

[6] is a continuous sensor that emits the energy consumption levels of the **Oven** in watts every second. When the **Oven** is not running, the sensor emits the value (close to) 0 every second. As the **Oven** is always kept at temperature, positive energy consumption does not necessarily mean that a part is currently processed.

Between all of these sensor values, a wide range of process events could be specified, leading to more or less fine-granular process models that describe the interaction between the **Conveyor**, the **Crane**, the **Oven**, and the **Stockpile**. Potential process events that we want to highlight for our running example are:

(a) The **Conveyor** moves a **part** into position, which can be deducted from [1].

(b) The **Crane** grabs a **part**, which can be deducted from certain values of [2], [3], and [4].

(c) A **part** is surface treated in the **Oven**. If [3], [4] have certain values, and [5] is 1, then the oven is currently being loaded. If [3], [4] have certain values, and [5] is 0, then the oven is processing a **part**.

(d) The **Crane** drops a **part** into the **Stockpile**, which can be deducted from certain values of [3] and [4].

2.3 *No Process Awareness*

This state of process awareness typically occurs in brownfield scenarios. IoT data collected from existing complex IoT environments is partially analyzed regarding certain domain-specific preordained, non-process-related questions. Whenever new

Table 1 Excerpt from an IoT data log with no process awareness

	Sensor ID	Sensor value	Sensor timestamp	Case	Process event ID	Event timestamp	Lifecycle transition	Label
1	1	512	1	–	–	–	–	–
2	2	0	1	–	–	–	–	–
3	1	0	2	–	–	–	–	–
4	2	1	2	–	–	–	–	–
5	3	100	3	–	–	–	–	–
6	4	0	3	–	–	–	–	–
7	...	...	4	–	–	–	–	–

questions arise, the data aggregation is refined to derive the required answers. This state is typically prone to breakages as the following properties might hold true:

- Unknown side effects: During the operation of the IoT environment, actuations might occur which are not foreseen in the data analysis. The software steering the actuations or the human's or environment's influence on the process might not be fully understood [62, 74].
- Breaking changes: Such side effects might come into existence whenever changes to the IoT environment are made. Changes not only include the addition of new actuators or sensors, but also the (software) upgrade of existing ones [1].

Consequentially, all changes have to be tracked and monitored, and all data analysis tasks have to be rechecked if they still deliver the required answers, or have to be adapted to do so. We can sum up the properties of this state as follows, while not all of them must apply to all scenarios [5, 29, 64, 73, 75, 76, 80, 97]:

- Hard-coded software
- No or only partial knowledge of the process
- Event correlation ambiguities: no clear assignment of events to individual actuations
- Partial monitoring and tracking of data
- Partial knowledge about the dependencies between sensors (i.e., sensor topology) and how they influence different scenarios
- Incomplete data, i.e., sensors might be missing to monitor relevant aspects of the process executions
- Inconsistent data resulting from different data granularities of different sensors

Table 1 shows an excerpt from a typical IoT data log referring to the example scenario described in Sect. 2.2 which only contains sensor readings with a minimal set of sensor attributes, sensor value, and timestamp of the sample. There is no process-related information available.

2.4 *Full Process Awareness*

This state of process awareness typically occurs in greenfield scenarios, when new IoT environments are planned and realized. Furthermore, it might occur in brownfield scenarios, when process management software is introduced as part of an automation effort, such as [59, 71]. Full process awareness is often characterized by [18, 31, 53, 56, 59, 75]:

- Full knowledge of the relevant processes in the form of a set of process models.
- Imperative control or at least full monitoring through a process engine.
- Sensors which are linked to process activities, i.e., sensors might contribute data to one or more activities.
- Complete data: if data is missing, the process is not specified well enough.
- Logs, which are automatically generated as part of the execution and containing events and process data at a suitable, homogeneous level of granularity [97]; complemented by IoT data [57].

In this case, data stores for individual sensors might still exist, but they are no longer used because all data is collected as part of the execution of process instances stored in an IoT-enriched process event log [31]. Table 2 contains the same sensor readings as Table 1, but these are now fully contextualized in the control flow of the process execution (cf. scenario process in Sect. 2.2) as indicated by the additional process attributes according to the XES standard [93]. In this example, the sensor events are clearly related to specific process events for one case, i.e., an executed process instance, namely, the workpiece "part 1." Furthermore, the following important concepts are exemplified in Table 2:

- One or multiple sensors and thus one or multiple sensor events might contribute to one process event.
- Row 1 and row 2 describe different events in the lifecycle of the same process task (start/complete of *move part*). Multiple combinations of sensor events might point to the same process task, i.e., in our example **1** is sufficient for start, but the combination of **1** and **2** with different values point at the and also is more robust (i.e., more sensors involved means better distinction).

Thus, it becomes clear that a process event is defined by a combination of sensors and the specific values they have at a given point in time.

But how are the sensor values collected? We differentiate between three options of realizing the data collection.

2.4.1 Option 1: Explicit Modeling of Sensor Data Collection

This way is about modeling the process in the most fine-grained and detailed way possible, including the explicit collection of sensor data as process tasks (activities) that are part of the process model [59]. This has the advantage that:

Table 2 Excerpt from an IoT-enriched event log with full process awareness related to control flow

	Sensor ID	Sensor value	Sensor timestamp	Case	Process event ID	Event timestamp	Lifecycle transition	Label
1	1	512	1	part 1	a	1	start	move part
2	1 , 2	0, 1	2, 2	part 1	a	2	complete	move part
3	2 , 3 , 4	1, 100, 0	2, 3, 3	part 1	b	3	start	grab part
4	…	…	…	part 1	…	…	…	…

- Every process engine can support it.
- Models are better to follow as everything is made explicit in an already existing modeling language (e.g., BPMN 2.0).

On the downside, the resulting models are complex and complicated. Furthermore, the data collection might be hindered by the expressiveness of the modeling language, e.g., BPMN 2.0 requires complex synchronization modeling if data relevant to multiple in-parallel running tasks has to be collected in separate process instances [89]. While this downside can be alleviated by modeling the process in a more abstract way (fewer tasks, higher abstraction, "more knowledge" in lower levels), this requires increased effort to maintain a good modeling quality and decreases the level of detail of monitoring.

2.4.2 Option 2: Implicit IoT Data Collection

In this approach, the process model contains additional information/annotations that allow data collection implicitly, independent of the traditional process execution, but still automatically. For example, if process activities contain information about linked sensors and how to collect data from them, this data collection can be enacted automatically in parallel to the normal process execution [42, 54].

The main advantage is that the process models stay simple, but on the downside some additional knowledge describing the connection to the IoT environment (e.g., in the form of annotations [42, 54]) has to be provided.

2.4.3 Option 3: Raw IoT Data Collection

In this approach, the process notion is completely ignored, and data is collected directly from all sensors. This is the most common way of data collection. In contrast to the two other collection methods, where the data can be used for compliance checking and process refinement, here the goal is to discover and create a new process model. The downside is that for getting the correct granularity of the process models, some of the steps in the framework proposed in this chapter (cf. Sect. 5) become much more reliant on domain expert input and validation [73].

3 Related Work

Augmenting business processes with IoT technologies brings several benefits, but does not come without challenges [17, 37, 70]. In this section, we discuss related research addressing IoT data processing as well as event extraction and event abstraction [19] in the context of IoT with the goal of increasing the level of process awareness associated with IoT data and, ultimately, to enable process mining [2].

One of the main benefits is the improved recognition of activities and processes from sensor data, e.g., with the help of multimodal approaches such as the one by Rebmann et al. [69] that combines motion and vision sensors with user feedback for detecting and disambiguating known activity types/tasks in real time. Other approaches employ camera-based object detection to track workpieces within a smart factory and correlate them to the execution of individual process activities [55]. Activity recognition is often combined with process mining for extracting knowledge to analyze and optimize the underlying processes [14, 60]. Applying process mining, in turn, requires bridging the gap between low-level sensor data and the event logs needed for process mining—a well-known issue, which gives rise to succeeding challenges [37].

The data used for process mining can reside in different data sources. *Event extraction* includes the identification of the event data and its extraction [19]. To leverage the opportunities IoT brings for improved recognition of (manual) activities and processes from sensor data [80], differences in granularity at which the data is recorded and at which it shall be analyzed [19] need to be bridged, also referred to as *event abstraction*. In turn, *event correlation* aims at associating event data extracted from data sources to *cases* of a process [12]. An overview of non-IoT-related event abstraction techniques in the field of process mining is provided in a literature review by van Zelst [96]. A paper by Diba et al. [19] provides an overview of different methods for event abstraction and correlation in the traditional process mining context.

3.1 Event Extraction

Identifying relevant data for process mining and extracting it from different sources is part of *event extraction* [19]. Existing approaches use database schema, process documents, domain models, event models, and domain knowledge for the identification and extraction of relevant data (cf. [19, 38]) and mostly focus on the extraction of event data from databases and traditional information systems. Recently, several works looked into the use of IoT as data source for process mining (e.g., [2, 7, 21, 23, 48, 50, 84]). Similarly, in our work we are moving towards IoT sensors as new data sources for event extraction and process correlation [76], which requires the sensor streams to be accessible and linked to processes [13, 65].

The IoT provides new sources of access to real-time data and thus opportunities to develop advanced monitoring approaches, e.g., for condition monitoring, predictive maintenance [36], or verification of process outcomes [74]. Large amounts of data, different data formats, sampling rates, and data quality are among the challenges that come with using the IoT as a new data source [2, 17, 40]. With our framework, we focus on using IoT data for event extraction, abstraction, and correlation where we relate raw IoT event streams to the execution of process activities [39]. Most works assume rather high levels of process knowledge when discovering activities and processes from sensor data, i.e., existing activity labels only have to be connected to the raw events [19, 59]. With almost no process knowledge and limited IoT topology knowledge, most approaches are not applicable.

3.2 Event Abstraction

Event abstraction in the context of BPM focuses on the *abstraction gap* between the granularity at which the data is recorded and at which it is analyzed [5, 19, 37]. When considering sensor data, the challenge of mapping fine-grained sensor data to more abstract process activities becomes more pronounced [18, 37, 96]. In literature [96], various approaches exist to bridge this abstraction gap using, e.g., complex event processing (CEP) [72, 76, 80] or machine learning (ML) [27, 39, 88].

Event abstraction is needed to map multiple fine-grained sensor data to coarse-grained process events [18, 96], e.g., with the help of CEP [80], which also supports data preprocessing and context enrichment [87], or with machine learning approaches such as clustering [18, 25, 27] or supervised learning [88]. Wanner et al. combine these two methods based on expert knowledge and observations in the context of a smart factory [94]. Then, to be suitable for the application of process mining techniques, such coarse-granular events need to be matched to the process activities in a process model or correlated to the process instances through *event-to-activity mappings* [3, 91]. We discuss work about the development of frameworks for IoT processes and event log generation from IoT events that mostly relates to our contribution, and refer to [2, 17, 80, 96] for a broader literature overview.

In [77], Senderovich et al. propose a knowledge-driven approach based on interactions mining to transform historical logs of sensor data into standardized event logs including information about process instances. In [61], Mannhardt et al. propose a supervised method for event abstraction based on activity patterns. These patterns are derived from low-level events using domain knowledge and then used to build an abstracted event log by exploiting existing alignment techniques. The authors remark the challenges stemming from the distributed, continuous, and ambiguous nature of IoT data, which makes event abstraction a necessary step for event log creation. Moreover, Koschmider et al. [48] discuss a modular framework for obtaining process event logs from sensor data, explicitly considering the data aggregations needed for event correlation, activity discovery, and event abstraction. The lack of explicit process information in the low-level IoT data may lead to *uncertainties* when being correlated to specific cases, as patterns within IoT data can often be associated with multiple activities and processes [48]. Ehrendorfer et al. present in [23] an approach for conformance checking and classification of manufacturing log data from a smart factory, thereby raising the analysis of low-level IoT data to a process-oriented perspective.

3.3 Conclusions from Related Work

From our investigations of related work at the intersection of BPM and IoT, we can conclude that the application of IoT technology in the domain of BPM has received increasing attention over the past years [17, 37]. Several approaches investigate aspects regarding IoT as new data source for enriching process event logs and for supporting process mining [2]. Our discussions in Sect. 2 have shown that detecting process-level events and activities from raw IoT data is a multistep process. Related work either focuses on using IoT data in early stages of activity detection (e.g., event extraction), on using specific sensors to detect specific types of activities independent of the process context, or on parts of the event abstraction [18]. In later stages of event and activity detection (e.g., event abstraction, event correlation, process mining), IoT data plays a less significant role. The goal of this work is to derive a list of challenges that have to be addressed when going the entire way to increase the level of process awareness of IoT data as defined in Sect. 2 from *no process awareness* to *full process awareness*. From these challenges, which are backed up by additional related work and literature, we intend to develop a holistic framework consisting of a generally applicable sequence of steps that address these challenges in order to increase process awareness of IoT data. Thereby, the framework should serve as a skeleton defining generic, abstract steps, and their expected inputs and outputs. The particular techniques and approaches discussed in the aforementioned related work can then be plugged in as concrete mechanisms to implement one or multiple steps of the framework.

4 Challenges

We have identified the following challenges regarding the automatic connection and derivation of process data and IoT data. These challenges are based on aforementioned related work and additional literature [8, 11, 37, 85] as well as on our own experience from implementing real-world use cases regarding the integration of BPM and IoT data in smart factories [42, 44, 56, 59, 75], smart healthcare [28, 82], and smart homes [71, 74].

The most relevant basis for this work is the BPM-IoT manifesto [37], which introduces 16 high-level challenges related to the interaction between the IoT and BPM. The manifesto presents these challenges along a similar spectrum as we have introduced in Sect. 2, going from rather low-level raw event data to higher-level process-related knowledge. We provide a more in-depth view on some specific challenges related to IoT and (business) process data posed in the manifesto. Specifically, we address aspects referring to (1) *placing sensors in a process-aware way*, (2) *connection of analytical processes with the IoT*, (3) *integrating the IoT with process correctness checks*, and most prominently (4) *bridging the gap between event-based and process-based systems* [37]. With that, we cover all major challenges that are related to collecting and analyzing raw event data from sensors and actuators with the goal of deriving higher-level knowledge. This knowledge is then used for tasks in the BPM context, e.g, the discovery of new processes, prediction and adaptation of processes, or enacting responses [37].

We have identified the following list of 10 challenges:

C0 Data delivered by IoT sensors is often incompletely described, e.g., the measurement procedure affecting the granularity and the units are missing, and sensors are not aware of their location [33, 78]. Moreover, missing knowledge about the granularity of sensors hinders the combination and aggregation of different sensors, which is crucial for conducting process mining on IoT data [8].

C1 Typically, IoT environments produce a plethora of data from heterogeneous sources [11], both regarding volume and variety [10], which can only be stored and analyzed with considerable computational effort [9, 40, 41].

C2 Data delivered by IoT sensors has different meaning in different scenarios, e.g., a certain temperature reading from one sensor might point towards an error in one scenario/process, while it signifies nominal operation for a second scenario/process [34, 42]. Thus, it is desirable that the implemented processes are stronger coupled with the corresponding data produced during execution (cf. Challenge 6 in [9]).

C3 Given a scenario and an IoT environment, the complete set of sensors and sensor information required for process analysis cannot be easily derived [11, 45], which is mainly due to the missing integration of process and IoT data within a common event log format or model [8, 31]. This results in potential ambiguities regarding the process–IoT data correlation [8, 29, 61, 76].

C4 Different sensors are relevant for different scenarios, both regarding if they are required at all and regarding their significance to the scenario/context [34, 51, 73, 90]. In addition, specific IoT data can be relevant in more than one scenario (cf. Challenge 3—*Scope of Relevance* in [8]).

C5 Data analysis is often dependent on visualization and human involvement [73, 95], which makes many of these interactive approaches infeasible, for example, for small lot sizes in manufacturing scenarios or for high-volume, data-intensive production processes [40]. This leads to data analysis being a highly demanding task and thus process refinements being difficult to achieve (cf. Challenge 9 in [9]).

C6 In semi-structured domains with high human involvement, it is challenging to get complete process/event and IoT data in a desirable and uniform granularity [5, 11, 28, 40]. Moreover, monitoring of human activities quickly raises privacy challenges [15] and is, in general, difficult to achieve [46].

C7 Refining existing process scenarios requires in-depth domain knowledge and is currently neither well-supported by a methodology nor by (semi)automatic decision support systems [37, 75, 90]. The automatic correlation of IoT data streams, process instances, and process models for the purpose of process refinement is currently not possible.

C8 In some scenarios, the explicit knowledge of processes might not be complete [11, 32, 37, 76]. Processes might be emerging and evolving [62, 74]. This makes predicting which IoT data is required for analysis difficult.

C9 Process outcome prediction, given explicitly known processes, is nontrivial given the high number of influencing IoT devices and data streams as well as physical surroundings (context), thus leading to emergent behavior [9, 37, 47].

With these 10 challenges we provide a comprehensive, though non-exhaustive, more detailed list and discussion—compared to the BPM-IoT manifesto [37]—of aspects that emerge when trying to collect and correlate BPM and IoT data to increase process awareness within the IoT. In the following section, we will propose a framework to address these challenges with the goal of moving from no process awareness to full process awareness within a given IoT data set or data stream.

5 Framework: From No Process Awareness to Full Process Awareness of IoT Data

In this section, we describe a data-driven process deduction framework consisting of eight steps to guide the analysis of a given IoT dataset from no process awareness to full process awareness in accordance with the definitions given in Sect. 2 and as depicted in Fig. 5. In the framework we provide a sequence of generally applicable steps and specify their interfaces including their expected inputs and outputs to increase process awareness of the datasets and intermediate artifacts. The goal is to (semi)automatically transform the raw IoT sensor streams and augment them with process models, knowledge of process events and IoT data, and their relationships. As a starting point, we assume that only the IoT data containing a list of IoT sensors and the respective raw sensor streams are given, i.e., there is no process awareness (cf. Sect. 2). For each step we provide a conceptual description and an instantiation of the framework based on the running example from smart manufacturing (cf. Sect. 2.2). We then discuss which of the challenges from Sect. 4 are addressed by the individual step. Note that the instantiations of the framework's individual steps are intended to show just one possible method or technique to achieve the desired outcome in this step, either relying on the domain expert or already on some (semi)automated means. Other methods or techniques may be possible to implement the specific steps, as well. The goal of the framework is to streamline the acquisition of process awareness in a series of generalized abstract steps with a high degree of automation potential. The conceptualization and implementation of all possible techniques and algorithms for each step is left for future investigations.

Throughout this section we will continue to use the example introduced in Sect. 2.2. As already mentioned, how the manufacturing process is executed/running is not important: it could be executed through a series of PLC programs that may

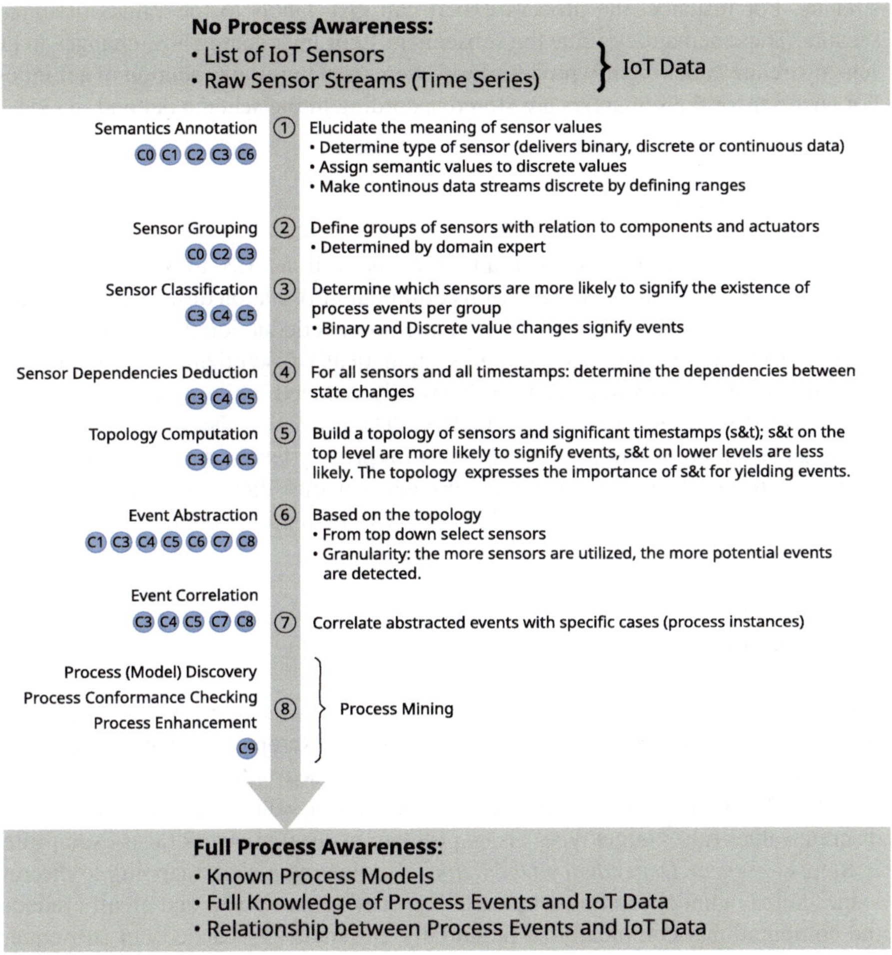

Fig. 5 Framework for (semi)automatic process deduction from IoT data, including references to challenges addressed by each step

or may not include the sensors, through a process engine enacting for an unknown process, or through a static Python program that controls all the machinery.

5.1 Semantic Annotation

5.1.1 Concept

In Step ① the domain expert elucidates semantics and characteristics of the sensors such as the type of the sensor values or relations after discretizing continuous streams. For instance, the process expert can give labels to the values of binary sensors. These semantics guide the subsequent steps in deducing how changes in the sensor stream values signify process-level events and thus the execution of activities. For each sensor the semantics are stored according to the schema defined in [31].

5.1.2 Challenges

Here we address challenges **C0** and **C2**, as well as **C3** to some extent, by enriching the set of sensor streams with additional information describing the sensor characteristics to enable the (semi)automatic process deduction through subsequent steps. For **C1**, we assume that through systematically discretizing continuous data, the amount of stored values can be drastically reduced while producing equal or better results with less computational effort. The privacy aspect in **C6** on the other hand can be supported by annotating data with privacy levels, which can be utilized further on to (a) always identify when privacy sensitive data is involved and (b) to regulate the visibility of such data. This step will be hard to conduct automatically, as it requires deep semantic knowledge.

5.1.3 Example

Table 3 is the result of the semantic annotations done by the domain expert for the set of 6 sensors from the smart factory introduced in the example in Fig. 4 (cf. Sect. 2.2). In our particular example we can see that all values typically have a semantic meaning. Not being able to map all continuous sensor values (i.e., “source type”) to discrete values (i.e., “target type”) is no problem in our framework (as is exemplified in Step 4—*Sensor Dependency Deduction*). But having a clear mapping to discrete values helps in understanding why certain process events occur and greatly reduces the computational complexity of potentially automatically carried out subsequent steps. From the example it also becomes clear that with a sufficiently trained AI or artificial general intelligence (AGI) this step could even be conducted automatically.

5.2 Sensor Grouping

5.2.1 Concept

In Step (2) the domain expert defines groups of sensors based on the setup of the physical CPS components including their sensors and actuators, which are henceforth also considered as *sensors* producing sensor events (cf. Fig. 1). This should be rather straightforward in traditional manufacturing scenarios (e.g., grouping for one production machine or one production cell) or medical settings as the grouping is usually done based on physical proximity of sensors and their belonging to one

Table 3 Sensor annotation, with the source type and target type, and the explanation of the values

					Value Mapping and Semantics				
ID	Component	Sensor	Source type	Target type	Source	→	Target	–	Semantics
1	Conveyor	Energy consumption	Continuous	Discrete	0	→	0	–	stopped
					1..1000	→	1	–	running
					>1000	→	2	–	stuck
2	Conveyor	Light barrier	Binary	Binary	0	→	0	–	unbroken
					1	→	1	–	triggered
3	Crane	Y movement	Continuous	Discrete	0	→	0	–	up
					100	→	1	–	down
					1..99	→	2	–	moving
4	Crane	X movement	Continuous	Discrete	0	→	0	–	conveyor pos
					30	→	1	–	oven pos
					60	→	2	–	stockpile pos
					1..29	→	3	–	moving
					31..59	→	3	–	moving
5	Oven	Door status	Binary	Binary	0	→	0	–	closed
					1	→	1	–	open
6	Oven	Energy consumption	Continuous	Discrete	0	→	0	–	off
					1..1000	→	1	–	heating
					>1000	→	2	–	error

specific CPS component (cf. Fig. 1). This grouping marks which sets of sensors are impacted by the same activities and should therefore be analyzed together to detect these activities. Groups that only contain one sensor may also exist.

5.2.2 Challenges

Here we address the challenges **C0**, **C2**, and **C3** in a similar way as in Step (1): by further adding information that guides the subsequent steps. Having a clear semantic relationship-based grouping of sensors will again help in increasing understandability of how process events could be deducted from sensor events, as process events are more likely to occur as a result of interactions *between* sensor groups (i.e., sensor groups are often equivalent to production cells).

5.2.3 Example

Sensor groups are formed by the domain expert through mostly considering the location of the sensors and the associated machine. For example, the sensor group *Buffer* contains different sensors which might encompass one production cell (cf. Table 4). For the purpose of this example we added an additional sensor 7,

Table 4 Grouping of sensors and assignment of new group identifiers

Sensor ID	Component	Sensor	Sensor group (G)	GID
1	Conveyor	Energy consumption	Buffer	B
2	Conveyor	Light barrier	Buffer	B
3	Crane	Y movement	Buffer	B
4	Crane	X movement	Buffer	B
7	Ambient	Temperature	Buffer	B
5	Oven	Door status	Oven	O
6	Oven	Energy consumption	Oven	O
3	Crane	Y movement	Stockpile	S
4	Crane	X movement	Stockpile	S
7	Ambient	Temperature	Stockpile	S

which is not part of the original running example (cf. Fig. 4). 7 measures the ambient temperature on the shop floor and in our case is considered important for the correct functioning of the conveyor and the crane. This exemplifies that:

- One sensor can be part of multiple groups
- Not all sensors must be part of or directly associated with a component

As can be seen, wherever groups (e.g., production cells) interact, sensors might be part of multiple logical groups (cf. ambient temperature, crane movement sensors in our example). Defining the correct granularity of the groups will not impact the overall result quality of subsequent analysis steps, but can contribute significantly to visualization and understandability of the results. For our running example we chose the conveyor (representing a buffer), the oven, and the stockpile as the logical groups, but other groups could also make sense. The oven group, for example, could also profit from containing values from crane sensors.

5.3 Sensor Classification

5.3.1 Concept

In Step (3) each sensor is ranked according to the significance regarding the existence of process-level events (e.g., the start or end of an activity) associated with it. Often, sensors with binary value changes are the most significant, followed by sensors with discrete value changes, followed by sensors with continuous value changes [73, 76].

While this ordering can be a good first choice, the classification should be confirmed by a domain expert or a more automated technique that analyzes sensor data for its relevance. The classification in this step is done per sensor

Table 5 Ranking of sensors per group (a lower ranking means a higher priority/significance of the sensor)

Sensor ID	GID	Sensor	Target type	Sensor ranking
1	B	Energy consumption	Discrete	2
2		Light barrier	Binary	1
3		Y movement	Discrete	2
4		X movement	Discrete	2
5	O	Door status	Binary	1
6		Energy consumption	Discrete	2
3		Y movement	Discrete	2
4		X movement	Discrete	2
3	S	Y movement	Discrete	1
4		X movement	Discrete	2

group, whereas inter-group relationships are analyzed in Step ④. The resulting classification ranks each sensor per group. A lower-ranking sensor (i.e., higher priority) is more likely to indicate its contribution to process-level events. Such a classification can be derived automatically by the information available from Steps ① and ② through analyzing the discretizations and the respective sensor value stream characteristics within a group.

5.3.2 Challenges

This step addresses challenges **C3**, **C4**, and **C5** by semiautomatically deriving a classification signifying the relevance of each sensor per group for the scenario and the data analysis.

5.3.3 Example

The information from the previous steps is analyzed to classify the sensors within a group resulting in Table 5.

As can be seen, we propose to rank sensors for each group that was defined in the previous step to simplify the ranking. For example, for the first group, the binary sensor has ranking (priority) 1, while discrete sensors are ranked 2 (lower priority).

Refined rules might include that the discreteness dimension (binary is a special case of discrete with 2 potential values) is further considered: e.g., X movement of the crane is considered less important because it has 4 discrete values in comparison to energy consumption or Y movement. Even more refined rules might consider the range of source values, e.g., X movement is more important because it only spans values from 1 to 59, as compared to the Y movement which covers values from 0 to 99, or energy consumption which covers source values from 0 to greater than 1000.

Finding a good set of rules will be subject of future research, i.e., multiple ranking algorithms with their own strength and weaknesses might emerge.

5.4 Sensor Dependency Deduction

5.4.1 Concept

In Step ④ dependencies across all sensors and sensor groups are deduced. The goal is to identify important timestamps, where multiple sensors show significant *joint changes*. While discrete sensors might just change their values a few times, continuous sensors typically change their values all the time. In order to identify the importance of a change in sensor values at a particular timestamp, as many as possible sensors, weighted by their importance as calculated in Step ③, have to contribute.

This analysis can be done manually by the domain expert, but it also shows high potential for automation based on techniques and metrics for sensor similarity and correlation calculation (e.g., using dynamic time warping [66], clustering, or more sophisticated machine learning approaches).

Nevertheless, the concept of detecting joint changes comes with a set of interesting questions:

- In which time frame do joint changes occur? If changes occur within, for example, 1 second, are they joint? How big does the time window need to be?
- Should we operate on the original data, or on the discretized data from Step ①?
- If a sensor that provides continuous data (and thus changes all the time) is part of every change, how can we minimize its influence or exclude its dependency to everything?

5.4.2 Challenges

This step addresses the same challenges **C3**, **C4**, and **C5** as Step ③ by revealing dependencies among sensors across groups. The biggest challenge in this step is that changes do not always happen at the same time, but some sensors might only show a delayed reaction, which makes the change detection for a group of sensors dependent on thresholds and thus computationally expensive.

5.4.3 Example

Sensor dependency deduction infers dependencies between all sensors across groups by means of finding joint changes. Figure 6a) shows the discretized changes of all sensors for the complete running example scenario, while Fig. 6b exemplifies the dependencies by showing how the values change for a set of sensors. Figure 6 also hints at potential solutions for the questions asked above. Joint changes rely on temporal proximity. We presume that by calculating solutions for multiple time frames (utilizing multiple thresholds), the granularity of dependencies and thus the granularity of potentially detected process events can be controlled. Working with discretized data makes the algorithm much faster, but by pursuing automatic discretization of original data (e.g., by utilizing symbolic aggregate approximation [86] to map to integer values), again the granularity of dependencies and thus the granularity of potentially detected process events can be controlled. Thus, a wide range of potential algorithms and optimizations for calculating dependencies seem possible.

For our running example we assume a naive algorithm and rely on the discretized data from Step (1). We assume changes happen exactly at the same time. Table 6 shows the changes for all sensors shown in Fig. 6b.

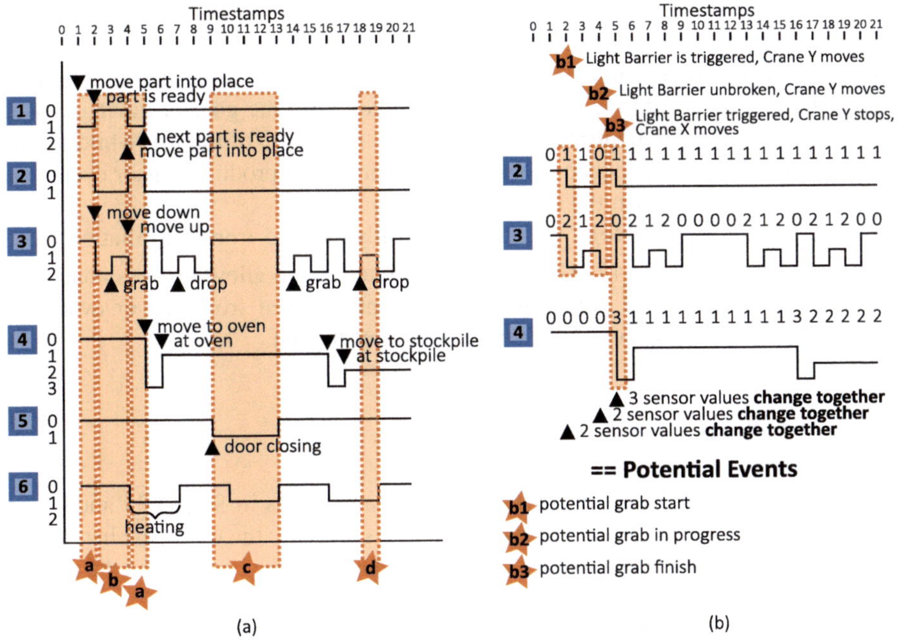

Fig. 6 Dependency deduction: sensor values changing together for all sensors in the running example

Table 6 Dependencies among sensors

Timestamp	Joint change
0	–
1	–
2	2, 3
3	3
4	2, 3
5	2, 3, 4
6	3, 4
...	...

While for Table 6 we only used sensors 2, 3, and 4, to get the all dependencies, all sensors and all timestamps have to be considered, leading to a large table of timestamps with linked sensor activity.

5.5 *Topology Computation*

5.5.1 Concept

In Step (5) a list of sensors that represents the sensor *topology*, i.e., the hierarchy and the dependencies among sensors and components and their significance, for a particular process is derived. Sensors are ordered with the goal of having sensors that are more likely to contribute process events on the top of the list, while sensors that are less likely to produce process events (or would produce sensor events at a much finer granularity) at the bottom of the list.

We envision the topology to be automatically derived from information calculated in Steps (3) and (4). This topology is intended to allow process analysts to fine-tune the number of potential process events derived from sensor events, by gradually including sensors from top to bottom of the list.

5.5.2 Challenges

Here we address the same challenges **C3**, **C4**, and **C5** as in Steps (3) and (4) by computing a topology of sensors in a list. Sensors closer to the top of the list indicate that value changes of these sensors are more relevant to the scenario and the analysis than sensors closer to the bottom.

Table 7 Sensor topology calculation

Timestamp t	α_t	Sensor ranking (SR_t)				Weighted sensor ranking (WSR_t)		
		2	3	4	Γ	2	3	4
0	0					0	0	0
1	0					0	0	0
2	2	1	2		B	2	1	0
3	1		$\frac{2+2}{2}$		B,O	0	0.5	0
4	2	1	2		B	2	1	0
5	3	1	2	2	B	3	1.5	1.5
6	2		2	2	O	0	1	1
					Σ	7	5	2.5
					#	3	5	3
					Σ/#	**2.30**	**1.00**	**0.83**

5.5.3 Example

A naive topology calculation for the event subset 2, 3, 4 and the first six timestamps (cf. Table 6, Step ④) is depicted in Table 7.

For each timestamp t the base importance α_t is set as the number of joint changes, as can be seen in Table 6, Step ④. The sensor ranking SR_t at timestamp t is taken from Table 5, Step ③. As sensors can occur in multiple sensor groups and can have a different importance in each group, the closest matching group for a combination of sensors is used as the source for the ranking. The sensor group id (GID) is denoted in column Γ. For example, for timestamp 2, Table 7 shows $\Gamma = \text{B}$, as sensors 2 and 3 only occur together in group B. For timestamp 3, with only sensor 3 changing, things are more difficult. Sensor 3 occurs in groups B and O. Thus, the weights for both groups have to be averaged (i.e., $\frac{2+1}{2}$). For our naive calculation, the weighted sensor ranking (WSR_t) for each timestamp t then amounts to:

$$WSR_t = \alpha_t \times \frac{1}{SR_t} \tag{1}$$

If a sensor ranking does not exist for a certain timestamp, its value becomes 0. To calculate the importance of a sensor in the topology, the sum (Σ) of all rankings per sensor is divided by the number of occurrences (#) of a sensor in all timestamps. This results in the final topology (i.e., sorting sensors in a list by importance) as shown in Table 8.

The sensor topology calculation is naive, as similarly to Step ④; future research into more complex scenarios will determine if alternative calculation algorithms (e.g., considering aspects of the source data) can yield better results.

Table 8 Naive sensor topology

Sensor ID	Importance	Timestamps
2	2.30	2, 4, 5
4	1.00	5, 6
3	0.83	2, 3, 4, 5, 6

5.6 Event Abstraction

5.6.1 Concept

Steps ①–⑤ served as event extraction and preparation of event abstraction. In Step ⑥ process-level events are abstracted from the information encoded in the topology calculated in Step ⑤ [19]. Due to their higher ranking and thus higher significance to indicate process-level events as explained in the previous steps, we assume that events from sensors at the top of the topology (list) represent process events related to process activities, which group all events from lower levels of the topology.

Just utilizing the top layer (sensors at the top position) of the topology will yield a coarse-grained process model to be mined in Step ⑧, which groups all existing events. By letting the domain expert specify that the first, second, or even deeper levels and only a subset of sensors at a particular level should be utilized to determine process activities and thus grouping of events, the process models can be manually tweaked to yield the desired granularity.

5.6.2 Challenges

Here we address the same challenges **C3**, **C4**, and **C5** as in the previous three steps by supporting the domain expert in selecting relevant sensors for analysis and event abstraction and automatically extracting the events based on the selection. **C1** can be addressed through having in-depth knowledge about the relationship between different sensors, leading to potential optimization regarding the required data velocity, but also relevance. In other words, data can potentially be discarded because it might not contribute to the data analysis tasks at hand. This step also allows us to control the granularity level of a process, thus for challenge **C6** allowing to stipulate a desired granularity of how human involvement in the process model must be represented. Challenges **C7** and **C8** are directly addressed by the development of this framework and realizing Steps ① to ⑥. Generating knowledge and refining knowledge in a structured and semiautomatic way with the goal to refine and evolve processes is the purpose of the proposed framework. For example, when the desired granularity of resulting processes is not fine enough, although we utilize the full topology generated in Step ⑤, we can deduct that we are missing sensors. On the other hand, the desired granularity of processes can be controlled by the mechanism introduced in Step ⑥.

5.6.3 Example

Beginning at the top of the topology shown in Table 8, timestamps 2, 4, and 5 are identified to be points in time when process events occurred.

For our specific sensor subset 2, 3, 4 and sensor events for the first six timestamps (cf. Table 6), this yields process events when the light barrier shows changes. These events can be extracted as shown in Table 9.

As can be seen, interpreting the context might require some further insights. When comparing to the large dependency deduction picture shown in Fig. 6b, it becomes clear that the three potential events can be interpreted as *pickup* b events, but the lifecycle transitions associated with the process events are unclear. Including sensor 1 could have helped with further differentiation, but it was not included in the sample calculation in Steps ④ and ⑤. Nonetheless, the interpretation (i.e., by a human domain expert) shown in Table 10 seems feasible.

It is important to note that the dependency deduction in Step ④ shows that for timestamp 5 many more sensors are involved. Thus, it is a stronger candidate for a process event. Also the semantic annotation in Step ① will provide further semantic information about the state of the involved sensors and thus further context for labeling of events and determining the process event lifecycle transition they represent.

Again, future research can show how all the structured information prepared in Steps ① to ⑥ can be incorporated in to either (1) a user interface to support domain experts or (2) allow for automatic deduction of process events.

5.7 *Event Correlation*

5.7.1 Concept

In Step ⑦ the extracted and abstracted events that were associated with specific process-level activities in Step ⑥ are correlated with a specific *case* (i.e.,

Table 9 Extracted process events

Event timestamp	Context
2	Group buffer; sensors 2 , 3
4	Group buffer; sensors 2 , 3
5	Group buffer; sensors 2 , 3 , 4

Table 10 Abstracted process events

Process event ID	Event timestamp	Lifecycle transition	Label
b1	2	Start	Grab part
b2	4	Unknown	Grab part
b3	5	Complete	Grab part

an executed process instance) to enable classic activity-centric process mining approaches [19]. This step may not be trivial due to the start and end of a case not being clearly identifiable from the IoT data log and multiple executions of process instances in parallel being possible. Thus, ambiguities may emerge when trying to correlate abstracted events from Step ⑥ with a specific case [29, 50]. Additional sensors being annotated to provide data about the start and end of a case or to track the execution of all activities belonging to one case (e.g., a camera [55]) might be necessary to support the correlation. The domain expert has to identify these sensors and specify their correlation with a case, e.g., what is the *object* that the execution of (a) process instance(s) is centered around. This could, for instance, be the *workpiece* in a production process or the *patient* in a treatment process, which would group all events into an executed case or an object-centric event log [6]. Moreover, context data from additional sensors and identified preceding process activities could be used to support the event correlation [12, 18].

5.7.2 Challenges

In this step we address challenges **C3**–**C5**, **C7**, and **C8** as process-related data is added here to complement and complete the available information from previous steps. By correlating the extracted and abstracted events with a specific case or object, we achieve the full contextualization of these events within the specific process scenarios.

5.7.3 Example

The domain expert performs the event correlation, i.e., adding the corresponding case identifiers (*Case IDs*) as additional attributes to the activities, by identifying all extracted events as pertaining to the production of the same workpiece and thus belonging to the same process instance (i.e., case). This step is straightforward under the assumption that no parallel process executions are possible and that all executed production activities relate to the same workpiece (i.e., lot size 1) [73]. The result of this step is similar to the excerpt of an IoT-enriched event log depicted in Table 2 where all IoT sensor readings are fully contextualized in the specific process and activity instances.

5.8 *Process Mining*

5.8.1 Concept

In Step ⑧ process mining techniques can be applied as soon as process activities are identified, which is achieved through *event abstraction* with a domain expert providing a mapping for meaningful labels of the abstracted activities in Step ⑥ and the correlation of these activities with specific cases in Step ⑦. Case-based, artifact-based, object-centric, or other process mining techniques can be utilized to mine the processes from the event logs—to discover process models, check conformance, and enhance processes [92]—and thus to achieve full process awareness of IoT data combined with process-related artifacts (e.g., process models).

5.8.2 Challenges

Challenge **C9** dealing with outcome prediction builds on all the data and relationships generated in Steps ① to ⑦. However, only if the resulting processes are known, potential outcomes can be identified, as they rely on the structure of the process and might only be realized at the very end of a process.

5.8.3 Example

Applying the inductive miner [49] on the event log shown excerpt in Table 10—extended with additional entries from all sensors shown in Fig. 6a—but only considering the events with "start" lifecycle, the final process model could look as depicted in Fig. 7. This depends on the process granularity chosen in Step ⑥.

5.9 *Discussion and Weaknesses*

As implied throughout this chapter, the successful application of this framework relies on a variety of future advances. We will discuss potential problems that have to be solved for each step in the proposed framework.

If during Step ① neither domain experts nor potential AI-based annotation mechanisms can provide further insight into the characteristics and meaning of sensor values, it will be very difficult to identify potential process events during

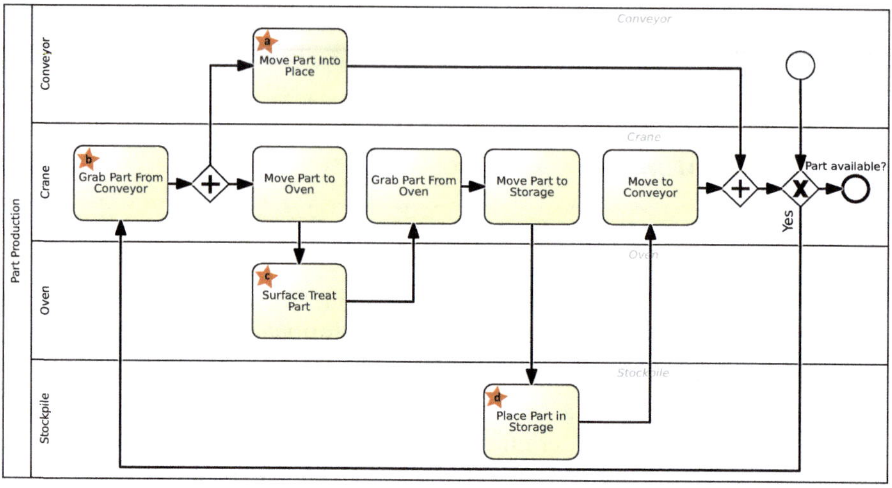

Fig. 7 Final process model of the production process in BPMN 2.0

Step ⑥. The same holds for Step ②, although the physical location of sensors, combined with knowledge about the purpose of sensors, might allow for automatic grouping.

Automatic discretization will have to be explored and can help with reducing human involvement while conducting Step ①.

We expect Steps ③ to ⑤ to run fully automatic without further human input, but the algorithms presented as part of this chapter need to be refined. Similar to the multitude of process mining algorithms that exist today, for Steps ③ to ⑤ we can see multiple refined algorithms dealing with potential shortcomings of topologies.

The abstraction and correlation in Steps ⑥ and ⑦ will potentially always be conducted by domain experts as they allow fine-tuning the granularity of the resulting process model(s).

One corner case that has to be mentioned is when the framework is applied to a scenario with only one sensor. Here Step ① is the only step that affects the outcome. If the sensor delivers continuous data, the discretization in Step ① will determine all process events, as the topology in Step ⑤ will contain only this sensor, and all sensor events will be considered to be process events.

Regarding the applicability of the framework in other domains (e.g., healthcare) we do not foresee any problems. As the approach is very basic, depending on just time series data (i.e., timestamps and associated values) from sensors, nothing prevents the application of the approach to data from other domains. For example, the approach for activity monitoring from IoT data in healthcare settings described in [28] exhibits the same characteristics within sensor data relevant for the framework presented in this work, namely, time series data and sensor groups.

High-volume sensor data will lead to a set of performance problems with the following steps being the most affected:

- Step ① will suffer from high computational cost for discretization, depending on the algorithm for automatic discretization, or the effort needed to transform data.
- Step ④ will suffer from performance problems, if a high number of sensors are included in the data, as all sensors have to be compared. The calculation of dependencies for different time windows further multiplies the computational costs.
- All other steps are considered to not be problematic.

Furthermore, the development of user interfaces to support domain experts during Steps ⑥ and ⑦ (and potentially Step ①) will be crucial to maximize the utility of the framework, as well as the deduction of process events from IoT events.

6 Conclusion and Future Work

Data engineers, data scientists, and domain experts have to be supported in collecting, structuring, and analyzing IoT data to extract process models and event logs in order to (1) allow potential enhancements to materialize faster, (2) minimize the errors occurring during the analysis, and (3) minimize the required domain knowledge.

In the context of these goals, we identified a set of 10 current research challenges that can be associated with deriving process-relevant data from complex IoT data. To support the goals and address these challenges, we presented a framework consisting of 8 generalized steps that can guide the semi-automated data analysis to produce process data and process models at different levels of granularity starting from raw IoT data (*no process awareness*) up to IoT data which is fully contextualized regarding a process (*full process awareness*). We envision future research to yield a wide variety of results for each step of the framework, including:

- Specialized tools for domain experts to annotate and group data artifacts
- Custom visualizations to show dependencies between data artifacts in real time
- Refined algorithms to calculate (hidden) dependencies between hardware artifacts and sensor data
- Refined data retention policies based on annotations and derived process intelligence

To demonstrate the potential of the framework we also introduced a sample scenario and exemplary IoT dataset to discuss how a semiautomatically generated process model based on the steps in the framework can be (a) of similar quality as a process model provided by domain experts, (b) used to point out problems in existing process models, and (c) utilized to refine existing processes, e.g., by pointing out the necessity to introduce additional sensors in a certain context. We think that this framework can serve as means to focus research towards the overall goal of allowing domain experts to make IoT datasets fully process aware without the involvement of data engineers and data scientists [22].

Acknowledgments This research was partly funded by the Austrian Research Promotion Agency (FFG) via the "Austrian Competence Center for Digital Production" (CDP) under the contract number 881843. This research was supported by the Pilot Factory Industry 4.0, Seestadtstrasse 27, Vienna, Austria. This work has received funding from the Swiss National Science Foundation under Grant No. IZSTZ0_208497 (*ProAmbitIon* project).

References

1. Ahmad, M.: Reliability models for the internet of things: a paradigm shift. In: 2014 IEEE International Symposium on Software Reliability Engineering Workshops, pp. 52–59 (2014)
2. Akhramovich, K., Serral, E., Cetina, C.: A systematic literature review on the application of process mining to industry 4.0. In: Knowledge and Information Systems, pp. 1–48 (2024)
3. Baier, T., Mendling, J., Weske, M.: Bridging abstraction layers in process mining. Inform. Syst. **46**, 123–139 (2014)
4. Bauer, M., Bui, N., De Loof, J., Magerkurth, C., Nettsträter, A., Stefa, J., Walewski, J.W.: IoT reference model. In Enabling Things to Talk: Designing IoT Solutions with the IoT Architectural Reference Model, pp. 113–162 (2013)
5. Beerepoot, I., et al.: The biggest business process management problems to solve before we die. Comput. Ind. **146**, 103837 (2023)
6. Berti, A., Koren, I., Adams, J.N., Park, G., Knopp, B., Graves, N., Rafiei, M., Liß, L., Unterberg, L.T.G., Zhang, Y., et al.: Ocel (object-centric event log) 2.0 specification (2024). arXiv preprint arXiv:2403.01975
7. Bertrand, Y., De Weerdt, J., Serral, E.: A bridging model for process mining and IoT. In: Munoz-Gama, J., Lu, X. (eds.) Process Mining Workshops, pp. 98–110. Springer, Cham (2022)
8. Bertrand, Y., Weerdt, J.D., Serral, E.: Assessing the suitability of traditional event log standards for IoT-enhanced event logs. In: Business Process Management Workshops. Springer, Berlin (2022)
9. Beverungen, D., Buijs, J.C.A.M., Becker, J., Di Ciccio, C., van der Aalst, W.M.P., et al.: Seven Paradoxes of Business Process Management in a Hyper-Connected World. BISE (2020)
10. Beyel, H.H., Makke, O., Gusikhin, O., van der Aalst, W.M.: Analyzing behavior in cyber-physical systems in connected vehicles: a case study. In: International Conference on Business Process Management, pp. 92–104. Springer, Berlin (2023)
11. Brauner, P., Dalibor, M., Jarke, M., Kunze, I., Koren, I., Lakemeyer, G., et al.: A computer science perspective on digital transformation in production. ACM Trans. Internet Things **3**(2), 1–32 (2022)
12. Brzychczy, E., Żuber, A., van der Aalst, W.: Process mining of mining processes: analyzing longwall coal excavation using event data. IEEE Trans. Syst. Man Cybern. Syst. **54**(5), 2723–2734 (2024)
13. Burattin, A., Eigenmann, M., Seiger, R., Weber, B.: MQTT-XES: real-time telemetry for process event data. In: BPM 2020 Best Dissertation Award, Doctoral Consortium, and Demonstration & Resources Track, pp. 97–101 (2020)
14. Cameranesi, M., Diamantini, C., Mircoli, A., Potena, D., Storti, E.: Extraction of user daily behavior from home sensors through process discovery. IEEE Internet Things J. **7**(9), 8440–8450 (2020)

15. Chanal, P.M., Kakkasageri, M.S.: Security and privacy in IoT: a survey. Wirel. Pers. Commun. **115**(2), 1667–1693 (2020)
16. Cook, A.A., Mısırlı, G., Fan, Z.: Anomaly detection for IoT time-series data: a survey. IEEE Internet Things J. **7**(7), 6481–6494 (2020)
17. De Luzi, F., Leotta, F., Marrella, A., Mecella, M.: On the interplay between business process management and internet-of-things: a systematic literature review. Bus. Inform. Syst. Eng. **67**(2), 265–288 (2025)
18. Di Federico, G., Burattin, A.: Cvamos–event abstraction using contextual information. Future Internet **15**(3), 113 (2023)
19. Diba, K., Batoulis, K., Weidlich, M., Weske, M.: Extraction, correlation, and abstraction of event data for process mining. Wiley Interdiscip. Rev. Data Mining Knowl. Discov. **10**(3), e1346 (2020)
20. Dumas, M., La Rosa, M., Mendling, J., Reijers, H.A., et al.: Fundamentals of Business Process Management, vol. 1. Springer, Berlin (2013)
21. van Eck, M.L., Sidorova, N., van der Aalst, W.M.P.: Enabling process mining on sensor data from smart products. In: 2016 IEEE Tenth International Conference on Research Challenges in Information Science (RCIS), pp. 1–12 (2016)
22. Ehrendorfer, M., Bertrand, Y., Malburg, L., Mangler, J., Grüger, J., Rinderle-Ma, S., Bergmann, R., Serral, E.: Internet of processes and things: a repository for IoT-enriched event logs in smart environments. In: International Conference on Business Process Management, Demonstration & Resources Forum, pp. 92–96. CEUR (2023)
23. Ehrendorfer, M., Fassmann, J.A., Mangler, J., Rinderle-Ma, S.: Conformance checking and classification of manufacturing log data. In: 21st CBI. vol. 1, pp. 569–577. IEEE, Piscataway (2019)
24. Ehrendorfer, M., Mangler, J., Rinderle-Ma, S.: Sensor data stream selection and aggregation for the ex post discovery of impact factors on process outcomes. In: 33rd CAiSE. pp. 29–37. Springer, Berlin (2021)
25. Ehrendorfer, M., Mangler, J., Rinderle-Ma, S.: Clustering raw sensor data in process logs to detect data streams. In: International Conference on Cooperative Information Systems, pp. 438–447. Springer, Berlin (2023)
26. Elsaleh, T., Bermudez-Edo, M., Enshaeifar, S., Acton, S.T., Rezvani, R., Barnaghi, P.: IoT-stream: a lightweight ontology for Internet of Things data streams. In: Global IoT Summit Proc., pp. 1–6. IEEE, Piscataway (2019)
27. Folino, F., Guarascio, M., Pontieri, L.: Mining predictive process models out of low-level multidimensional logs. In: International Conference on Advanced Information Systems Engineering, pp. 533–547. Springer, Berlin (2014)
28. Franceschetti, M., Seiger, R., González, M.J.G., Garcia-Ceja, E., Flores, L.A.R., García-Bañuelos, L., Weber, B.: Proambition: online process conformance checking with ambiguities driven by the internet of things. In: CAiSE Research Projects Exhibition, pp. 52–59 (2023)
29. Franceschetti, M., Seiger, R., López, H.A., Burattin, A., García-Bañuelos, L., Weber, B.: A characterisation of ambiguity in bpm. In: International Conference on Conceptual Modeling, pp. 277–295. Springer, Berlin (2023)
30. Franceschetti, M., Seiger, R., Weber, B.: An event-centric metamodel for IoT-driven process monitoring and conformance checking. In: International Conference on Business Process Management, pp. 131–143. Springer, Berlin (2023)
31. Grüger, J., Malburg, L., Mangler, J., Bertrand, Y., Rinderle-Ma, S., Bergmann, R., Asensio, E.S.: Sensorstream: an XES extension for enriching event logs with IoT-sensor data (2022). https://arxiv.org/abs/2206.11392
32. Gruhn, V., Han, Y., Hesenius, M., Reichert, M., Wang, G., Yu, J., Zhang, L.: BRIBOT: towards a service-based methodology for bridging business processes and IoT big data. In: Service-Oriented Computing, vol. 13121, pp. 597–611. Springer, Berlin (2021)
33. Hirmer, P., Wieland, M., Breitenbücher, U., Mitschang, B.: Automated sensor registration, binding and sensor data provisioning. In: CAiSE Forum, pp. 81–88 (2016)

34. Hirmer, P., Wieland, M., Schwarz, H., Mitschang, B., Breitenbücher, U., Leymann, F.: SITRS—a situation recognition service based on modeling and executing situation templates. In: Proceedings of the 9th Symposium and Summer School on Service-Oriented Computing, pp. 113–127 (2015)
35. Hoffmann, M., Malburg, L., Bergmann, R.: ProGAN: toward a framework for process monitoring and flexibility by change via generative adversarial networks. In: BPM Workshops, LNBIP, vol. 436, pp. 43–55. Springer, Berlin (2022)
36. Hoppenstedt, B., Pryss, R., Stelzer, B., Meyer-Brötz, F., Kammerer, K., Treß, A., Reichert, M.: Techniques and emerging trends for state of the art equipment maintenance systems–a bibliometric analysis. Appl. Sci. **8**(6), 916 (2018)
37. Janiesch, C., Koschmider, A., Mecella, M., Weber, B., Burattin, A., Di Ciccio, C., Fortino, G., Gal, A., Kannengiesser, U., Leotta, F., et al.: The Internet of Things meets business process management: a manifesto. IEEE Syst. Man Cybern. Mag. **6**(4), 34–44 (2020)
38. Jans, M., Soffer, P., Jouck, T.: Building a valuable event log for process mining: an experimental exploration of a guided process. Ent. Inf. Syst. **13**(5), 601–630 (2019)
39. Janssen, D., Mannhardt, F., Koschmider, A., van Zelst, S.J.: Process model discovery from sensor event data. In: Process Mining Workshops, pp. 69–81 (2021)
40. Kammerer, K., Pryss, R., Hoppenstedt, B., Sommer, K., Reichert, M.: Process-driven and flow-based processing of industrial sensor data. Sensors **20**(18), 5245 (2020)
41. Küfner, T., Schönig, S., Jasinski, R., Ermer, A.: Vertical data continuity with lean edge analytics for industry 4.0 production. Comput. Ind. **125**, 103389 (2021)
42. Kirikkayis, Y., Gallik, F., Reichert, M.: IoTDM4BPMN: an IoT-enhanced decision making framework for BPMN 2.0. In: 2022 International Conference on Service Science (ICSS), pp. 88–95 (2022)
43. Kirikkayis, Y., Gallik, F., Seiger, R., Reichert, M.: Integrating IoT-driven events into business processes. In: International Conference on Advanced Information Systems Engineering, pp. 86–94. Springer, Berlin (2023)
44. Kirikkayis, Y., Gallik, F., Winter, M., Reichert, M.: BPMNE4IoT: a framework for modeling, executing and monitoring IoT-driven processes. Future Internet **15**(3), 90 (2023)
45. Klein, P., Malburg, L., Bergmann, R.: Ftonto: A domain ontology for a fischertechnik simulation production factory by reusing existing ontologies. In: Lernen. Wissen. Daten. Analysen. (LWDA), Berlin, pp. 253–264. CEUR Workshops Proceedings (2019)
46. Knoch, S., Ponpathirkoottam, S., Schwartz, T.: video-to-model: unsupervised trace extraction from videos for process discovery and conformance checking in manual assembly. In: 18th BPM, LNCS, vol. 12168, pp. 291–308. Springer, Berlin (2020)
47. Kopetz, H., Bondavalli, A., Brancati, F., Frömel, B., Höftberger, O., Iacob, S.: Emergence in cyber-physical systems-of-systems (CPSoSs), pp. 73–96. Springer, Cham (2016)
48. Koschmider, A., Janssen, D., Mannhardt, F.: Framework for process discovery from sensor data. In: EMISA, pp. 32–38 (2020)
49. Leemans, S.J.: Robust Process Mining with Guarantees: Process Discovery, Conformance Checking and Enhancement, vol. 440. Springer, Berlin (2022)
50. Li, C.Y., Joshi, A., Tam, N.T., Lau, S.S.F., Huang, J., Shinde, T., van der Aalst, W.M.: Rectify sensor data in IoT: a case study on enabling process mining for logistic process in an air cargo terminal. In: International Conference on Cooperative Information Systems, pp. 293–310. Springer, Berlin (2023)
51. Maamar, Z., Kajan, E., Guidara, I., Moctar-M'Baba, L., Sellami, M.: Bridging the gap between business processes and IoT. In: 24th Symposium on International Database Engineering and Applications, pp. 1–10. ACM, New York (2020)
52. Malburg, L., Bergmann, R.: Towards adaptive workflow management by case-based reasoning and automated planning. In: 30th ICCBR Workshops Proceedings. CEUR-WS.org (2022). Accepted for Publication
53. Malburg, L., Grüger, J., Bergmann, R.: An IoT-enriched event log for process mining in smart factories (2022). CoRR abs/2209.02702

54. Malburg., L., Klein., P., Bergmann., R.: Semantic web services for AI-research with physical factory simulation models in industry 4.0. In: Proceedings of the International Conference on Innovative Intelligent Industrial Production and Logistics—IN4PL, pp. 32–43. INSTICC, SciTePress (2020)
55. Malburg, L., Rieder, M.P., Seiger, R., Klein, P., Bergmann, R.: Object detection for smart factory processes by machine learning. Procedia Comput. Sci. **184**, 581–588 (2021)
56. Malburg, L., Seiger, R., Bergmann, R., Weber, B.: Using physical factory simulation models for business process management research. In: Business Process Management Workshops, pp. 95–107. Springer, Berlin (2020)
57. Mangler, J., Grüger, J., Malburg, L., Ehrendorfer, M., Bertrand, Y., Benzin, J.V., Rinderle-Ma, S., Serral Asensio, E., Bergmann, R.: Datastream XES extension: embedding IoT sensor data into extensible event stream logs. Future Internet **15**(3), 109 (2023)
58. Mangler, J., Rinderle-Ma, S.: CPEE—cloud process execution engine. BPM Demo Sessions **1295** (2014)
59. Mangler, J., Pauker, F., Rinderle-Ma, S., Ehrendorfer, M.: centurio.work—industry 4.0 integration assessment and evolution. In: 17th BPM Conference, pp. 106–117 (2019)
60. Mannhardt, F., Bovo, R., Oliveira, M.F., Julier, S.: A taxonomy for combining activity recognition and process discovery in industrial environments. In: Intelligent Data Engineering and Automated Learning (IDEAL 2018), Madrid, Spain. LNCS, vol. 11315, pp. 84–93. Springer, Cham (2018)
61. Mannhardt, F., de Leoni, M., Reijers, H.A., van der Aalst, W.M.P., Toussaint, P.J.: From low-level events to activities—a pattern-based approach. In: International Conference on Business Process Management (BPM), Rio de Janeiro, Brazil. LNCS, vol. 9850, pp. 125–141. Springer, Cham (2016)
62. Marrella, A., Mecella, M., Sardiña, S.: Supporting adaptiveness of cyber-physical processes through action-based formalisms. AI Commun. **31**(1), 47–74 (2018)
63. Messner, M., Pauker, F., Mauthner, G., Frühwirth, T., Mangler, J.: Closed loop cycle time feedback to optimize high-mix/low-volume production planning. Procedia CIRP **81**, 689–694 (2019)
64. Monostori, L.: Cyber-physical production systems: roots, expectations and R&D challenges. Procedia Cirp **17**, 9–13 (2014)
65. Mottola, L., Picco, G.P., Oppermann, F.J., Eriksson, J., Finne, N., Fuchs, H., Gaglione, A., Karnouskos, S., Montero, P.M., Oertel, N., Römer, K., Spieß, P., Tranquillini, S., Voigt, T.: makesense: simplifying the integration of wireless sensor networks into business processes. IEEE Trans. Software Eng. **45**(6), 576–596 (2019)
66. Müller, M.: Dynamic time warping. In: Information Retrieval for Music and Motion, pp. 69–84 (2007)
67. Pauker, F., Mangler, J., Rinderle-Ma, S., Ehrendorfer, M.: Industry 4.0 integration assessment and evolution at EVVA GmbH: process-driven automation through centurio. work. In: BPM Cases, vol. 2, pp. 81–91. Springer, Berlin (2021)
68. Pauker, F., Mangler, J., Rinderle-Ma, S., Pollak, C.: centurio.work—modular secure manufacturing orchestration. In: 16th International Conference on Business Process Management, pp. 164–171 (2018)
69. Rebmann, A., Emrich, A., Fettke, P.: Enabling the discovery of manual processes using a multi-modal activity recognition approach. In: Business Process Management Workshops. LNBIP, vol. 362, pp. 130–141. Springer, Berlin (2020)
70. Rinderle-Ma, S., Mangler, J.: Process automation and process mining in manufacturing. In: Polyvyanyy, A., Wynn, M.T., Van Looy, A., Reichert, M. (eds.) Business Process Management, pp. 3–14. Springer, Cham (2021)
71. Seiger, R., Assmann, U., Huber, S.: A case study for workflow-based automation in the internet of things. In: 2018 IEEE International Conference on Software Architecture Companion (ICSA-C), pp. 11–18 (2018)
72. Seiger, R., Franceschetti, M., Weber, B.: Data-driven generation of services for IoT-based online activity detection. In: International Conference on Service-Oriented Computing, pp. 186–194. Springer, Berlin (2023)

73. Seiger, R., Franceschetti, M., Weber, B.: An interactive method for detection of process activity executions from IoT data. Future Internet **15**(2), 77 (2023)
74. Seiger, R., Huber, S., Heisig, P., Aßmann, U.: Toward a framework for self-adaptive workflows in cyber-physical systems. Softw. Syst. Model. **18**(2), 1117–1134 (2019)
75. Seiger, R., Malburg, L., Weber, B., Bergmann, R.: Integrating process management and event processing in smart factories: a systems architecture and use cases. J. Manuf. Syst. **63**, 575–592 (2022)
76. Seiger, R., Zerbato, F., Burattin, A., García-Bañuelos, L., Weber, B.: Towards IoT-driven process event log generation for conformance checking in smart factories. In: 24th International EDOC Workshop, pp. 20–26. IEEE, Piscataway (2020)
77. Senderovich, A., Rogge-Solti, A., Gal, A., Mendling, J., Mandelbaum, A.: The road from sensor data to process instances via interaction mining. In: International Conference on Advanced Information Systems Engineering (CAiSE), Stockholm, Sweden. LNCS, vol. 9097, pp. 257–273. Springer, Cham (2016)
78. Shi, F., Li, Q., Zhu, T., Ning, H.: A survey of data semantization in Internet of Things. Sensors **18**(1) (2018)
79. Sisinni, E., Saifullah, A., Han, S., Jennehag, U., Gidlund, M.: Industrial Internet of Things: challenges, opportunities, and directions. IEEE Trans. Ind. Inf. **14**(11), 4724–4734 (2018)
80. Soffer, P., Hinze, A., Koschmider, A., Ziekow, H., Di Ciccio, C., Koldehofe, B., Kopp, O., Jacobsen, A., Sürmeli, J., Song, W.: From event streams to process models and back: challenges and opportunities. Inf. Sys. **81**, 181–200 (2019)
81. Stertz, F., Mangler, J., Rinderle-Ma, S.: Data-driven improvement of online conformance checking. In: 24th EDOC, pp. 187–196. IEEE, Piscataway (2020)
82. Stertz, F., Mangler, J., Rinderle-Ma, S.: Balancing patient care and paperwork automatic task enactment and comprehensive documentation in treatment processes. In: Enterprise Modelling and Information Systems Architectures, vol. 15, p. 11 (2020)
83. Stertz, F., Mangler, J., Rinderle-Ma, S.: Temporal conformance checking at runtime based on time-infused process models (2020). arXiv:2008.07262
84. Stertz, F., Rinderle-Ma, S., Mangler, J.: Analyzing process concept drifts based on sensor event streams during runtime. In: Fahland, D., Ghidini, C., Becker, J., Dumas, M. (eds.) Business Process Management, pp. 202–219. Springer, Cham (2020)
85. Stoiber, C., Schönig, S.: Event-driven business process management enhancing IoT—a systematic literature review and development of research agenda. In: Innovation Through Information Systems, vol. 48, pp. 645–661. Springer, Berlin (2021)
86. Sun, Y., Li, J., Liu, J., Sun, B., Chow, C.: An improvement of symbolic aggregate approximation distance measure for time series. Neurocomputing **138**, 189–198 (2014)
87. Sztyler, T., Carmona, J., Völker, J., Stuckenschmidt, H.: Self-tracking Reloaded: Applying Process Mining to Personalized Health Care from Labeled Sensor Data, pp. 160–180. Springer, Berlin, Heidelberg (2016)
88. Tax, N., Sidorova, N., Haakma, R., van der Aalst, W.M.: Event abstraction for process mining using supervised learning techniques. In: Proceedings of SAI Intelligent Systems Conference, pp. 251–269. Springer, Berlin (2016)
89. Traganos, K., Spijkers, D., Grefen, P., Vanderfeesten, I.: Dynamic process synchronization using bpmn 2.0 to support buffering and (un)bundling in manufacturing. In: Fahland, D., Ghidini, C., Becker, J., Dumas, M. (eds.) Business Process Management Forum, pp. 18–34. Springer, Cham (2020)
90. Valderas, P., Torres, V., Serral, E.: Towards an Interdisciplinary Development of IoT-Enhanced Business Processes. BISE (2022)
91. Van Der Aa, H., Leopold, H., Reijers, H.A.: Efficient process conformance checking on the basis of uncertain event-to-activity mappings. IEEE Trans. Knowl. Data Eng. **32**(5), 927–940 (2019)
92. Van Der Aalst, W., Adriansyah, A., De Medeiros, A.K.A., Arcieri, F., Baier, T., Blickle, T., Bose, J.C., van den Brand, P., Brandtjen, R., Buijs, J., et al.: Process mining manifesto. In: Business Process Management, pp. 169–194. Springer, Berlin (2011)

93. Verbeek, E.: IEEE standard for extensible event stream (XES) for achieving interoperability in event logs and event streams. IEEE Std 1849-2016, pp. 1–50 (2016)
94. Wanner, J., Herm, L.V., Janiesch, C.: Countering the fear of black-boxed ai in maintenance: towards a smart colleague. In: Proceedings of the 2019 Pre-ICIS SIGDSA Symposium (2019)
95. Weyers, F., Seiger, R., Weber, B.: Method to identify process activities by visualizing sensor events. In: International Conference on Business Process Management, pp. 76–88. Springer, Berlin (2022)
96. van Zelst, S.J., Mannhardt, F., de Leoni, M., Koschmider, A.: Event abstraction in process mining: literature review and taxonomy. Granular Comput. **6**(3), 719–736 (2021)
97. Zerbato, F., Seiger, R., Di Federico, G., Burattin, A., Weber, B.: Granularity in process mining: can we fix it? In: International Workshop on BPM Problems to Solve Before We Die, pp. 40–44. CEUR-WS (2021)

Human-Centric Processes: Where IoT and People Meet

Andrea Burattin, Giovanni Meroni, Gemma Di Federico, Hugo A. López, Andrey Rivkin, and Ekkart Kindler

1 Introduction and Background

Traditional workflow management systems (WfMSs) made it easier for Humans to participate in business processes and to manage and control them. To this end, Humans were required to explicitly interact with computers through dedicated user interfaces. The workflow reference model [32] identified two main user interfaces of WfMSs for this purpose: the work list shows available work items to the Humans, who have the authority and capability to work on them. When available, the Human can then pick a work item from the work list, and this way start the respective Activity; the Activity itself is associated with an Application, which will automatically be started by the WfMS, when Human picks a work item; and the user interface of this Application is then used to do the actual work on the Activity.

The idea of the Internet of Things (IoT) is that Humans no longer need to explicitly provide information and interact with a computer, but that IoT devices would collect that information automatically [4]. In this way, Humans can knowingly or unknowingly be part of business processes without actively or explicitly interacting with computers or communication devices; their participation could be by some IoT devices providing them with some information and by IoT devices perceiving the human behavior; Humans could just "be themselves" and part of some processes. We call such processes human-centric processes (HCPs).

Figure 1 gives an overview of the interplay between humans, IoT, and business processes. Both humans and IoT devices are immersed in an environment, which they both can observe and interact with. In addition, framing the actions of observing and interacting with the environment in the context of BPM makes possible to

A. Burattin (✉) · G. Meroni · G. Di Federico · H. A. López · A. Rivkin · E. Kindler
Technical University of Denmark, Kgs. Lyngby, Denmark
e-mail: andbur@dtu.dk; giom@dtu.dk; gdfe@dtu.dk; hulo@dtu.dk; ariv@dtu.dk; ekki@dtu.dk

G. Fortino, M. Mecella (eds.), *Internet of Things Meets Business Process Management*, Internet of Things, https://doi.org/10.1007/978-3-031-90746-3_3

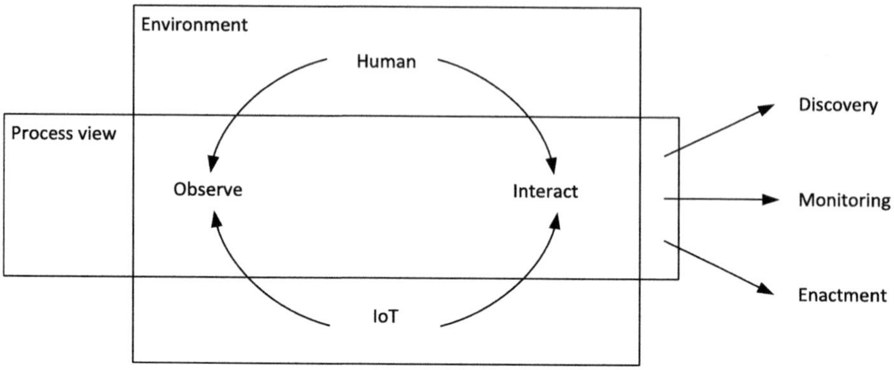

Fig. 1 Overview of the interplay between humans, IoT, and business processes

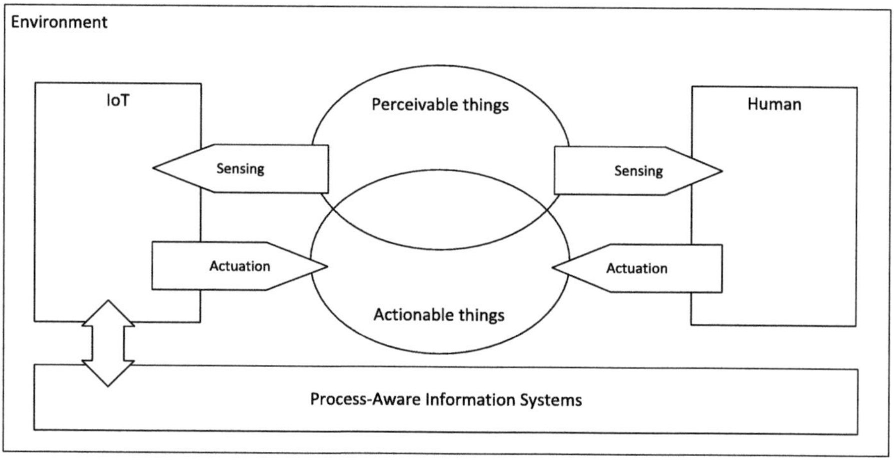

Fig. 2 Actionable overview

use existing tools and techniques in this discipline. For instance, the observations being collected can be used by process discovery algorithms to determine the causal relations of events. Such observations can also be used by process monitoring tools to identify deviations between the expected execution of a process and the real one. Interactions can be coordinated by a Process-Aware Information System (PAIS) so to ensure the enactment of a process to happen as specified in a process model.

This overview can then be refined as shown in Fig. 2. In particular, we distinguish the environment into a set of perceivable things (e.g., the weather) and a set of actionable things (e.g., the ideas that emerged during a meeting). It is worth noting that the two sets can overlap, as things can be both actionable and perceivable (e.g., a self-driving car). IoT devices, similar to humans, can then sense perceivable things and actuate actionable things through sensors and actuators, respectively. On top of

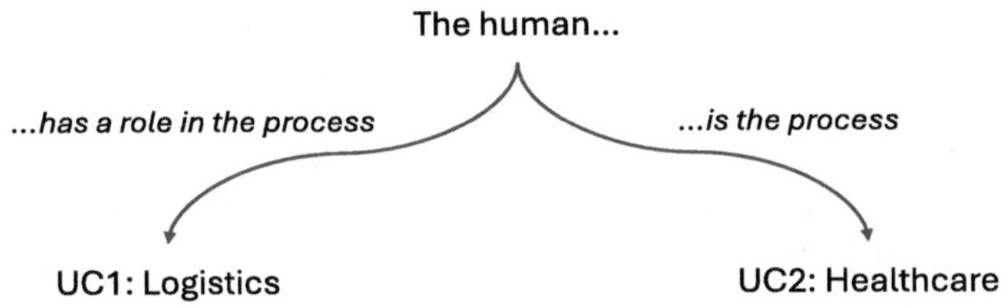

Fig. 3 Visual representation of the two classes of use cases reported in this chapter

that, IoT devices can communicate with a PAIS to provide information on the status of a process and to request a specific action to be taken.

It is worth noting that humans do not communicate directly with the PAIS. Instead, they do so by sensing and actuating the environment. In this view, IoT becomes the main interface between the PAIS and the human. It is also worth noting that, in IoT, we also include any device capable of interfacing with the human and the PAIS through the environment. For example, an IoT device could simply be a light bulb to tell the human to take a specific action, or a switch to be pushed to inform the PAIS that a task has been completed.

In this chapter, we discuss how IoT can be incorporated into the conventional BPM and PAIS landscape to make processes human-centric, based on the original idea of WfMS from the 1990s. We do that in two steps. We provide a conceptual model for the relationship between IoT and BPM, and then we provide two use cases.

When focusing on the relationship between IoT and BPM, it is important to mention that one key aspect of HCPs is flexibility: IoT provides an interface between human actors and the BPM system, which means that the need to have fixed rules that organize the work of human actors is not a hard requirement anymore; instead, it is expected that human actors perform their activities and the system reacts to them.

In the use cases part, we present different ways for processes to be human-centric (cf. Fig. 3). Specifically, the first use case, which is in the logistics domain, heavily relies on humans as they have a very active role in the process. In the second use case, focusing on healthcare, the human *is* the actual process under investigation. The two use cases complement each other and offer a comprehensive overview of human-centric processes.

Related Works To reflect on the various relationships between IoT, process management and mining, and, ultimately, human-centricity, we briefly discuss some recent works in these areas that may be relevant to the subjects covered in this chapter.

Conceptual Models for BPM in Relation to IoT Our investigation can start from the basics of conceptual models for BPM (and its extensions), including *why* we need such models (e.g., UML diagrams are crucial when implementing WF management systems/PAIS).

A paper by Leotta et al. [42] discusses a BPM engine architecture capable of capturing and processing exogenous, unexpected events using IoT sensors so that executed processes can have higher degrees of adaptivity and reactivity. In Sect. 3, the paper discusses a conceptual architecture of their framework, highlighting how to automatically adapt processes at run-time when unanticipated exceptions occur in IoT-based environments and how to avoid explicit recovery specifications. Becker et al., in [7], propose a conceptual model to describe the digital shadows of manufacturing processes. The conceptual model, in this case, includes traces generated from multiple sources, as well as different types of process models. The role of the human is limited to a data source, not considering its influence on the operation of the process. Frigo and Hirmer [30], by Frigo et al., introduces the notion of *building blocks* and *building block implementations* for IoT processes. Building blocks are general structures that can be later refined for physical devices, network protocols, or software applications, among others. Such building blocks will contain a software artifact defining its operation. Building blocks are later orchestrated by a process management engine. However, no notion of humans in the loop is presented in this work. Berges et al., in [8], present an ontology that captures how the sensors relate to indicators about the performance of an industrial machine. This ontology focuses on the sensing aspects and does not consider the human in the loop. In [60], Vila et al. describe the Connectivity Management Tool Semantics (CMTS), an ontology aimed at allowing digital tools for monitoring physical infrastructures through the use of IoT devices: gateways, nodes, and sensors. The focus is more on the sensing aspects, and there is no mention of the implications of human participants in the process. Differently from the previous works, [3] introduces a conceptual architecture that handles both self-awareness and IoT sensors. This architecture considers both sensing and actuating phases, as well as the role of models in the reasoning and execution of processes. The impact of humans is not explicitly mapped, although one can see that its impact could be mapped into external sources in an environment.

Human-Centricity in BPM One of the recent surveys [61] identifies the human-centricity in IoT as "an approach that can meet the needs of communities served by IoT technology and that helps to find out what technically works best for humans and society first and foremost." While, as the same survey suggests, there is no agreement in the scientific communities studying IoT what "human-centricity" actually stands for, the BPM community has considered human-centricity in two major contexts. First, human-in-the-loop is a standard approach in workflow automation scenarios where domain experts can intervene at specific stages to resolve errors, make decisions, and guide the process execution [34]. Second, approaches that allow the development of systems and solutions that are proactive and can anticipate or accommodate the needs of human actors have been extensively studied by the BPM community [19]. Finally, Lee et al, in [41], discuss ethical aspects of human-sensing frameworks, including those driven by IoT.

IoT and BPM and Process Mining It has been acknowledged that the integration of the Internet of Things with business process management can enhance business processes by increasing flexibility, efficiency, and responsiveness by collecting and acting upon real-time data with the help of IoT sensors and actuators. In fact, the recent "BPM meets IoT" manifesto [35] suggests that BPM and IoT domains can mutually benefit from each other: BPM can rely on IoT as a data-centric and data-intensive technology that can improve business processes, both at design- and run-time; IoT can use well-defined processes and event data to improve automated decision-making.

Following the aforementioned manifesto, [51] studies an industrial use case showing how process adaptation and process mining—a sub-discipline of BPM that focuses on interpreting event data to understand and improve business processes—can be applied in the context of Industry 4.0, whereas [16] presents a systematic literature review on the adoption of IoT-aware BPM solutions as well as their maturity through the prism of challenges identified in the manifesto.

Recent research in both BPM and process mining reveals a growing interest in IoT adoption in these fields. [11] proposes a conceptual model that mixes the event-data realm with data coming from physical IoT devices. The model can then be used in developing new and/or adopting existing process mining techniques for scenarios involving IoT. Diamantini et al., in [20], propose an ontology for process-aware scenarios in the Industrial Internet of Things, where raw sensor data is augmented with details about process activities and the physical production environment.

Some of the works also propose extensions of existing conventional process modeling formalisms and notations with specific IoT elements, where some of the approaches propose well-defined conceptual extensions of BPMN with location-aware elements [53], while others propose concrete extensions of the BPMN 2.0 metamodel [39, 40] and show how such extensions can be employed in complex decision-making tasks driven by DMN [38]. Other works focus on more generic frameworks for adopting IoT in process-aware systems (e.g., [31] proposes a framework for process-aware cyber-physical systems allowing the integration of IoT environments and human actors into higher-level process-oriented systems).

Moving closer to process mining topics, [12] proposes an approach for multi-perspective trace clustering that is applied to logs enriched with time series sensor data; [29] addresses the topics of monitoring and conformance checking in IoT-driven PAIS and suggests a conceptual event-centric metamodel that only assumes observability of low-level events from IoT devices that can be easily linked to a concrete process. Some of the recent works propose IoT-aware extensions of event data by either building on top of existing formats like XES [45] or by proposing new, conceptually well-defined solutions [9, 11] with their further adoption and validation in practice [10].

2 Human-Centric Conceptual Model of BPM Using IoT

In this section, we discuss the concepts of business process management, IoT and their relation, and how IoT helps make business processes more human-centric. The result is a conceptual model, which then is used to present the opportunities and challenges of IoT and BPM to help humans to more seamlessly use, work, and interact with Process-Aware Information Systems (PAISs).

Traditional workflow management systems (WfMSs) as reflected by the Workflow Reference Model of the Workflow Management Coalition (WfMC) [32, 33] suggested two user interfaces for Humans (end users), to participate in and to interact with workflows: the work list, where the Humans can manage and pick their work items, and the Applications where the actual work on a work item is done. In Process-Aware Information Systems [21], interactions of Humans became much more flexible. And with IoT, the Human might not even need to explicitly interact with information systems anymore to participate in a process. Using IoT devices, the behavior of the human or changes in attributes or conditions of involved goods or equipment can be automatically detected, and events relevant to the execution and progress in a given process can automatically monitored and aggregated, which then drive the process.

We start with introducing the core concepts of business processes, following the workflow reference model of the WfMC [32] and our own AMFIBIA approach [5]. A business process, which we call Process for short, consists of different Tasks that need to be done to achieve a certain goal. The process typically defines in which order these Tasks need to be done. Each different execution (or instance) of such a Process is called a Case of that Process. And each time performing a Task in a Case of that Process is called an Activity. The Process and its Tasks are the conceptual ideas of the business process. When this idea of a process is made explicit in a more or less formal and graphical way, this is called the model of the process. The Cases and activities are the manifestations of this idea; and, when there is a process model, each Case is considered an execution of this process model in the real world.

Figure 4 shows these concepts and their relation as a UML class diagram—an excerpt of and slightly adapted from the AMFIBIA model [5]. On top of these core concepts, Fig. 4 shows the Humans who are involved in Cases and activities; the counterpart of the concrete Humans participating in a process in the process model are the Roles, which define who is required or allowed to do a certain Task. This is part of the organizational aspect of a business process and can be much more involved than in our simple model. At last, the actual work on an activity is done with the help of some application, such as general-purpose software like Word or dedicated software for a specific task. The dashed dependency between the Human and the Application indicates this interaction, and the application might make all kinds of changes to data and documents in the process, which, however, are not in the focus of this paper and, therefore, are not included in the conceptual model discussed here.

Note that Fig. 4 reflects the paradigm of traditional business process and workflow management, where each Activity belongs to exactly one single Case and is associated with one Task of a specific Process. In the paradigm of object-centric processes [58], this view is relaxed, so that the same activity can belong to different cases (or, in the object-centric terminology, objects) and reflect tasks of different processes. For this paper, however, we stay with the traditional paradigm of "case-

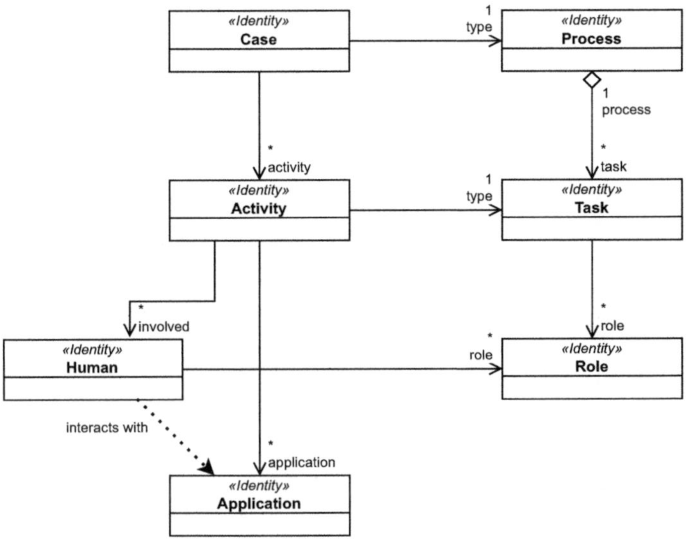

Fig. 4 Core concepts of BPM

centric processes" in order not to blur the conceptual model of BPM and IoT with the additional paradigm shift toward object-centric processes, which is not too relevant to our discussion.

Figure 5 extends the core concepts of BPM from Fig. 4 and brings them together with IoT to help make BPM more human-centric. First of all, this conceptual model acknowledges that modern PAIS will not only have Applications that humans can interact with but might also use IoT devices for letting humans interact with the system. Therefore, Applications are generalized to any kind of information technology (IT), which can be classical Applications but also IoT devices as an explicit part of the PAIS.

And the Human might interact with the environment and cause all kinds of effects in the environment through IoT devices. In turn, the environment (which is not an explicit part of the PAIS) could detect all kinds of events and changes by IoT devices. A part of what is going on in the environment or internally in the PAIS might be recorded as Events. These Events will typically not have a direct meaning in the PAIS, which is why we might also call them raw events. But, by filtering and combing some of the raw Events into aggregated events, these aggregated events could bear some importance or meaning for the processes running in the PAIS; such aggregated events, we call BPM events. For example, a human adding some items or goods into the basket in a supermarket might start a new case of a shopping process.

With this conceptual model, we want to put attention to two components that are needed for this: event monitoring and configuration for collecting data or, actually, Events from the environment, and event aggregation and configuration for aggregating these raw Events to higher-level BPM events. Today, the collection and

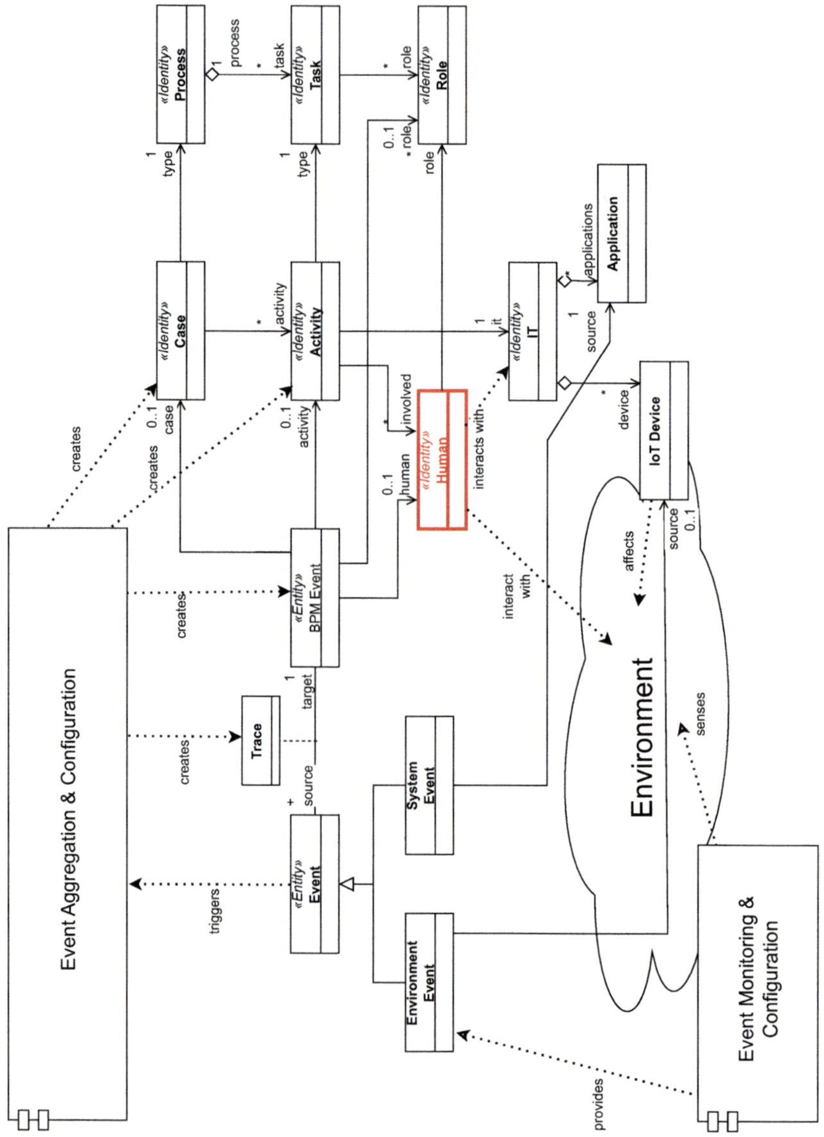

Fig. 5 BPM IoT conceptual model (complete)

aggregation are mostly done programmatically in some specific setting. However, we believe that this could and should be done by configurations as an integral part of setting up PAIS without programming, much like workflow models were used for enacting workflows without programming them anymore.

The classes in the conceptual model in Figs. 4 and 5 do not have any attributes, since these are not too relevant to the discussion in this paper. But we included the nature and characteristics of some classes. The most important characteristics are indicated in the diagram by the two stereotypes «Identity»and «Entity».

The concepts with the stereotype «Identity» indicate that these objects have an identity that can somehow be traced and identified across different systems, and they are permanently stored somewhere in the PAIS. This applies to most of the BPM concepts, where the processes, tasks, cases, and activities and some of their relevant attributes and history need to be stored permanently for legal and documentation purposes. In particular, for activities, these attributes could be the start and end time.

By contrast, the concepts with stereotype «Entity» are concepts where the particular attributes would depend on the domain in which the processes are running. This concerns the payload of Events. But all kinds of Event objects will have a timestamp. An additional characteristic of «Entity» is that their amount might be so high that not all of them could or should be stored in a PAIS in the long run. So these might not be stored for longer periods. These objects might have all kinds of attributes that cannot be defined upfront since they depend on the domain. Therefore, there needs to be a way to handle the attributes of these objects more dynamically, such as entities with attribute-value pairs. This is the reason why we call it «Entity» stereotype; but we do not dwell on this since the model we present here is just a conceptual model and not a data model. For now, events (or objects of using stereotype «Entity») can have any kind of property and they have a timestamp.

Figure 6 shows which parts of the conceptual model relate to the areas of BPM (blue) and IoT (orange), and it shows that the human (red) is at the intersection of both. The dashed lines show areas where we currently do not see or want to assign to which area they belong. This, in particular, applies to the area that we indicated by the components for the event aggregation and its configuration, which is responsible for combining and aggregating the raw Events from IoT to some meaningful BPM events. As mentioned above, this part is often programmed specifically for some dedicated application. This, however, would need to be programmed by software developers who have a very good understanding of the domain and the processes, or it would require complex event processing (CEP). But we believe that this could and should be done by domain experts themselves (who typically are not software developers), e.g. by defining rules for combining and aggregating raw events to meaningful events, which we call event monitoring and event aggregation configurations. And these configurations should be part of modeling the business process.

The way how these configurations could be done in an appropriate way for domain experts who are not software engineers and programmers or IoT experts is still a research issue, which is of great importance for making processes more

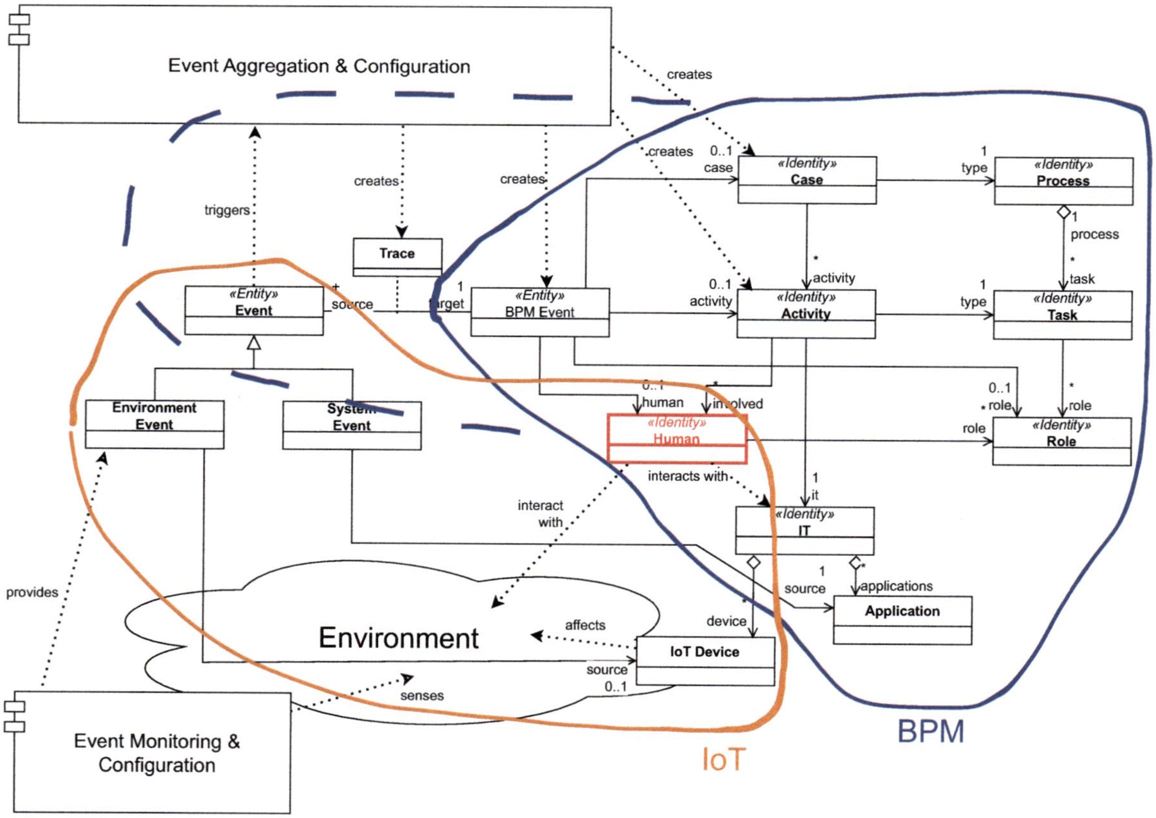

Fig. 6 BPM IoT conceptual model (areas)

human-centric—right at the intersection of BPM, IoT, and CEP. This also applies how to adequately integrate IoT devices to PAIS just by configuration.

3 Use Case UC1: Detecting Deviations in Logistics Processes at Runtime

3.1 Motivations

Business processes belonging to the logistics domain are multi-party, that is, they require multiple organizations to cooperate for the Process to be successfully executed. This implies that no organization has full control over the whole Process, but only on the portion that the organization is responsible for. For example, the BPMN model shown in the upper portion of Fig. 8 represents a (simplified) logistics process, where some goods are loaded into a shipping container, which is then shipped, and finally the goods are unloaded. In this process, the manufacturer would be responsible for loading the goods. However, (s)he would not be able to control how the container is shipped and the goods unloaded, since these tasks are performed by the shipper and the customer, respectively. In particular, (s)he cannot guarantee that the container will never be overheated.

In this setting, being able to continuously monitor how such processes are executed, and to detect when a Case deviates from the expected execution, becomes critical. However, monitoring processes that are distributed among multiple participants poses significant challenges. Existing monitoring solutions typically rely on BPM events that—as discussed in Sect. 2—are generated by a Business Process Management System (BPMS), which is usually confined within organizational boundaries. Thus, monitoring collaborative processes may require federating such systems, a daunting task, particularly for short-term collaborations.

Another significant challenge in process monitoring is ensuring the accuracy, timeliness, and reliability of Events related to process execution. While monitoring automated processes through a single software component like a BPMS may involve collecting and analyzing execution logs generated by these systems, the presence of manual Tasks complicates obtaining reliable monitoring information. In this setting, the Human responsible for executing the Activity related to that Task has to interact with an Application to notify that the Activity was started and completed. As such, this person may forget to send, delay, or forge this information, leading to incorrect monitoring results. In addition, doing so disrupts that Human from his/her working routine.

Lastly, continuously and autonomously detecting deviations from the Process model is far from trivial. Conformance-checking techniques typically operate postmortem with complete event logs, providing monitoring information only after process completion. Conversely, compliance-checking techniques can operate in real time with partial event logs, but they often only indicate whether a constraint

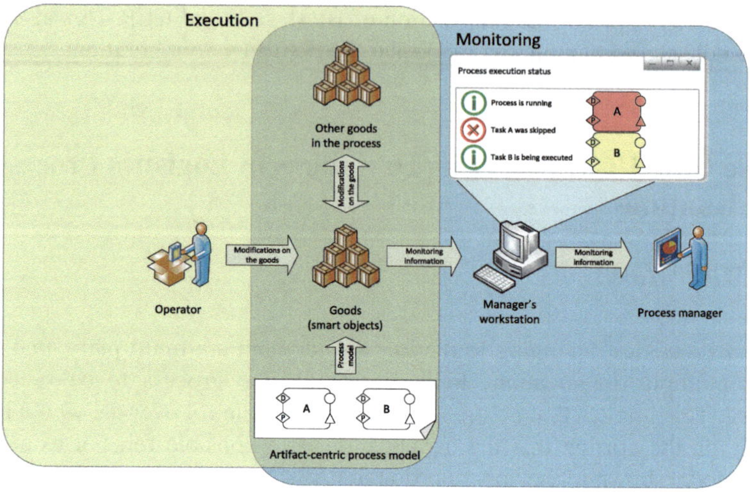

Fig. 7 Key idea behind artifact-driven process monitoring. (Figure from [48])

was met or violated without pinpointing the exact discrepancies. Breaking down the model into multiple constraints may alleviate this issue, but it also significantly increases the complexity of compliance checks if all potential discrepancies must be accounted for.

3.2 *Key Idea*

To address the aforementioned challenges, artifact-driven monitoring was introduced [48]. As shown in Fig. 7, the key idea behind this monitoring technique is that, by observing the changes in the artifacts—physical or virtual objects involved in a process—one can infer how the Process is executed. Specifically, artifact-driven monitoring adopts the IoT paradigm to make physical artifacts taking part in a process "smart." Being equipped with sensors, computing devices (such as single board computers), and communication interfaces, physical artifacts become IoT devices aware of their own conditions and of the environment in which they reside. Also, they can autonomously exchange this information with other IoT devices.

The main benefits of artifact-driven monitoring are the following:

- Manual notifications are no longer required. Manual Tasks inherently alter the state of one or more artifacts. For instance, delivering a package alters its position. Consequently, if the conditions of these artifacts are automatically monitored by IoT devices, they will produce external Events. If an Event aggregation between Events and BPM events has been defined, operators overseeing such activities no longer need to interact with an Application to manually report on their execution details.

- Ease of monitoring collaborative processes: Physical artifacts are in close contact with the processes they are involved in, even if such processes cross the boundaries of a single organization. Hence, they can independently collect information relevant for that Process, without the need for this information to go through Applications belonging to external organizations.
- Deviations in the process model can be promptly identified: Artifact-driven monitoring adopts an artifact-centric view of the process, treating Task dependencies as descriptive rather than prescriptive. This approach allows to achieve runtime conformance checking and, in case a deviation is detected, to identify which portion of the process is affected.

3.3 Modelling Language

Most business processes are modeled with imperative models, such as Business Process Modeling Notation (BPMN) process diagrams. While such models are fine to document how the Process is executed and to instruct a BPMS, they do not allow flexibility in process monitoring. Indeed, all control flow dependencies are expected to always hold. In addition, such languages lack the possibility to express conditions on data to determine when activities are started or completed. For example, the BPMN model in the upper part of Fig. 8 shows a simplified delivery process consisting of three manual Tasks (Load goods, Ship container, and Unload goods). This model would not allow for a container to be shipped without the goods being loaded first. However, being all three tasks manual, such behavior may happen and it would not be captured by that model.

To address these issues, the Extended-GSM (E-GSM) language has been proposed to model the Process to monitor [49]. E-GSM is an extension of the Guard-Stage-Milestone (GSM) language, originally designed for automating the execution of knowledge-intensive process models [14]. In GSM, Tasks and process portions are represented by stages. Stages can be atomic to represent atomic Tasks, or they can hierarchically nest other stages.

To determine under which conditions an Activity starts and completes its execution, the stage representing the corresponding Task is decorated with one or more data flow guards and with one or more milestones. Data flow guards and milestones are Event-Condition-Action (ECA) rules, requiring an Event and a Boolean expression to be specified. When the specified Event happens, the ECA rule is said to be fired. If an ECA rule is fired and its Boolean expression evaluates to true, the ECA rule is said to hold.

Stages can assume one of the following states: *unopened*, *opened*, or *closed*. When the monitoring starts, all stages are marked as *unopened*, denoting that they have not been executed. When a data flow guard holds, if the stage to which that data flow guard is attached is *unopened* or *closed*, that stage becomes *opened*. This indicates that the process element represented by that stage is running. When a milestone holds, it becomes achieved. Also, if the stage to which the milestone is

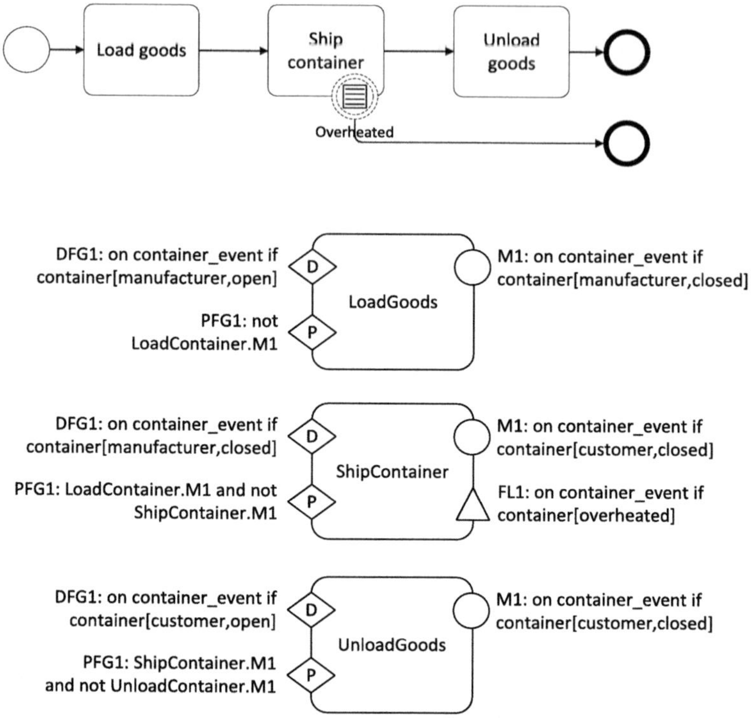

Fig. 8 BPMN model of a (simplified) logistics process (upper part) and equivalent E-GSM model for monitoring (lower part)

attached is *opened*, that stage becomes *closed*. This indicates that the execution of the process element represented by that stage is completed. In addition, if that stage has any child stage that is *opened*, these stages become *closed* as well.

Thank to data flow guards and milestones, it is possible to define conditions on attributes of artifacts, thus allowing to automatically infer when activities are executed.

E-GSM extends GSM by introducing two additional constructs: process flow guards and fault loggers. To specify the control flow dependencies that are expected to hold for a Task, the stage representing that Task is decorated with a process flow guard. A process flow guard is a Boolean expression, which predicates on the other elements in an E-GSM model. Such expression is evaluated when at least one of the data flow guards of the attached stage holds, before the stage becomes *opened*.

To specify under which conditions, independently from control flow deviations, an Activity would fail its execution, the stage representing the corresponding Task is decorated with a fault logger. A fault logger is an ECA rule that is evaluated while a stage is *opened*.

To keep track on whether the execution of an Activity deviates from the expected flow, E-GSM classifies stages as *onTime*, *outOfOrder*, or *skipped*. Initially, all stages are marked as *onTime*. Upon satisfaction of a data flow guard of a stage, if the process flow guard of that stage evaluates to true, the stage remains *onTime*. Otherwise, the stage is marked as *outOfOrder*. Moreover, if the process flow guard requires another stage to be *opened* or *closed*, and that stage is *unopened*, then that other stage is marked as *skipped*.

To detect if the execution of an Activity is unsuccessful, E-GSM also classifies stages as *regular* or *faulty*. Initially, all stages are marked as *regular*. If, while a stage is *opened*, one of its fault loggers holds, that stage is marked as *faulty*.

It is important to note that process flow guards and fault loggers do not enforce control flow dependencies. A stage can still be *opened* even if its process flow guard evaluates to false, as long as at least one of its data flow guards holds. Similarly, a stage can remain *opened* even if one of its fault loggers hold. This capability facilitates continuous and autonomous process monitoring. Also, ECA rules effectively turn Events into BPM events. Thus, they play the role of Event aggregations.

The lower part of Fig. 8 shows an E-GSM model capable of monitoring the logistics process previously introduced, whose BPMN model is shown in the upper part of that figure. In particular, the atomic stages LoadGoods, ShipContainer, and UnloadGoods represent the homonymous tasks in the BPMN model. Data flow guards and milestones on that stage are triggered by and predicate on Events produced by the IoT device attached to the container being shipped, which is capable of providing the position, the condition of the doors (open or closed), and whether it is overheated or not. For example, LoadGoods.DFG1 requires the container to be at the manufacturer with the doors open to cause the stage LoadGoods to open, indicating that Activity Load goods has started. Control flow dependencies are represented by process flow guards. For example, ShipContainer.PFG1 requires LoadContainer to be closed and ShipContainer not to be closed, indicating that it must be executed after activity Load container and no more than once. However, if the container is shipped without loading the goods, this behavior can still be captured by the model. In this case, ShipContainer.PFG1 will not hold. Thus, ShipContainer will become outOfOrder and LoadContainer skipped. The non-interrupting boundary event Overheated in the BPMN model is represented by the fault logger ShipContainer.FL1.

3.4 *Software Architecture*

To implement an artifact-driven monitoring platform, the reference architecture shown in Fig. 9 has been proposed, organized into four main modules:

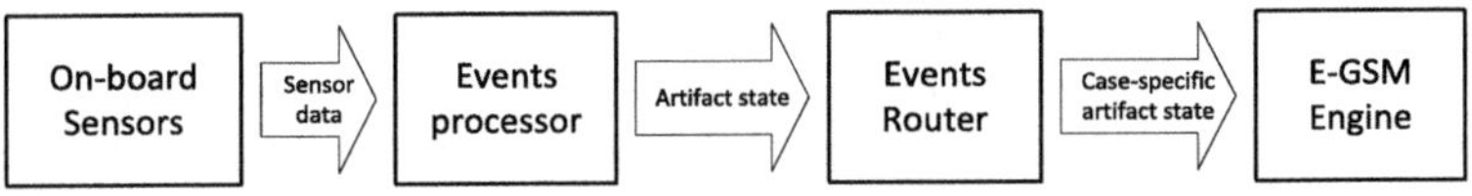

Fig. 9 Reference architecture of an artifact-driven monitoring platform

- On-board Sensors Gateway: This module operates on each IoT device, responsible for periodically gathering Events from the environment.
- Events Processor: This module receives Events from the On-board Sensors Gateway, analyzes them, and determines the artifact's state. Depending on complexity and computing capabilities, it may run locally on IoT devices or remotely in an on-premises or cloud environment.
- Events Router: This module is responsible for keeping track of which IoT device participates in which Case. In this way, it will forward only Events relevant for a specific Case to the corresponding E-GSM instance.
- EGSM Engine: Housing the EGSM model of the Process to monitor, this module detects when each Activity is executed and when the a Case deviates with respet to the planned Process. Upon receiving a new Event from the Events Router, it triggers ECA rules in the EGSM model corresponding to that Event.

This architecture was implemented in the SMARTifact platform [50], with specific technologies: Node-RED for the Events Processor, MQTT protocol for the Events Router, and Node.js for the EGSM Engine. The platform's modest computing requirements allowed deployment on an Intel Galileo single-board computer.

4 Use Case UC2: Understanding Deviations in Daily Routine for Healthcare Purposes

In the domain of Human behavior, recurrent behaviors form what we commonly call *routines*. These routines, marked by repetitive actions, offer a sense of predictability and regularity in individuals' lives [36]. A person's daily routine encompasses the Tasks of everyday life. Understanding these patterns sheds light on preferences and habits and enables us to observe how routines shift and develop over time. Delving into the intricacies of daily routines is crucial for a comprehensive grasp of human behavior.

A business Process can be depicted through a process model, which offers a representation of the flow of Tasks and decisions. This model provides a structured and abstract overview of the process, aiding in the comprehension and analysis of its execution.

Once we establish the ability to represent daily routines as Process models, process mining techniques [13, 57, 59], typically employed for analyzing processes, can be repurposed to explore human behavior. Essentially, process mining involves analyzing and extracting valuable insights from data, operating under the assumption that any process with recordable events can potentially benefit from its application. This adaptability makes process mining a potent and widely applicable tool for gaining deeper insights into various processes.

In healthcare, utilizing individual patient data enables more personalized medical care, with the potential of leading to enhanced patient well-being. For example, individuals dealing with neurological conditions face numerous challenges in their daily lives. This is also due to insufficient tracking of cognitive and behavioral changes [37, 55]. Investigating these challenges is crucial for clinical interventions, as a thorough understanding of deficiencies can improve therapeutic effectiveness. Since physiological functions are not easily quantifiable, information about a person's daily routine can help identify relevant health conditions. For example, let's consider a person suffering from dementia. If such a person is used to sticking to a structured routine (e.g., during the evening, night, or when waking up) and this routine gets disrupted in an unprecedented way, it could be a signal of the worsening of dementia.

To benefit from process mining in the healthcare context, however, several challenges need to be addressed [52] first. Among these, when the human *is* the process (cf. Fig. 3), modeling human behavior as a process model and capturing data referring to daily activities are the most important ones.

4.1 Modeling Human Behavior

A process describing human behavior differs significantly from a business process in several aspects [25]. Specifically, human behavior blends intentions, perceptions, and states. Routines, on the other hand, represent intermediate structures composed of aggregated tangible actions resulting from human behavior—that is, the actual sequence of activities undertaken during a day. For our investigation, we consider routines as a proxy for human behavior. Human behavior can be understood as a sequence of routines. A routine consists of a series of activities, which, at a lower level of abstraction, are sequences of observable units. The same routine can happen multiple times, thus resulting in several Cases. As illustrated in Fig. 10, the structure of human behavior is hierarchical. A routine may comprise other routines, and an activity may consist of other activities.

When modeling human behavior and routines, psychological and external factors must also be considered, as they create a close interdependence between behavior and context [6]. Behavior is then guided by the person who decides how and when to perform certain actions, but it is also contingent on specific circumstances (or

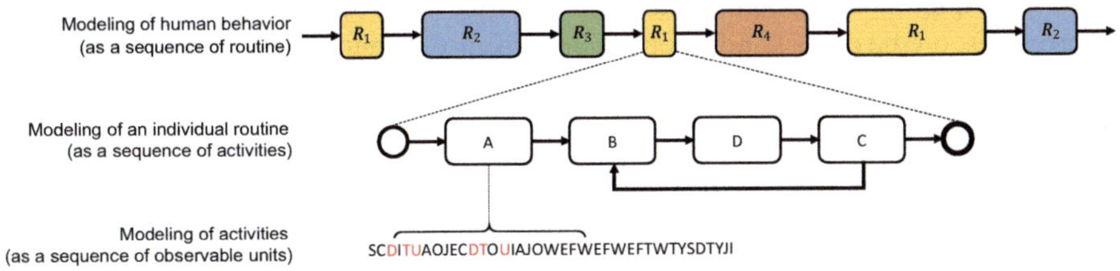

Fig. 10 A simplified hierarchical structure of human behavior. (Figure from [25])

context). Thus, based on these variables, individuals choose how and which routine to implement.

A comprehensive list of requirements to model routines and how these are matched by existing process modeling languages is available in the literature [18, 25]. However, it is important to mention that no single language is capable of capturing all shades and peculiarities of human behavior. A common approach, therefore, is to use a combination of multiple languages to describe the behavior from different angles. For example, to describe the general structure of the routine, an imperative language can be used (such as Petri nets or BPMN); local constraints among activities are better captured by declarative language (such as DECLARE or DCR), and finally, statistics (such as frequency and absolute time distributions) can help to capture aspects otherwise neglected by the control-flow alone.

4.2 Capturing Human Behavior

Ambient-assisted living environments leverage technologies and services to support individuals with specific needs in living longer and independently in their natural surroundings [54]. These environments can capture individuals' daily routines using technologies such as Internet of Things (IoT devices) systems, which collect data on daily activities [6]. Process mining techniques can automatically derive a process model representing the typical daily routine of an individual from this data [22, 27]. Such a model has the potential to offer a comprehensive representation of the human behavior manifested in the routine. Additionally, the person's behavior can be continually monitored using IoT systems and compared with the reference model to identify and explain deviations in execution. Detecting variations in behavior can serve as a tool for identifying critical health conditions and facilitating interventions.

Event and Activity Abstraction In order to benefit from the ambient-assisted living data, it is necessary to abstract these into more tangible activities, which can be fed to process mining algorithms via Event aggregations. From the literature, we can distinguish two primary approaches to accomplishing such an abstraction: model-based and trace-based abstractions.

Model-based approaches seek to recognize activities by representing them through process models, such as "local process models" expressed using Petri nets. Examples of these approaches are reported in [43, 46, 47, 56]. Trace-based approaches, on the other hand, focus on identifying activities at the trace level: these try to transform a trace into another trace, where events are now at a higher level of abstraction [15, 23, 24, 44].

Routine Discovery and Conformance In the literature, we can find that two main approaches have been used to discover routines and verify the adherence of new behavior with existing routines. These have implications in terms of how the discovery is performed as well as how the conformance is computed.

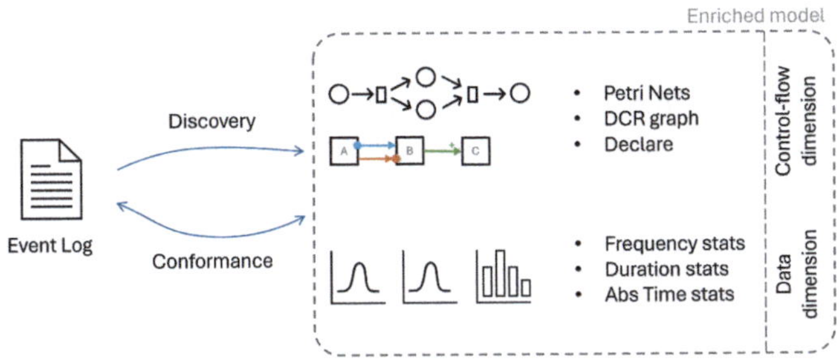

Fig. 11 Multidimensional approach. (Figure from [18])

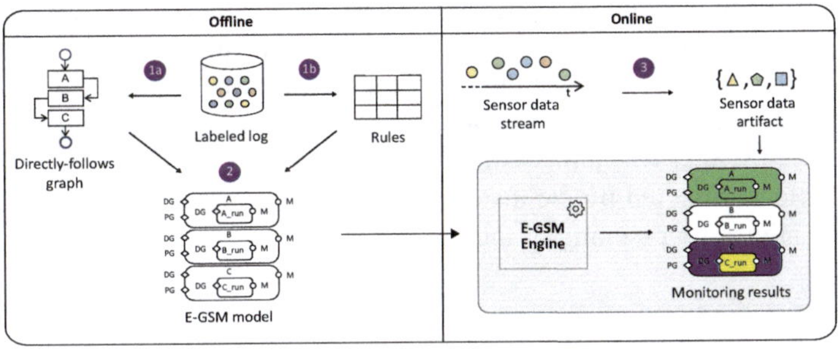

Fig. 12 Overview of the single-model approach based on E-GSM. (Figure from [27])

Since human behavior is multifaceted, leveraging only the control flow is insufficient. To comprehensively understand personal behavior processes, a possible approach is to adopt a multidimensional process discovery and conformance approach [22]. This not only considers control flow but also incorporates a data dimension. Figure 11 gives a graphical intuition of the idea: it starts with an abstracted event log, where activities are already aggregated at the process level. The control flow is then captured using both declarative and imperative languages, whereas the data flow is captured by computing statistics. Adequate algorithms are used for each notation to compute the conformance of these models. For example, alignment algorithms are used for Petri nets [13]. Rule checkers are employed for declarative models [17]. For the statistical aspects (i.e., frequencies), the normal distribution function is employed. By considering the mean frequency as the reference value, it is possible to interpret the probability density function as the likelihood of the mean frequency value in the trace being close to the reference. This likelihood is then returned as the fitness value. All these fitness values are aggregated to a final conformance value.

An alternative approach to multidimensional models is to consider models that embed both data conditions and the process aspects, such as E-GSM, which has been introduced in Sect. 3.3. In Fig. 12, an overview of a possible approach is provided, consisting of two phases: the offline phase (where the model is derived) and the online phase (which includes streaming conformance). The peculiarity of this approach is that the identification of activities and conditions is fully automated. To compute the conformance, the existing E-GSM engine evaluates the conditions on the different activities and extracts conformance properties on them, such as activities executed when expected (on time), activities that are not compliant with control flow constraints (out of order), or activities that are skipped.

4.3 Evaluation

Results on synthetic [28] as well as real data [26] have shown that it is possible to leverage the presented techniques to discover deviations in human behavior when a process representation is leveraged. These results, combined with visual analytics techniques [1, 2], have the potential to empower nonexpert users to better understand human behavior. While tests on patients affected by dementia are still very limited [18, Sec. 10-2], this family of techniques shows the potential to closely monitor patients affected by the disease, with the ultimate goal of spotting as early as possible signs of worsening of their disease.

5 Conclusion

In this chapter, we focused on the challenges and opportunities that the IoT would offer in the context of human-centric processes, that is, business processes heavily involving and centered around interaction with humans. To this aim, we discussed related work in this area, and we designed a conceptual model that captures the interactions between the IoT, BPM, and humans. Finally, we then presented two different use cases of IoT-enabled human-centric processes discussing how these can be framed within the proposed model.

The first use case relied on the IoT to monitor the behavior of humans involved in a process and their impact on the correct execution of the process. The second use case relied on IoT to observe the daily routines that humans perform—which can be seen as business processes—and to detect changes over time. Despite being two very different use cases, with humans having, respectively, an active and passive role in the process, the proposed model was capable of describing them.

Glossary

Activity An activity is the execution of some Task of a Process in a specific Case.

Application In traditional workflow management, the applications are assigned to certain Tasks or the activities, which will be used by a Human actor to perform the tasks. And via these applications Humans will make changes to the state of the case and the data and documents in it (where we do not dwell on concepts involving the information aspects of a Process in the context of this paper).

BPM event Some events from the environment in aggregated might have meaning in the BPM Process and can be used tp as input and to control the execution of a Case. These are called BPM events, which can be associated with a Case, a Activity, a Human or Role.

Case A case is single instance of the execution of a Process.

Event All things that occur in environment of a workflow system or PAIS can be perceived by a events. A single raw event, might not have a meaning from a Process perspective. But, a combination of them might have a meaning be useful to control the Process.

Event aggregation Event aggregation derives higher-level BPM events from low-level or is established the raw Events. The provenance of this high-level BPM event and raw Events it was derived from can represented by a trace.

Human In a Process activities are performed by or need the participation or authorization of human actors (notwithstanding the fact that some activities can be performed automatically, here we focus on the human actors). In the process model, there is typically an organizational aspect or perspective, which defines an organization model, and defines Roles, assigns humans to these roles, and defines which roles a human needs to have to perform a certain Task.

IoT device In modern PAIS, activities can be performed not only by Applications, but also by IoT devices, which can be an explicit part of the PAIS; or and implicit part by using Events from environment.

Process A process is a business process executed in some organization for achieving certain goals. The process is actually the conceptual idea (often called the model), how each individual execution of this process would look. A process consists of the different Tasks that need to be done to achieve the goal; and the processes typically defines in which order these tasks need to be done. Each different execution (or instance) of this same process is called a Case of that Process.

Role In a Process, their different roles and these roles defines which activities can be performed by which Humans.

Task A defines task are the different units of work that need be done to execute a Process. The instance of the execution within a specific Case is called an Activity.

References

1. Alman, A., Arleo, A., Beerepoot, I., Burattin, A., Ciccio, C.D., Resinas, M.: Tiramisù: a recipe for visual sensemaking of multi-faceted process information. In: ICPM Workshop Proceedings, pp. 19–31. Springer, Berlin (2023)
2. Alman, A., Arleo, A., Beerepoot, I., Burattin, A., Ciccio, C.D., Resinas, M.: Tiramisù: making sense of multi-faceted process information through time and space. J. Intell. Inform. Syst. (2024).

3. Arnaiz, D., Vila, M., Alarcón, E., Moll, F., Sancho, M.R., Teniente, E.: Relating Context and Self Awareness in the Internet of Things. In: Cooperative Information Systems, pp. 384–402. Springer, Berlin (2024)
4. Ashton, K.: That "Internet of Things" thing. RFID J. **22**(7), 97–114 (2009)
5. Axenath, B., Kindler, E., Rubin, V.: AMFIBIA: a meta-model for the integration of business process modelling aspects. Int. J. Bus. Process Integration Manag. **2**(2), 120–131 (2007)
6. Banovic, N., Buzali, T., Chevalier, F., Mankoff, J., Dey, A.K.: Modeling and understanding human routine behavior. In: Proceedings of the 2016 CHI Conference on Human Factors in Computing Systems, pp. 248–260. (2016)
7. Becker, F., Bibow, P., Dalibor, M., Gannouni, A., Hahn, V., Hopmann, C., Jarke, M., Koren, I., Kröger, M., Lipp, J., Maibaum, J., Michael, J., Rumpe, B., Sapel, P., Schäfer, N., Schmitz, G.J., Schuh, G., Wortmann, A.: A conceptual model for digital shadows in industry and its application. In: Conceptual Modeling (pp. 271–281). Springer, Berlin (2021)
8. Berges, I., Ramírez-Durán, V.J., Illarramendi, A.: Facilitating data exploration in industry 4.0. In: Advances in Conceptual Modeling, Lecture Notes in Computer Science, pp. 125–134. Springer, Berlin (2019)
9. Bertrand, Y., Veneruso, S., Leotta, F., Mecella, M., Serral, E.: NICE: the Native IoT-centric event log model for process mining. In: ICPM Workshop Proceedings, pp. 32–44. Springer, Berlin (2023)
10. Bertrand, Y., Weerdt, J., Serral, E.: An expert-validated bridging model for IoT process mining. Bus. Inform. Syst. Eng. **66**(6), 737–756 (2024)
11. Bertrand, Y., Weerdt, J.D., Serral, E.: A bridging model for process mining and IoT. In: ICPM Workshop Proceedings, volume 433 of Lecture Notes in Business Information Processing, pp. 98–110. Springer, Berlin (2021)
12. Bertrand, Y., Weerdt, J.D., Serral, E.: A novel multi-perspective trace clustering technique for IoT-enhanced processes: a case study in smart manufacturing. In: Proceedings of BPM, pp. 395–412. Springer, Berlin (2023).
13. Carmona, J., van Dongen, B., Solti, A., Weidlich, M.: Conformance Checking. Springer, Berlin (2018)
14. Damaggio, E., Hull, R., Vaculín, R.: On the equivalence of incremental and fixpoint semantics for business artifacts with guard-stage-milestone lifecycles. Inform. Syst. **38**(4), 561–584 (2013)
15. de Leoni, M., Dündar, S.: Event-log abstraction using batch session identification and clustering. In: Proceedings of the ACM SAC, pp. 36–44 (2020)
16. De Luzi, F., Leotta, F., Marrella, A., Mecella, M.: On the interplay between business process management and Internet-of-Things. Bus. Inform. Syst. Eng. **67**(2), 265–288 (2025)
17. Debois, S., Hildebrandt, T.T., Laursen, P.H., Ulrik, K.R.: Declarative process mining for DCR graphs. In: Proceedings of the Symposium on Applied Computing, pp. 759–764 (2017)
18. Di Federico, G.: A Process Mining Framework to Analyze Variability in Human Behavior. Ph.D. Thesis, Technical University of Denmark, 2024
19. Di Francescomarino, C., Ghidini, C.: Predictive process monitoring. In: Process Mining Handbook, pp. 320–346. Springer, Berlin (2022)
20. Diamantini, C., Mircoli, A., Potena, D., Storti, E.: Process-aware IIoT knowledge graph: A semantic model for industrial IoT integration and analytics. Future Gener. Comput. Syst. **139**, 224–238 (2023)
21. Dumas, M., van der Aalst, W.M.P., ter Hofstede, A.H.M. (eds.): Process-Aware Information Systems: Bridging People and Software Through Process Technology. Wiley, London (2005)
22. Federico, G.D., Burattin, A.: Do you behave always the same? A process mining approach. In: ICPM Workshop Proceedings, pp. 5–17. Springer, Berlin (2022)
23. Federico, G.D., Burattin, A.: vAMoS: eVent Abstraction via Motifs Search. In: BPM Workshop Proceedings, pp. 101–112. Springer, Berlin (2022)
24. Federico, G.D., Burattin, A.: CvAMoS—Event Abstraction Using Contextual Information. Future Internet **15**(3), 113 (2023)

25. Federico, G.D., Burattin, A., Montali, M.: Human behavior as a process model: which language to use? In: Proceedings of the Italian Forum on BPM, volume 2952 of CEUR Workshop Proceedings, pp. 18–25. CEUR-WS.org, 2021
26. Federico, G.D., Fernández-Llatas, C., Ahmadi, Z., Shirali, M., Burattin, A.: Identifying variation in personal daily routine through process mining: a case study. In: ICPM Workshop Proceedings (PODS4H) (2023)
27. Federico, G.D., Meroni, G., Burattin, A.: Viola: detecting violations of behaviors from streams of sensor data. In: BPM Workshop Proceedings, pp. 118–130. Springer, Berlin (2023)
28. Federico, G.D., Nikolajsen, E.R., Azam, M., Burattin, A.: Linac: a smart environment simulator of human activities. In: ICPM Workshop Proceedings, pp. 60–72. Springer, Berlin (2021)
29. Franceschetti, M., Seiger, R., Weber, B.: An event-centric metamodel for IoT-driven process monitoring and conformance checking. In: BPM Workshop Proceedings, pp. 131–143. Springer, Berlin (2023)
30. Frigo, M., Hirmer, P.: A toolbox for the Internet of Things—easing the setup of IoT applications. In: ER Forum/Posters/Demos Proceedings (2020)
31. Guldner, A., Hoffmann, M., Lohr, C., Machhamer, R., Malburg, L., Morgen, M., Rodermund, S.C., Schäfer, F., Schaupeter, L., Schneider, J., Theusch, F., Bergmann, R., Dartmann, G., Kuhn, N., Naumann, S., Timm, I.J., Vette-Steinkamp, M., Weyers, B.: A framework for AI-based self-adaptive cyber-physical process systems. Inform. Technol. **65**(3), 113–128 (2023)
32. Hollingsworth, D.: The workflow reference model. Technical Report TC00-1003, The Workflow Management Coalition (WfMC), 1995
33. Hollingsworth, D.: The workflow reference model: 10 years on. Technical report, The Workflow Management Coalition (WfMC), 2004
34. Holzmüller-Laue, S., Göde, B., Thurow, K.: A BPM-based approach of human task support within automated workflows. In: Human-Computer Systems Interaction: Backgrounds and Applications 3, pp. 117–128. Springer, Berlin (2014)
35. Janiesch, C., Koschmider, A., Mecella, M., Weber, B., Burattin, A., Di Ciccio, C., Fortino, G., Gal, A., Kannengiesser, U., Leotta, F., Mannhardt, F., Marrella, A., Mendling, J., Oberweis, A., Reichert, M., Rinderle-Ma, S., Serral, E., Song, W., Su, J., Torres, V., Weidlich, M., Weske, M., Zhang, L.: The internet of things meets business process management: a manifesto. IEEE Syst. Man Cybern. Mag. **6**(4), 34–44 (2020)
36. Katz, S.: Assessing self-maintenance: activities of daily living, mobility, and instrumental activities of daily living. J. Am. Geriatr. Soc. **31**(12), 721–727 (1983)
37. Kaye, J., Mattek, N., Dodge, H.H., Campbell, I., Hayes, T., Austin, D., Hatt, W., Wild, K., Jimison, H., Pavel, M.: Unobtrusive measurement of daily computer use to detect mild cognitive impairment. Alzheimer's Dementia **10**(1), 10–17 (2014)
38. Kirikkayis, Y., Gallik, F., Seiger, R., Reichert, M.: Integrating IoT-driven events into business processes. In: Proceedings of CAiSE Forum, pp. 86–94. Springer, Berlin (2023)
39. Kirikkayis, Y., Gallik, F., Winter, M., Reichert, M.: BPMNE4IoT: a framework for modeling, executing and monitoring IoT-driven processes. Future Internet **15**(3), 90 (2023)
40. Kirikkayis, Y., Winter, M., Reichert, M.: A user study on modeling IoT-aware processes with BPMN 2.0. Information **15**(4), 229 (2024)
41. Lee, S.U., Paik, H., Kanhere, S.S.: Human-sensing platforms and ethical considerations throughout their data life-cycles. IEEE Internet Things Mag. **7**(1), 66–73 (2024)
42. Leotta, F., Marrella, A., Mecella, M.: IoT for BPMers. Challenges, case studies and successful applications. In: Proceedings of BPM, pp. 16–22. Springer, Berlin (2019)
43. Leotta, F., Mecella, M., Sora, D.: Visual process maps: a visualization tool for discovering habits in smart homes. J. Ambient Intell. Hum. Comput. **11**(5), 1997–2025 (2020)
44. Lu, X., Fahland, D., Andrews, R., Suriadi, S., Wynn, M.T., ter Hofstede, A.H., van der Aalst, W.M.: Semi-supervised log pattern detection and exploration using event concurrence and contextual information. In: Proceedings of OTM, pp. 154–174. Springer, Berlin (2017)
45. Mangler, J., Grüger, J., Malburg, L., Ehrendorfer, M., Bertrand, Y., Benzin, J., Rinderle-Ma, S., Asensio, E.S., Bergmann, R.: Datastream XES extension: embedding IoT sensor data into extensible event stream logs. Future Internet **15**(3), 109 (2023)

46. Mannhardt, F., Leoni, M.D., Reijers, H.A., Van Der Aalst, W.M., Toussaint, P.J.: From low-level events to activities-a pattern-based approach. In: Proceedings of BPM, pp. 125–141. Springer, Berlin (2016)
47. Mannhardt, F., Tax, N.: Unsupervised event abstraction using pattern abstraction and local process models (2017). arXiv:1704.03520
48. Meroni, G.: Artifact-Driven Business Process Monitoring - A Novel Approach to Transparently Monitor Business Processes, Supported by Methods, Tools, and Real-World Applications, volume 368 of Lecture Notes in Business Information Processing. Springer, Berlin (2019)
49. Meroni, G., Baresi, L., Montali, M., Plebani, P.: Multi-party business process compliance monitoring through IoT-enabled artifacts. Inf. Syst. **73**, 61–78 (2018)
50. Meroni, G., Garda, S.: Artifact-driven process monitoring at scale. In: Proceedings of ICSOC, pp. 3–12. Springer, Berlin (2023)
51. Monti, F., Mathew, J.G., Leotta, F., Koschmider, A., Mecella, M.: On the application of process management and process mining to industry 4.0. Softw. Syst. Model. **23**(6), 1407–1419 (2024)
52. Munoz-Gama, J., Martin, N., Fernandez-Llatas, C., Johnson, O.A., Sepúlveda, M., Helm, E., Galvez-Yanjari, V., Rojas, E., Martinez-Millana, A., Aloini, D., et al.: Process mining for healthcare: characteristics and challenges. J. Biomed. Inform. **127**, 103994 (2022)
53. Poss, L., Dietz, L., Schönig, S.: LABPMN: location-aware business process modeling and notation. In: Proceedings of CoopIS, pp. 198–216. Springer, Berlin (2023)
54. Queirós, A., Silva, A., Alvarelhão, J., Rocha, N.P., Teixeira, A.: Usability, accessibility and ambient-assisted living: a systematic literature review. Universal Access Inform. Soc. **14**, 57–66 (2015)
55. Rosen, H.J.: Anosognosia in neurodegenerative disease. Neurocase **17**(3), 231–241 (2011)
56. Tax, N., Sidorova, N., Haakma, R., van der Aalst, W.M.: Mining local process models. J. Innov. Digit. Ecosyst. **3**(2), 183–196 (2016)
57. van der Aalst, W.M.P.: Process Mining: Data Science in Action, 2nd edn. Springer, Berlin (2016)
58. van der Aalst, W.M., Berti, A.: Discovering object-centric Petri nets. Fundamenta Informaticae **175**(1–4), 1–40 (2020)
59. van der Aalst, W.M., Carmona, J. (eds.): Process Mining Handbook. Springer, Berlin (2022)
60. Vila, M., Sancho, M.R., Teniente, E., Vilajosana, X.: Semantics for Connectivity Management in IoT Sensing. In: Proceedings of Conceptual Modeling, pp. 297–311. Springer, Berlin (2021)
61. Ystgaard, K.F., Atzori, L., Palma, D., Heegaard, P.E., Bertheussen, L.E., Jensen, M.R., Moor, K.D.: Review of the theory, principles, and design requirements of human-centric Internet of things (IoT). J. Ambient Intell. Humaniz. Comput. **14**(3), 2827–2859 (2023)

Behaviour-Centred Design of IoT Systems

Udo Kannengiesser, Christian Stary, Richard Heininger, and Thomas Ernst Jost

1 Introduction

The pervasive use of IoT technologies has spawned increased research interest in understanding the opportunities and challenges of integrating IoT with BPM systems. A great deal of research is concerned with issues of modelling, such as investigating business process notations regarding their coverage of the characteristics and functionalities of IoT systems [4]. In this chapter, we take a step back and examine the connection between BPM and IoT on a conceptual level. In particular, we focus on the overall design process through which IoT models and business process models are created and transformed. Such an understanding of model use is a precondition for determining the entities which process models, including their meta-models, are to cover [28].

Building a unified view of design that encompasses BPM and IoT is challenging. Over the years, each field has developed its own understanding and terminology of design, which makes establishing common ground difficult. Even within the BPM domain there is no uniform understanding of this notion. For example, business process design is often equated with business process modelling—a view which has been criticised as too narrow [20]. The approach we take in this chapter is to apply a well-established, ontological framework of design, which is independent of any particular domain or discipline, to the business process-driven design of IoT systems. It represents designing as involving eight fundamental transformations of design issues that are partitioned into a problem space and a solution space. It is shown that BPM concerns can be located mainly in the problem space and IoT

U. Kannengiesser (✉) · C. Stary · R. Heininger · T. E. Jost
Institute of Business Informatics - Communications Engineering, Johannes Kepler University, Linz, Austria
e-mail: udo.kannengiesser@jku.at; christian.stary@jku.at; richard.heininger@jku.at; thomas.jost@jku.at

G. Fortino, M. Mecella (eds.), *Internet of Things Meets Business Process Management*, Internet of Things, https://doi.org/10.1007/978-3-031-90746-3_4

concerns mainly in the solution space. Based on the framework, the key concept connecting the two spaces is that of behaviour, which can be represented in the form of process models. Hence, we refer to our view of business process-driven IoT design as "behaviour-centred design".

The chapter is structured as follows: The ontological basis for behaviour-centred design of IoT systems—the function-behaviour-structure (FBS) ontology of design—is introduced in Sect. 2. Its application to BPM and IoT is presented in Sect. 3. The particular role of behaviour, as a boundary object between BPM and IoT domains, is shown in Sect. 4. A case study is reported in Sect. 5, illustrating the concepts of behaviour-centred design. The results, limitations and potential for future work are discussed in Sect. 6. The chapter is concluded in Sect. 7.

2 The Function-Behaviour-Structure (FBS) Ontology of Design

The function-behaviour-structure (FBS) ontology is a uniform representation of designing, covering any designed entity or system and any domain of the design. With over 4,700 citations on Google Scholar for the two foundational papers [9, 10], the FBS ontology is one of the most highly referenced models of designing. It is unique in its meta-cognitive perspective and its focus on the micro-level processes in designing. It has been applied to a wide range of domains and instances of designing, including the design of IoT manufacturing systems [16].

Function (F) is defined as the teleology (purpose) of the system, i.e. "what it is for". For example, functions of an IoT system in logistics may be to support transportation, packing and commissioning. Components of the system may have more specific functions such as gripping and releasing goods.

Behaviour (B) is defined as the attributes derived or expected to be derived from the system's structure and its interactions with the environment, i.e. "what it does". It can often be described as the inputs, outputs and transformations of the entity (e.g. using a process model), as well as associated characteristics (e.g. speed, cost, accuracy).

For instance, the behaviour of a packing system includes a sequence of steps (i.e. the business logic or "process") that transform unpacked products into ones that are packed and ready for delivery. Its behaviour also includes various performance characteristics attributed to the steps, such as their execution time, energy consumption and material consumption.

Structure (S) is defined as the system's components and their relationships, i.e. "what it is (composed of)". Depending on the domain of the design, there can be different types of structure. For example, in the physical domain, structure includes geometry, topology and material. In the software domain, it includes software architecture and code.

Consider packing to be accomplished in a cyber-physical system (CPS). Its physical components may include a transportation robot with a LIDAR and pickup arm, a box, Internet-of-Things sensors and a storage shelf. The digital components are an intelligent navigation system and the pickup and packing software. The individual components and their connections similarly have a physical structure (e.g. the material of the box, the physical form of the robot system, etc.) and a software structure (e.g. the control program of the robot).

Function is ascribed to a system by teleological reasoning that associates human goals with observable or measurable behaviour. Behaviour is causally connected to structure, based on physical laws or other types of causal knowledge, including heuristics. There is no direct connection between function and structure [6].

For instance, take a sensor system that collects temperature data and transmits it to another system. This behaviour is determined by its physical components and a control program. Yet, its function, e.g. whether the sensor system is used for packing or for smart home control, is not determined by structure. It is dependent on the human goals associated with the system.

Most design literature posits that, ultimately, designing is a transformation of desired purposes (i.e. function) into a design solution (i.e. structure) that (hopefully) achieves them. In the FBS ontology, this corresponds to a transformation of function into behaviour and then from behaviour into structure. Behaviour serves as an intermediate state in this transformation, playing a dual role: Firstly, it sets expectations regarding the required performance of the system, based on the desired functions. Secondly, it represents the “actual” performance of the system, based on the structure generated and its interaction with the environment. The FBS ontology accounts for this dual role of behaviour by defining two subclasses: expected behaviour (Be) and behaviour derived from structure (Bs). A great deal of designing is concerned with comparing and aligning Be and Bs (often involving several cycles of producing and analysing different design alternatives), so designs are produced that are both desired and feasible.

Based on these considerations, the FBS ontology defines the following transformations:

(1) F $\rightarrow$ Be (formulation)
(2) Be $\rightarrow$ S (synthesis)
(3) S $\rightarrow$ Bs (analysis)
(4) Be $\leftrightarrow$ Bs (evaluation)

For example, in IoT system design the purpose of transporting goods (F) is transformed into a process (Be), including performance attributes such as speed, that is expected to achieve that purpose. The structure (S) of a specific IoT system is then synthesized in terms of a set of IoT components and their control software. In order to test the IoT system, its actual behaviour (Bs) is then analysed, e.g. by testing a prototype or by simulation. It is then compared with the expectations about the process and its performance (Be), as a basis for deciding whether the designed IoT system achieves its purpose.

The FBS ontology introduces two further categories to represent the external input and output of designing: requirements (R) and descriptions (D). Requirements are often provided by (internal or external) clients. They are transformed into functions and are subsumed in formulation (1): R $\rightarrow$ F and F $\rightarrow$ Be. For example, the automated transportation of goods may be articulated as a requirement (R) by the management of a company, which is planning to move a production step to a different location in its factory. This requirement is transformed into function (F), when it is taken into consideration by the designers of the (IoT-based) transportation system.

Descriptions (D) are the output of designing. They include representations of the structure (S) of the designed system, for purposes of supporting downstream activities, such as manufacturing or purchasing the system components and embedding control code.

(5) S $\rightarrow$ D (documentation)

For example, the structure of an IoT system may be transformed into a list of system components, CAD models specifying their physical form and the code to be embedded in IoT controllers. These models and representations are provided to manufacturers, suppliers and software specialists.

Designing occurs in an open world in which new ideas and issues emerge, frequently leading to focus shifts and reformulations of design problems. The FBS ontology accounts for this by additional transformations that allow for changes in structure, expected behaviour and function after design solutions have been produced:

(6) S $\rightarrow$ S' (reformulation type 1)
(7) S $\rightarrow$ Be' (reformulation type 2)
(8) S $\rightarrow$ F' (reformulation type 3)

For example, in reformulation type 1, the current design of the IoT system may be changed to include an additional sensor component (S) that was not originally considered, such as a vibration sensor for automatic adjustments of the transportation speed to prevent goods from getting thrown off the vehicle on uneven ground. An example for reformulation type 2 is when the (to-be) process model (Be) is augmented by an (exception-handling) operation for recovering goods that have fallen off the vehicle. Reformulation type 3 occurs when, for example, the purpose (F) of the IoT system shifts to including not only transportation but also commissioning of goods. This may be the result of associating a new goal to the vehicle's ability of identifying and recovering fallen goods.

Reformulations are likely to lead to subsequent modifications of function, behaviour and structure, via any of the transformations described so far. They can occur throughout the design process, up until its final stages.

The eight fundamental transformations compose what is commonly called the FBS framework, shown in Fig. 1. The numbers represent labels for the different classes of transformations; they do not imply any sequence of execution. For example, observing the behaviour of an existing system (i.e. analysis: S $\rightarrow$ Bs) may

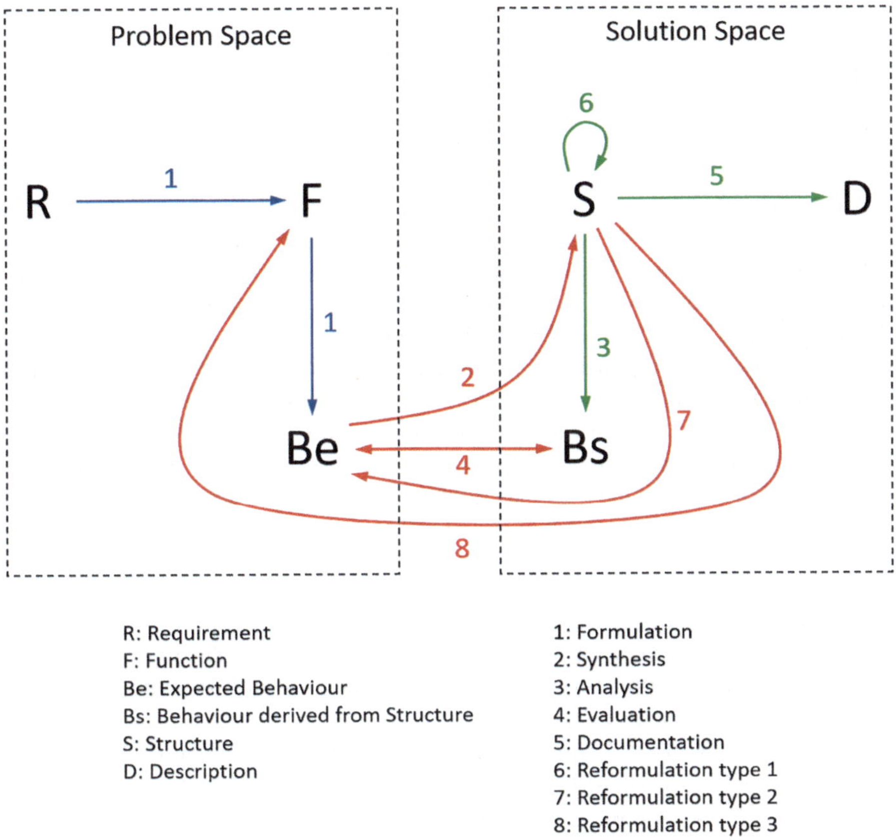

Fig. 1 The FBS framework (after [9] and [10]), augmented with representations of problem and solution spaces (after [11]). Transformations can occur within the problem space (blue arrows), within the solution space (green arrows) and across problem and solution spaces (red arrows).

be used as a starting point for redesigning that system. This is often the case in BPM, where the elicitation of an as-is process (Bs) can trigger a cycle of improvements and optimizations.

3 BPM and IoT: Separation Between Problem and Solution Space

In Fig. 1, two spaces are superimposed on the FBS framework: the problem space, composed of R, F and Be, and the solution space, composed of Bs, S and D. Some of the eight fundamental transformations occur entirely within the problem space (blue arrows), within the solution space (green arrows) and across problem and solution spaces (red arrows) [11].

Problem and solution spaces in design may correspond to separate disciplines. This is often the case for the design of highly specialised, technical systems to address problems in business-oriented domains. One of these domains includes IoT-supported business processes. Here, BPM experts typically focus on the problem space that comprises purposes, processes and KPIs related to the business operations. This is based on the common view in BPM that the "process" is the central design concern. While a literature search for "business process management" and "design" returns a large number of publications, most of them share the view that the artefact being designed is the business process rather than the underlying resources (systems) responsible for that process. This view seems valid in cases where the IT systems remain the same across different processes, for example, an existing workflow engine that is integrated with standard business systems such as ERP and CRM. Here, most of the intellectual effort is usually spent on process modelling, as indicated by the high personnel cost in many business process modelling projects [24]. However, for developing new technical solutions, such as IoT-based transportation, it would seem quite odd to say "let us design a new transportation process", if the main innovation is the IoT system being developed for automating the transportation. The FBS framework shows very clearly that designing does not stop with generating expected behaviour (Be) (which is what a "designed" business process corresponds to). There is always structure (S) needed to bring about the desired behaviour.

To be fair, it should be mentioned that large parts of the general BPM life cycle [7] are in fact concerned with structure (S) and behaviour derived from structure (Bs). Examples include the phases of process discovery (involves understanding an existing process [Bs]), process implementation (establishes the resources [S] responsible for process execution) and process monitoring and controlling (includes observing the process [Bs] executed by the resources [S]). Process implementation, in particular, makes use of to-be process models (Be) produced in the redesign phase. For example, BPMN models can be transformed into a BPEL representation that is then executed by a workflow engine. However, when a process is to be implemented using IoT, transformations into executable IoT control code are usually not supported. Most BPM for IoT approaches treat IoT tasks as black boxes encapsulated by services [25].

While BPM focuses on the problem space, the IoT domain concentrates on the solution space, as it is mainly concerned with the various hard- and software components (S), their interplay and their technical performance (Bs). A lack of function and (expected) behaviour models for IoT systems has most explicitly been noted in adaptive manufacturing [22]. Based on the scarcity of such models, the design of IoT systems is mostly viewed as a configuration problem, consisting of selecting and associating pre-existing components and assessing the resulting compositions

[2]. For example, the approach proposed by [13] for developing IoT manufacturing systems consists of four steps: (1) selecting hardware components, (2) structuring their information, (3) creating digital representations (of the components) and (4) "orchestrating" their interplay. Here, the first three steps are concerned with structure (S) and the fourth step is concerned with behaviour derived from structure (Bs). Expected behaviour (Be), which would be necessary to provide criteria for validating configuration choices, is not addressed in this approach.

In their approach to designing and operating digital twins as part of cyber-physical systems [14] propose layering behaviour models bridging the gap between problem and solution space. By component abstraction and a common communication mechanism at varying levels of granularity, the problem space can be refined to operational CPS requirements, retaining the controllability of a CPS for implementation and operation. Work on architectural design of IoT systems described in Bassi et al.[1] uses the notions of "functional views" and "domain models", which can be seen as including function (F) expected behaviour (Be). However, these views and models focus on implementation concerns, e.g. what kind of services are provided and where they are hosted. They are described using UML models, which are not easily understandable by domain experts. Approaches like IoTsecM [8] extend UML/SysML for dedicated purposes, such as security requirement modelling, and address the problem space in a development project. IoTsecM supports the specification of security requirements with the goal to bridge the gap for IoT developers between analysis and design.

The heterogeneity of components of IoT application has led to the introduction of a respective model-based systems engineering methodology [5]. Focusing on the design phase, SysML4IoT captures hardware devices and software components by using stereotypes. An IoT domain model supports the communication between stakeholders along development. The system model is conceived as a composition of views based on viewpoints. Stakeholders need to become aware of their concerns, i.e. the problem space, to specify these viewpoints properly. To support architecture specification of IoT systems, [23] proposed an IoT style for SysADL, a SysML-based language. The SysADL+IoT style was shown to be easy to use and sufficiently expressive to represent IoT-specific needs.

Efforts to integrate domain-specific models of function and expected behaviour into representations of IoT systems have emerged only recently, especially in the domain of manufacturing [22]. Here, the notion of required capabilities or skills [18], such as "drilling" and "gripping", is viewed as corresponding to function (representing design intent) or expected behaviour (representing desired operations). Skills facilitate the configuration and reconfiguration of production systems by providing reusable descriptors for the capabilities of individual components, thus allowing identifying and selecting components in response to current needs of the system. Various approaches have been developed for representing skills, including semantic technologies and graphical models [17, 30].

4 Behaviour as a Boundary Object Between Design Disciplines

The focus of BPM and IoT on separate spaces in design is not necessarily problematic. However, there needs to be sufficient communication and coordination between BPM and IoT experts to bring the two spaces together for effective overall IoT system design. This reduces the risk that processes (Be) are specified by business stakeholders that turn out to be technically infeasible and that IoT systems (S) are designed by technical stakeholders that turn out to insufficiently address business problems. It is therefore important to allow for effective, bidirectional communication between the two stakeholder groups.

As shown in the FBS framework, behaviour has a pivotal role in connecting the problem space with the solution space, as it is directly involved in most cross-space transformations: synthesis (transformation 2 in Fig. 1), evaluation (transformation 4) and reformulation type 2 (transformation 7). In addition, behaviour is indirectly involved in reformulation type 3 (transformation 8), as reformulated functions are based on a transformation of structure via behaviour. Therefore, most of the communication between BPM and IoT experts can be expected to involve representations of behaviour. Such representations, which are shared among different actors and communities of practice, are commonly termed boundary objects [3].

Behaviour has an additional role in multidisciplinary design, such as mechatronic system design that comprises mechanical engineering, electronics and software engineering. Here, the overall system design is decomposed into different subsystems, each of which is refined and implemented by experts from separate disciplines [31]. The results of integrating the partial solutions and subsystems are then assessed against the expected behaviour (Be) specified for the overall system design. Therefore, behaviour is also a boundary object between experts concerned with different, domain/discipline-specific subspaces of the solution space.

Based on these considerations, our notion of behaviour-centred IoT system design highlights the importance of behaviour representations such as process models in the communication and coordination between different domain experts in design.

5 Illustrative Case Study

To illustrate the ideas discussed in the previous sections, we conducted a preliminary case study. For this case study, we considered student projects that were conducted as part of a course in Johannes Kepler University's Bachelor's Program in Business Informatics during the winter semester of 2022/2023. As part of this particular course, students tackle a project in groups by applying their acquired knowledge from both business and technical domains. The projects are usually carried out in the final two semesters of study, as they require a solid understanding of fundamentals including software engineering and programming. A project contains three groups, each of which consisting of four to six students, that are tasked to design and implement an IoT system, starting with the specification of the underlying business processes and ending with the technical realization in the form of a prototype. A common scenario for these systems was provided to them as a design requirement, and the three groups collaborated to develop an IoT system realizing the scenario. Two of the authors of this chapter served as project supervisors and clients who provided the required scenario. This section further outlines our case study, describes the IoT scenario and analyses the execution of the project through the lens of the FBS framework.

The scenario was set in the logistics domain, for which there were no dedicated experts with specialised knowledge about the problem space available. However, the notion of behaviour-centred design could still be examined based on the role of behaviour as a boundary object between technical domains focusing on different solution subspaces. The scenario was sufficiently simple that final-year students in business informatics could understand it without requiring particular domain knowledge.

5.1 Case Study Approach

The students were first provided with a textual scenario description of a business process that is to be enabled/supported by an IoT systems with multiple subsystems and components. We constructed this scenario based on previous research, aiming at depicting a realistic application of IoT technology in an industrial context, in this case, smart logistics. As it required multiple interacting systems, there was sufficient complexity while still keeping it at an appropriate size so that selected components can be realized as part of a student project. Based on this scenario, the student groups were tasked with picking one of the involved IoT systems each, which they subsequently designed and developed in accordance with a development approach that started with the business processes involved and then shifted to the concrete technical realization using hardware and software.

5.2 Scenario Description

The scenario presented to the students concerned a smart logistics case as depicted in Fig. 2, focusing on supporting transport preparation through different interacting IoT and robotic systems (the original scenario that our version was based on can be found in Ref. [15]). The envisioned scenario was described as follows: "A good to be prepared for further transport arrives at an inbound location, where a sensor system detects its presence and notifies a four-legged robot. This robot moves to the inbound location and picks up the good using an attached arm unit and transports it

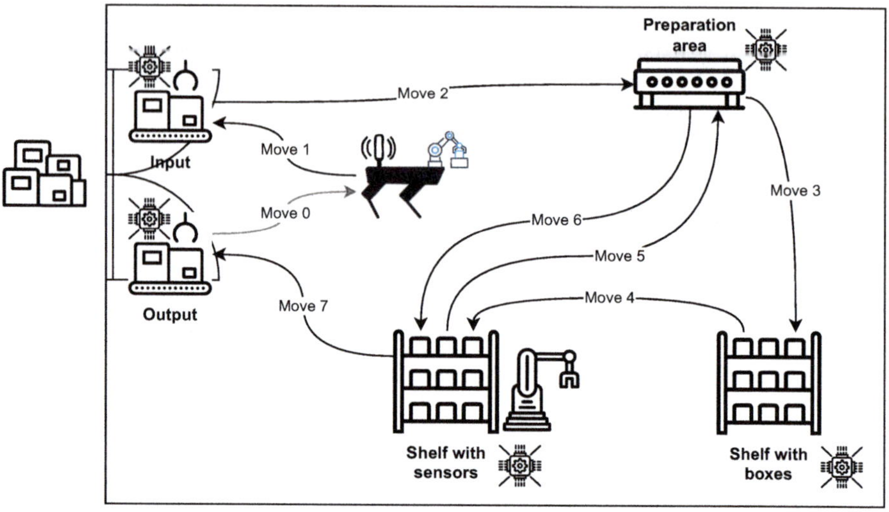

Fig. 2 Smart logistics scenario

to a preparation area. A label attached to the good provides information on its type and the details of the transportation assignment, e.g., how it should be prepared. Then, the robot moves on to outfit an intelligent transport box for the good, situated in a dedicated shelf. To this end, it moves to another shelf and fetches different sensors units, depending on the details of the transportation assignment, and installs them in the transport box. Both shelves should also be IoT systems, keeping track of the stored boxes and sensor units, and being able to inform the robot upon request, if the required items are currently stored. If an item is not present, this would lead to an interruption of the process and the notification of a logistics centre worker to investigate and resolve the issue. The shelves should also be capable of conveying the correct positions of the required items (e.g., through pick-by-light, or by sending of position coordinates). Once an intelligent transport box is outfitted with all the required sensor units, the robot can retrieve the good and place it inside. The box should then be transported to an outbound location, where it can be retrieved by the party responsible for transport."

The student groups were then tasked with designing the key IoT components and their interactions that should realize the smart packing process. In the examined projects, they chose the smart shelf (for housing the intelligent transport boxes), the general room infrastructure and a four-legged robot system ("Spot": https://bostondynamics.com/products/spot/).

5.3 Design Project Execution

In this subsection, we describe the steps that the students took while working on the project. We accompanied them in our roles as supervisor and client over a period of 15 weeks. After presenting the scenario, the students formed three groups of four to six members and chose high-level task blocks, which are described in the following.

Group A (*Spot*) chose to explore the Spot API to identify functionalities relevant to the use case, such as navigation to waypoints and object identification. The associated tasks were to learn about the API documentation, document potential features and create a shortlist of candidates for testing. In addition, experiments had to be designed to test these functionalities (e.g., moving the robot from point A to B) and the results documented and analysed. As part of the task block, a program also had to be developed to implement part of the behaviour of the use case.

Group B (*infrastructure*) opted for the package of tasks related to setting up the required software environment. This included the selection, installation and configuration of the software on a computer provided by the institute, requiring detailed documentation of each step. An important part of the package was also the integration of custom messaging middleware, ideally MQTT, to enable the sending and receiving of messages. Finally, the middleware had to be tested using a suitable model, for which a step-by-step guide had to be created.

Group C (*smart room/smart shelf*) chose to program and integrate the hardware using Node-RED for visual low-code programming. The task block involved the group familiarizing themselves with the programming tool and the hardware available, including microcontrollers and sensors. In addition, those parts of the scenario related to IoT components had to be implemented and documented in detail.

The project structure comprised the three basic phases of mechatronic systems engineering [31]: (1) system design, (2) domain-specific design and (3) system integration.

In the *system design* phase, all three student groups modelled the expected behaviour of the IoT system based on their understanding of the required scenario, using BPMN modelling. After two weeks of preparation, the groups including the supervisors/clients aligned their models in a meeting to produce a commonly agreed model. The individual models of groups A and C were almost identical and included the global, end-to-end process. It was only group B that focused on their own task block by modelling only the BPMN Pool that was relevant for them. It was relatively easy to reconcile the different perspectives of the problem space in this initial design phase. The various perspectives and requirements were discussed, and as a result, an aligned BPMN collaboration diagram was agreed upon to serve as a common reference model of the expected behaviour of the IoT system. An excerpt of the diagram is shown in Fig. 3. The activities in this phase can be mapped onto the FBS framework as shown in Fig. 4: The scenario provided to the students represents requirements (R) that are transformed into the students' conceptualisations of the IoT system's functionality (F) and expected behaviour (Be) (Fig. 4a). A high-level

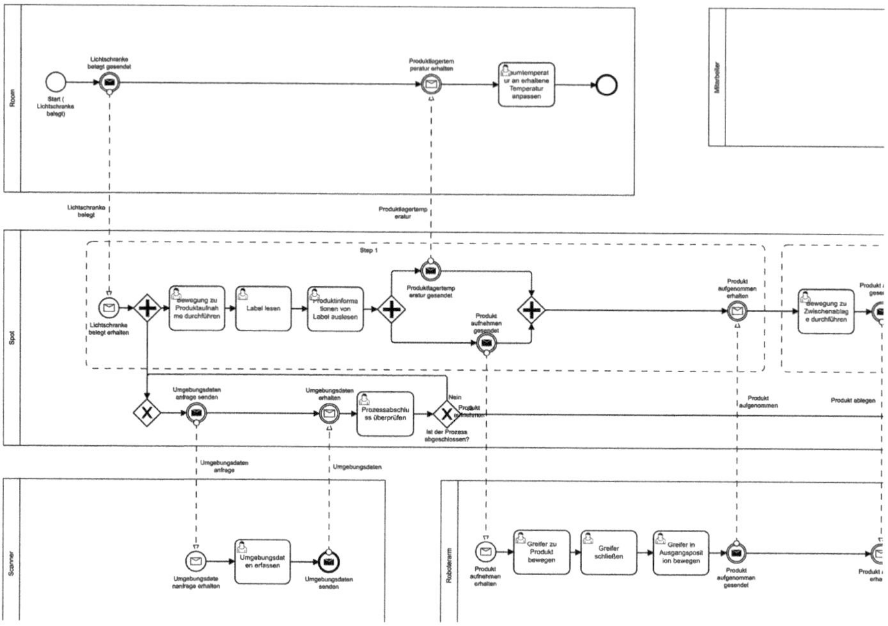

Fig. 3 Excerpt of the BPMN diagram (with labels in German language) representing the expected behaviour (Be) of the IoT system

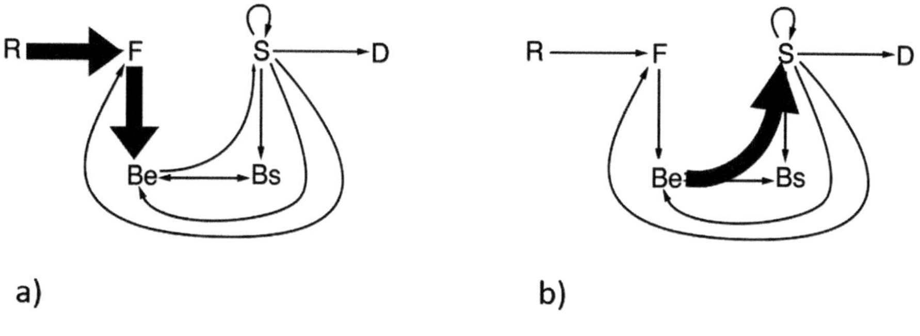

Fig. 4 Design activities during the *system design* phase: (**a**) Based on the scenario (R) provided to the students, the system functionality (F) was interpreted and process models representing the expected system behaviour (Be) were produced; (**b**) basic system components (S) were identified as responsible for realising the desired process (Be)

structure (S) is produced for realising that behaviour (Fig. 4b), in terms of the basic components of the system (such as robot, shelves, etc.) that are added as Pools in the BPMN diagram.

In the *domain-specific design* phase, each student group started working independently on their own solution subspace. This is shown in Fig. 5a. In fortnightly meetings, the progress of the work was presented by every group and possible

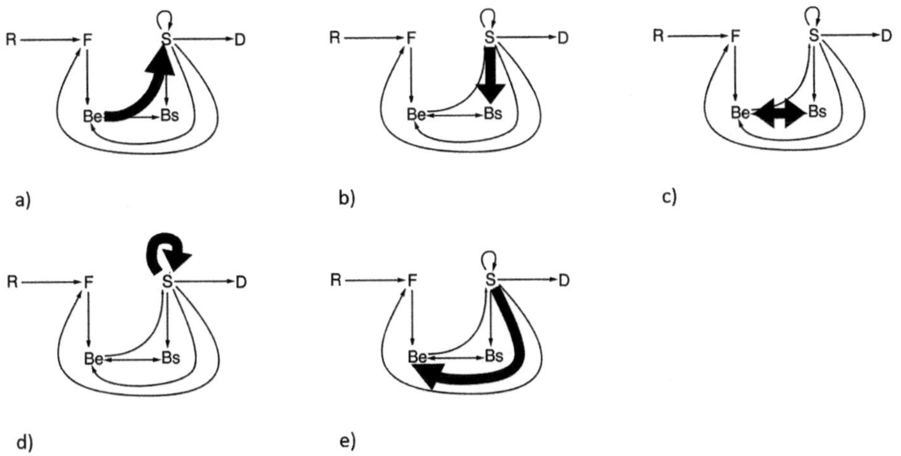

Fig. 5 Design activities during the *domain-specific design* phase: (**a**) Every group produced (partial) solutions (S) based on the process model (Be); (**b**) prototype testing produced "actual" component behaviour (Bs); (**c**) "actual" (Bs) and expected behaviour (Be) were compared; (**d**) an initial solution (S) for robot navigation was modified; (**e**) the process model (Be) was changed to better guide implementation of IoT components

effects on the other groups were analysed and evaluated (Fig. 5b, c). The groups were exclusively occupied with their respective technologies; therefore, nothing had to be adjusted in this phase of the project. They developed and tested their prototypes in order to satisfy their part of the scenario's requirements. Sometimes they had to take a step back and try a different solution because the initial ideas were misleading. For example, group A initially tried to navigate Spot using coordinates. However, the group then decided to use more static Autowalk missions for navigation because the other solution was too complex for the project. This modification of the solution structure (S) is represented in Fig. 5d. Another modification resulted from a problem faced by group C: Here, the expected behaviour had originally been specified at a level too high to easily derive the IoT control software. Therefore, group C had to make small adjustments to the BPMN model. In the FBS framework, this corresponds to reformulated expected behaviour (Be), as shown in Fig. 5e. Other groups were not affected by the changes in the model, as they were all located within the Pool owned by group C.

In the *system integration* phase, there was increased coordination effort between the groups. For example, group B developed Java skeleton code for the communication of IoT devices with the MQTT broker, which was then presented in a project meeting. This solution was analysed and evaluated by the other groups (Fig. 6a, b). It was generally accepted by all groups but required some adaptations of the software code that had already been developed by group C. This corresponds to reformulated structure (S) in the FBS framework, as shown in Fig. 6c. Various properties of the integrated IoT system also led to the discovery of new design opportunities: Group C implemented a temperature and humidity sensor in one of their IoT devices,

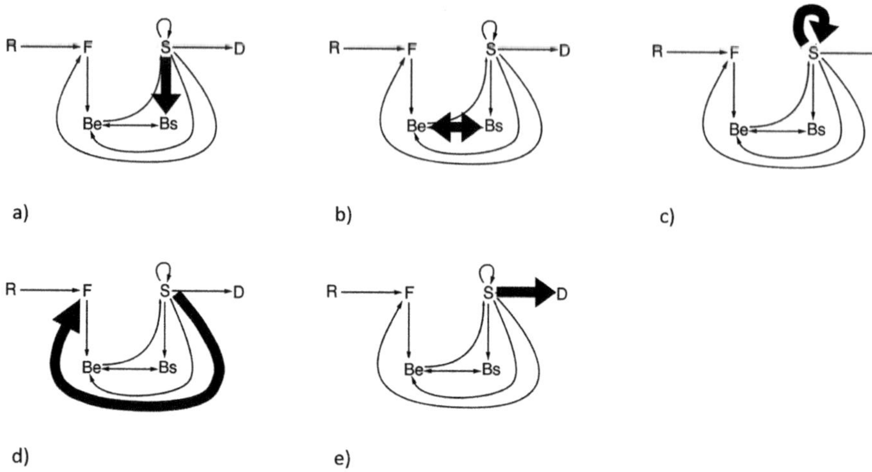

Fig. 6 Design activities during the *system integration* phase: (**a**) A proposed MQTT communication solution was analysed and (**b**) evaluated for integration with other components; (**c**) the integration of the code required changes in other components; (**d**) the new, partial function (F) of room data monitoring was introduced; (**e**) software code was documented (D)

which led to the inclusion of room data monitoring as a new design objective, corresponding to reformulated function (F) (Fig. 6d). The *system integration* phase ended with the documentation of the students' work in a project report including annotated source codes (Fig. 6e).

6 Discussion

The analysis of the use case through the lens of the FBS framework has provided first indications of the validity of the concepts regarding the notion of behaviour-centred design of IoT systems. In particular, it has shown the applicability of the FBS framework to business process-driven IoT system design and the central role of behaviour as a boundary object to coordinate work on the different subspaces of the design space.

The behaviour representation articulated as a process model served as the students' central point of reference when they needed to coordinate their work with other groups, most notably in the initial phase and the final phase of the project. It has been observed, however, that structure (S) became an additional boundary object during system integration, as the different IoT components had to be connected by software interfaces that the students needed to discuss. Conceptually speaking, the subspaces of the solution space were not completely disjoint. Such overlaps in design expertise can be argued to occur quite frequently in mechatronic system design, given the complex dependencies between different experts' viewpoints [29].

There are a few limitations related to the case study. Firstly, it focused only on certain software aspects of IoT system design. It did not cover the mechanical design of the IoT system. All of the IoT devices needed for the given scenario were provided and considered mainly as constraints for software design decisions. Secondly, all students involved in the project were already trained in BPMN modelling, as it is a mandatory study module in their curriculum. Yet, in many industry projects such shared knowledge of formal process modelling cannot be assumed to be sufficiently available for producing effective, error-free models [21]. Thirdly, as the project was originally carried out for other research purposes, the data collected during the project is insufficient for more rigorous, quantitative analyses of behaviour-centred design approaches including an assessment of the quality of design outcomes.

Our mapping of business process-driven IoT system design onto the FBS ontology opens up a number of research opportunities. There are increasing indications that designing can be viewed as a fundamental human activity that transcends different design disciplines, which has been shown using protocol studies coded using the FBS schema [12]. Similar studies could be used to empirically study behaviour-centred design of IoT systems, using different approaches, notations, tools, levels of expertise, project settings, etc.

The insights gained from our work may also provide a new perspective to inform research in process modelling notations for IoT system design. We mentioned earlier that all project participants in our case study had prior modelling experience in BPMN. What kind of notations do we need if this is not the case? Alternative approaches may be explored that could help novices produce process models more easily and with fewer modelling errors [19]. In addition, individual characteristics of process models, such as the notion of modularity, may be further investigated. The BPMN diagram produced by the students used only Pools, rather than Lanes within a Pool. This encapsulation of different parts of the end-to-end process has increased the modularity of the model, which was likely to be the reason for the rather easy alignment of process perspectives of the different groups (in *system design*) and the isolation of changes in the Pool of group C without affecting the other Pools (in *domain-specific design*).

7 Conclusion

For building a unified view of design that encompasses BPM and IoT, we have applied a well-established, ontological framework of design. The framework represents designing as involving eight fundamental transformations of design issues targeting the problem and its solution in separate spaces. BPM supports informed problem elaboration in the respective space, while IoT supports the development of working solutions in the respective space. The focus on behaviour represented in the form of process models bridges the gap between different subspaces as a boundary object. Consequently, behaviour-driven process modelling meets both design issues:

understanding and refining the problem and linking possible IoT solutions. Both become increasingly important when propagating the Internet of Behaviours and Web3 (cf. [26, 27]), due to the increase of complexity and variability of system designs.

References

1. Bassi, A., Bauer, M., Fiedler, M., Kramp, T., Van Kranenburg, R., Lange, S. and Meissner, S.: Enabling things to talk: designing IoT solutions with the IoT architectural reference model. Springer, Berlin (2013). https://doi.org/10.1007/978-3-642-40403-0
2. Brown, D.C.: Defining configuring. Artif. Intell. Eng. Design Anal. Manuf. **12**, 301–305 (1998). https://doi.org/10.1017/S0890060498124034
3. Carlile, P.R.: A pragmatic view of knowledge and boundaries: Boundary objects in new product development. Organ. Sci. **13**, 442–455 (2002). https://doi.org/10.1287/orsc.13.4.442.2953
4. Compagnucci, I., Corradini, F., Fornari, F., Polini, A., Re, B., Tiezzi, F.: A systematic literature review on IoT-aware business process modeling views, requirements and notations. Softw. Syst. Model. **22**, 969–1004 (2023). ISSN: 1619-1366. https://doi.org/10.1007/s10270-022-01049-2
5. Costa, B., Pires, P.F. and Delicato, F.C.: Modeling IoT applications with sysml4iot. In: 2016 42th Euromicro Conference on Software Engineering and Advanced Applications (SEAA), pp. 157–164. IEEE, Piscataway (2016)
6. De Kleer, J., Brown, J.S.: A qualitative physics based on confluences. Artif. Intell. **24**, 7–83 (1984). https://doi.org/10.1016/0004-3702(84)90037-7
7. Dumas, M., La Rosa, M., Mendling, J., Reijers, H.A.: Fundamentals of Business Process Management. Springer, Berlin (2013). https://doi.org/10.1007/978-3-642-33143-5
8. Escamilla-Ambrosio, P.J., Robles-Ramírez, D.A., Tryfonas, T., Rodríguez-Mota, A., Gallegos-García, G., Salinas-Rosales, M.: IoTsecM: a UML/SysML extension for internet of things security modeling. IEEE Access **9**, 154112–154135 (2021)
9. Gero, J.S.: Design prototypes: a knowledge representation schema for design. AI Mag. **11**, 26–36 (1990). https://doi.org/10.1609/aimag.v11i4.854
10. Gero, J.S., Kannengiesser, U.: The situated function—behaviour—structure framework. Design Stud. **25**, 373–391 (2004). https://doi.org/10.1016/j.destud.2003.10.010
11. Gero, J.S., Milovanovic, J.: Creation and characterization of design spaces. In: DRS2022 (2022). https://doi.org/10.21606/drs.2022.265
12. Gero, J.S., Kannengiesser, U., Pourmohamadi, M.: Commonalities across designing: Empirical results. In: Design Computing and Cognition '12, pp. 265–281 (2014). https://doi.org/10.1007/978-94-017-9112-0_15
13. Heidel, R., Hoffmeister, M., Hankel, M., Döbrich, U.: Basiswissen RAMI 4.0: Referenzarchitekturmodell mit Industrie4.0-Komponente. Beuth Verlag GmbH (2017)
14. Jost, T.E., Heininger, R., Stary, C.: Layering abstractions for design-integrated engineering of cyber-physical systems. In: Handbook of Digital Twins, pp. 221–237. CRC Press, Boca Raton (2024)
15. Jost, T.E., Stary, C., Heininger, R.: Geo-spatial context provision for digital twin generation. Appl. Sci. **12**(21), 10988 (2022). ISSN: 2076–3417. https://doi.org/10.3390/app122110988. (Visited on 11/21/2023)
16. Kannengiesser, U., Weichhart, G.: Designing and executing interoperable IoT manufacturing systems. In: Enterprise Interoperability: Interoperability for Agility, Resilience and Plasticity of Collaborations: I-ESA'14 Proceedings, pp. 15–20 (2015). https://doi.org/10.1002/9781119081418.ch2

17. Köcher, A., Hildebrandt, C., Caesar, B., Bakakeu, J., Peschke, J., Scholz, A. and Fay, A.: Automating the development of machine skills and their semantic description. In: 2020 25th IEEE International Conference on Emerging Technologies and Factory Automation (ETFA), vol. 1, pp. 1013–1018 (2020). https://doi.org/10.1109/ETFA46521.2020.9211933
18. Malakuti, S., Bock, J., Weser, M., Venet, P., Zimmermann, P., Wiegand, M., Grothoff, J., Wagner, C., Bayha, A.: Challenges in skill-based engineering of industrial automation systems. In: 2018 IEEE 23rd International Conference on Emerging Technologies and Factory Automation (ETFA), vol. 1, pp. 67–74 (2018). https://doi.org/10.1109/ETFA.2018.8502635
19. Moattar, H., Bandara, W., Kannengiesser, U., Rosemann, M.: Control flow versus communication: comparing two approaches to process modelling. Bus. Process Manag. J. **28**, 372–397 (2022). https://doi.org/10.1108/BPMJ-01-2021-0015
20. Reijers, H.A.: Business process management: the evolution of a discipline. Comput. Ind. **126** (2021). https://doi.org/10.1016/j.compind.2021.103404
21. Roy, S., Sajeev, A.S.M., Bihary, S., Ranjan, A.: Empirical study of error patterns in industrial business process models. IEEE Trans. Services Comput. **7**, 140–153 (2014). https://doi.org/10.1109/TSC.2013.10
22. Sanderson, D., Chaplin, J.C., Ratchev, S.: A function-behaviour-structure design methodology for adaptive production systems. Int. J. Adv. Manuf. Technol. **105**, 3731–3742 (2019). https://doi.org/10.1007/s00170-019-03823-x
23. Santos, L., Silva, E., Batista, T., Leite, J., Cavalcante, E., Oquendo, F.: Evaluating a SysML-based graphical notation for modeling Internet of Things system architectures. In: 2020 IEEE 6th World Forum on Internet of Things (WF-IoT), pp. 1–6. IEEE, Piscataway (2020)
24. Schmidt, R., Möhring, M., Sandkuhl, K., Zimmermann, A.: Social paradigms for reducing process modeling cost. In: Innovation in Medicine and Healthcare Systems, and Multimedia. Smart Innovation, Systems and Technologies, vol. 145, pp. 393–405 (2019). https://doi.org/10.1007/978-981-13-8566-7_37
25. Schönig, S., Ackermann, L., Jablonski, S., Ermer, A.: IoT meets BPM: a bidirectional communication architecture for IoT-aware process execution. Softw. Syst. Model. **19**, 1443–1459 (2020). https://doi.org/10.1007/s10270-020-00785-7
26. Stary, C.: Digital process twins as intelligent design technology for engineering metaverse/XR applications. Sustainability **15**(22), 16062 (2023)
27. Stary, C.: The Internet-of-Behavior as organizational transformation space with choreographic intelligence. In: Subject-Oriented Business Process Management. The Digital Workplace–Nucleus of Transformation: 12th International Conference, S-BPM ONE 2020, Bremen, Germany, December 2-3, 2020, Proceedings 12, pp. 113–132. Springer, Berlin (2020)
28. Thalheim, B.: The theory of conceptual models, the theory of conceptual modelling and foundations of conceptual modelling. In: Handbook of Conceptual Modeling, pp. 543–577 (2011). https://doi.org/10.1007/978-3-642-15865-0_17
29. Törngren, M., Qamar, A., Biehl, M., Loiret, F., El-Khoury, J.: Integrating viewpoints in the development of mechatronic products. Mechatronics **24**, 745–762 (2014). https://doi.org/10.1016/j.mechatronics.2013.11.013
30. Trautner, T., Ayatollahi, I., Strutzenberger, D., Frühwirth, T., Pauker, F., Kittl, B.: Behavioral modeling of manufacturing skills in OPC UA for automated execution by an independent cell controller. In: Procedia CIRP 99 (2021). 14th CIRP Conference on Intelligent Computation in Manufacturing Engineering, 15–17 July 2020, pp. 633–638. https://doi.org/10.1016/j.procir.2021.03.085
31. VDI. VDI 2206—Design methodology for Mechatronic Systems. Verein Deutscher Ingenieure (2004)

LACII: Enhancing Decision-Making in Business Processes with the IIoT: A Cognitive-Load-Aware Design Approach

Maximilian Nebel, Seyyid Ahmed Ciftci, and Christian Janiesch

1 Introduction

The advent of the fourth industrial revolution has led to significant changes in the nature of work in production. The integration of sensors into production machinery, the decline in the cost of powerful computing and the availability of miniaturized embedded systems, and the growing popularity of artificial intelligence (AI) has motivated companies to introduce cyber-physical systems (CPS) [1]. CPS employ the Industrial Internet of Things (IIoT) to connect devices through a network, for example, the Internet, allowing communication between network participants [2]. The participants are represented by both human workers and a variety of machinery and robots that can act as autonomous agents in the system following their own agenda [3, 4]. To guarantee that all participants in the system are working towards a common goal, managers must act as authorities.

The managers of these IIoT systems themselves pursue several goals when planning business processes. In doing so, they make trade-offs and try to harmonize these sometimes very different goals. Usually, their long-term objective is to generate profit through their business operations to sustain the success of the operation. This can only be achieved by designing business processes efficiently [5].

Additionally, they are responsible for the well-being of their employees. In particular, given the scarcity of skilled workers and the difficulty that companies face in retaining and acquiring them, it is not feasible for a company to place excessive pressure on its employees. Instead, the manager should also pay attention to the specific needs of the employees. This may, for example, include the removal of

M. Nebel · S. A. Ciftci · C. Janiesch (✉)
TU Dortmund University, Dortmund, Germany
e-mail: maximilian.nebel@tu-dortmund.de; seyyid.ciftci@tu-dortmund.de; christian.janiesch@tu-dortmund.de

G. Fortino, M. Mecella (eds.), *Internet of Things Meets Business Process Management*, Internet of Things, https://doi.org/10.1007/978-3-031-90746-3_5

repetitive and tedious tasks from the workload of employees or ensuring their job security [6].

In IIoT-enabled production facilities, communication typically occurs via the data recorded by the sensors on the machines. In the event of an anomaly in the data, a decision on the most appropriate course of action may be required. If this complex decision cannot be made autonomously by the machine, for example, due to its severity, the data is made available to the human-in-the-loop [4, 7]. As the quantity of data accessible in real time within the CPS continues to expand, the role of the human-in-the-loop is becoming increasingly stressful, as the volume of necessary decisions also increases. Further, in deciding on anomalies in real-time production data, they rely on the information of various algorithms and must address the dilemma of determining when to react to anomalies detected while attempting to minimize false alarms and avoid overlooking smaller anomalies [8]. Such decisions frequently have to be made within a narrow time frame, with minimal room for error [9]. Errors in judgement or carelessness have the potential to result in costly accidents and production downtime due to damage. Consequently, it is of critical importance that decision-makers are kept in a cognitive state that allows them to make optimal decisions for the production processes.

To ensure the well-being of employees in an IIoT environment, it is necessary to consider the signs of stress and cognitive load in a manner comparable to that of machine anomalies. Cognitive load and task load are two well-known constructs in psychology that are involved in the well-being of humans. Increases in cognitive load and task load can result in stress [10, 11] and lead consequently to poorer performance. Consequently, a feasible approach is to include humans as an object of consideration in a similar way to machines as the physiological reactions of the human participant can be measured. Subsequently, the human's physiological data is automatically analyzed and decisions can be made to protect the human from exhaustion due to stress or cognitive load.

Hence, the objective of this study is to propose a design approach of a *"load-aware co-pilot for IIoT"* (LACII) that acts as a computer-in-the-loop by observing the human-in-the-loop, learning, and intervening if the human decision-maker exhibits signs of cognitive load or task load. The goal of *LACII* is to enable decision-makers to engage in productive work while maintaining their well-being. Accordingly, our study targets to answer the following research question (RQ).

RQ: *How should LACII, a load-aware computer-in-the-loop co-pilot, be designed that aims to increase productivity as well as the well-being of the human-in-the-loop decision-maker?*

To address our RQ, we based our design process on Peffers' Design Science Research Methodology (DSRM) [12]. First, we present the necessary background with the connection of IIoT and business process management (BPM) in Sect. 2. Here, we also present the background of cognitive load, task load, and related concepts. Section 3 summarizes our adoption of the DSRM, followed by presenting our design requirements in Sect. 4. The conceptual architecture is presented in Sect. 5 also containing the design principles, the architecture, and the interaction patterns to discuss human-AI-interaction between the human decision-makers and

LACII. We complement this with an illustration in Sect. 6 and considerations on practical implications in Sect. 7. The chapter concludes with the discussion and conclusion in Sect. 8.

2 Background

2.1 Technological Background

BPM and IIoT BPM is an interdisciplinary field that seeks to enhance organizational processes, for instance, by avoiding faults to reduce costs [5]. The potential for synergy between IIoT and BPM is significant, with the potential to enhance the overall performance of an organization [3]. In an IIoT production environment, a high volume of data is exchanged in real time. To enable successful processes, work assignments and process data must be monitored during operation. In addition, the system must keep track of affected resources and their capacities in order to allocate them to upcoming process instances [5]. However, individual process participants lack awareness of the agendas and goals of other process participants in the absence of BPM. BPM can facilitate task coordination, for example, by accounting for critical paths of interfering processes. Likewise, BPM directly benefits from the integration of sensors and the IIoT, as this greatly simplifies communication between process participants [3].

Sensors Sensors are an integral part of IIoT. They generate data by measuring physical objects, thereby enabling the detection of changes in the physical world and facilitating automation [13]. The sensors utilized in machinery can be classified into several categories, including proximity sensors, acoustic sensors, and motion sensors [13, 14]. The specific type of sensor employed depends on the characteristics of the machinery and the intended product. Machine sensors are categorized into direct and indirect sensors. Direct sensors measure properties that directly reflect the state of the machine (for example, proximity sensors), whereas indirect sensors measure values that can correlate with the state of the machine (for example, vibration sensors) [14]. Physiological sensors to track the users' physical reactions are widely discussed in literature; see, for example, [15–18]. What all these discussions have in common is that, among other data, they suggest heart rate as a measurement. This is because the heart rate can be measured objectively and can serve as a proxy for emotional reactions such as stress [18]. When using a smart watch or similar devices, it is beneficial to measure heart rate at the nondominant hand, to minimize external influences due to hand movements [15].

In addition, various other sensors can be used to measure physiological reactions. Eye trackers built into webcams or glasses [15] are commonly used devices to predict cognitive load based on blinking and eye movement [16, 19].

In addition to physiological sensors, other measurements such as the dust or carbon dioxide level in the working environment can be measured, as these can have an influence on the well-being of employees as well [20]. Regardless of which sensor is used, a computer-in-the-loop in the CPS has to analyze the emitted data in real time to find evidence for increases in cognitive load.

Anomaly Detection Anomalies are patterns in data that differ from a well-defined normal behavior [21]. The goal of anomaly detection is therefore to detect those patterns to enable reactions. In the case of this research, an exemplary anomaly could be a lasting increase in the decision-maker's heart rate, which indicates increases in cognitive load and requires the decision-maker to take a break. In reality, it can be very challenging to identify anomalies. This is due to various reasons. For example, it is not always trivial to even define what the normal behavior of the data should be [21]. For instance, different decision-makers could react differently to specific tasks, because they may like or dislike the given task. With the sensors measuring data continuously, real-time anomaly detection on data streams becomes more prevalent [7]. This is, because organizations aim to detect anomalies as soon as possible to allow fast reactions or even predict faults before they appear and possibly damage the machine or product [7, 8].

The design of LACII benefits directly from the above. Due to the subconscious physical reactions of the decision-maker, physiological sensors enable direct communication between the human-in-the-loop and LACII without interfering with the decision-making process. LACII aims to predict the cognitive load of the decision-maker based on their reactions.

2.2 *Psychological Constructs*

Cognitive Load The concept of cognitive load theory originates from educational psychology [10]. Initially, cognitive load theory tried to describe how cognitive capacities are used in learning and how they can be measured. These capacities include long-term memory and working memory. It is important to understand that an environment that focuses on reducing the cognitive load of learners can lead to increased well-being and learning efficiency [10]. The observations of cognitive load theory are applied in various areas and are also used in information systems.

Usually cognitive load is described to consist of three components linked to different subsets of learning. The capacities involved in understanding the complexity needed to enable task fulfillment is named intrinsic cognitive load. Extraneous cognitive load is described by the cognitive capacities used to filter and process irrelevant information to understand the instructions. Last, germane cognitive load is described by the cognitive capacities that are used to connect existing knowledge with the new task [16]. In learning, for each of the three types, strategies have already been developed as to how they can be avoided or minimized through design decisions in accordance with cognitive load theory [16].

In the context of this work, the design should be able to respond to an increase in cognitive load to keep the worker's performance high, but also to relieve the worker if necessary. However, responding to an increase in cognitive load is only possible if the changes are measurable. Literature identified mental load, mental effort, and performance as the three dimensions in cognitive load that can be measured [22, 23]. Mental load is the fraction of cognitive load used by the task and environmental demands; the cognitive capacity used is represented in the mental effort. Increases in mental demand and mental effort result in lower performance in the task [23]. Our work focuses on user performance, but also on the physical responses that are seen as a symptom of cognitive load.

In many cases, the performance can be measured by diverse objective factors relevant to the task, for example, time or precision. But performance alone is not a reliable source for the change in cognitive load, because it is affected by different factors. To measure cognitive load following a task, Klepsch proposes to use self-report questionnaires. After a learning task is done, the participant is asked to answer a questionnaire. In each question participants have to rate their answer on a Likert scale to calculate the cognitive load during the learning task [16]. In addition and less invasive and interrupting, physiological measures are used to track the increases in cognitive load. Different kinds of tools have been used trying to capture cognitive load increases, for example, heart rate monitors [17] or eye tracking [19]. These approaches assume increases in cognitive load because it may be correlated to stress and its symptoms, which then manifests itself in physiological reactions [24].

In addition, the approach to measure cognitive load with dual-task studies is widely used in literature [25, 26]. This approach measures an increase in cognitive load through an involvement of a secondary task that affects the performance of the primary task [26].

Task Load Task load is a concept related to cognitive load. It examines the capacities that individuals possess to complete tasks within a certain time frame [11]. Similar to cognitive load, questionnaires can be employed to measure task load. Typically, researchers rely on the NASA Task Load Index (NASA-TLX) [27, 28].

The relevant dimensions observed by the NASA-TLX include mental demand, physical demand, and temporal demand. If a task is frustrating, it is represented by the perceived mental demand. The perceived physical demand is linked to the effort a person has during task fulfilment. With a high perceived temporal demand the user's performance is influenced [27].

In the context of task load measurement studies, participants are required to complete the NASA-TLX questionnaire following the fulfilment of the assigned tasks [28]. With regard to LACII, this entails that in order to gain insight into the task load of the decision-maker, it is essential to request information on the three aforementioned concepts. In addition to the surveys and the physiological data, the task history, including the time required, can also be analyzed to allow conclusions to be drawn about the task load.

Flow At last, flow can be considered to complement cognitive load and task load. We see flow as a counteracting factor to cognitive load. Flow is defined as the comprehensive sensation experienced when individuals engage in activities with complete absorption [29]. That is, people can become so engrossed in their task that they forget the time, for example, or hardly feel any cognitive strain. It is considered by researchers and practitioners to be crucial in the work environment. This phenomenon is closely associated with positive outcomes in the workplace, including improved well-being and task performance [29].

3 Methodology

We design LACII using the DSRM by Peffers [12]. This methodology comprises six reiterable consecutive steps: identify problem and motivate, derive objectives of solution, design and development, demonstration, evaluation, and communication (cf. Fig. 1). In our study, we chose the problem-centered entry point. In other words, we start with the problem identification. The following step is to derive the objectives and design the artifact next. Demonstration and evaluation are combined and discussed in an illustration. The communication is done via this publication.

We identified the phenomenon that researchers usually try to measure cognitive load and task load to rate how demanding a certain task is. Less common, they discuss how to minimize cognitive load. This is why we decided to formalize the problem how LACII should be designed. For this, we formulate design requirements for decision-makers interacting with LACII.

The objective definition was done by developing design principles based on the design requirements. Chandra et al. [30] define design principles on the basis of Sein et al. [31] and Hevner and Chatterjee [32] as "knowledge about instances of a class of artifacts"[30, 31] that "[. . .] can also be formed as a rule or a standard of conduct"[30, 32]. The proposed structure formulating design principles reads as follows: "Provide the system with [material property—in terms of form and function] in order for users to [activity of user/group of users—in terms of action], given that [boundary conditions–user group's characteristics nor implementation settings]"[30].

The outcome of the design phase is the framework presented in Sects. 5.2 and 5.3. It comprises the architecture illustrating the implementation of the artifact into an IIoT production presented in the FMC notation. Moreover, the framework provides

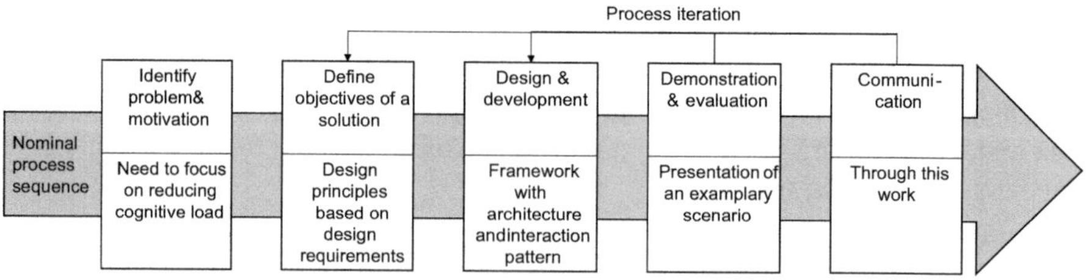

Fig. 1 Nominal process sequence according to Peffers [12]

interaction patterns used to plan out how individual decision-makers engage with LACII utilizing the Business Process Model and Notation (BPMN) 2.0. The design is demonstrated and evaluated by discussing the exemplary scenario in Sects. 7 and 6.

4 Design Requirements

In order to define design principles for the LACII architecture and the corresponding process patterns, we formalize design requirements first. The goal of these design requirements is to build the basis for the latter considerations taken in designing an artifact. The formulation of the design requirements should enable an evaluation of the developed design solution [33]. Here, the design requirements are extracted from problems and statements identified in related literature.

DR1: React to Increased Cognitive Load The main objective of the artifact is to reduce the cognitive load of decision-makers. The artifact therefore has to be able to react, when cognitive load increases. In order to be able to react to changes in real time, LACII needs continuous measurements of the cognitive load [34]. Symptoms of increased cognitive load are evident in the physiological responses of the decision-makers. They can be observed with diverse physiological sensors [15–17], eye tracking [19], dual-task settings [16, 25], and environmental sensors [18, 20]. Additionally, flow is observable with comparable sensor-driven methods [29]. The sensors then are linked to LACII, to allow reactions.

DR2: Adapt Workload Responding to detected increases in cognitive load and to prevent poorer task performance [10, 11], LACII must react accordingly, for example, by delegating a break for the decision-maker or by decreasing the frequency in which new tasks are presented. Similarly, persons are able to recognize their own increased cognitive load and task load and can report [16, 27, 28]. The artifact should therefore enable the decision-makers to self-report cognitive load or to request cognitive-load-based breaks.

DR3: Consider Decision-Makers' Experience Different tasks and the number of them place different demands on people [27, 28]. When observing the different task performances, LACII should be able to learn. For example, if the workers' performance improves, LACII assumes learning. With gained experience, the decision-maker needs less mental capacity to perform the same task in the future [10]. Experienced workers perform differently and overall better than beginners. LACII should detect and learn to assign tasks fitting to the strengths of an individual.

DR4: Support Human-Machine Interaction As described in DR3, it is important that the system supports the individual strengths of the decision-makers. In order to achieve this, the form of human-machine interaction must be adapted to the personality of the decision-makers. If LACII detects cognitive load, this can be communicated in various forms. It is crucial that the system chooses a way and formulation that is easy for people to understand [34] and motivates them to continue working with LACII. If a delegation is addressed to a human by the AI, this may result in a feeling of control loss for some people [35]. It is therefore necessary that these persons are able to select a form of interaction that reduces the level of autonomy of LACII [36] and instead places control with the decision-maker.

DR5: Allow Precise Performance Measurements It is also important to ensure that the decision-maker's interaction with the system is not influenced by the Hawthorne effect [37]. That is, in order to ensure the reliability of statements about the performance of the system during testing and operation, it is important to minimize the influence of self-perception on one's own observations [34]. Therefore, it is necessary to integrate measuring instruments such as sensors in such a way that they do not unnecessarily change the familiar workplace and as a consequence change the users' performance. One method of placing sensors on people without causing discomfort is the use of smart jewelry [38]. Contactless environmental sensors can also be employed [18, 20].

5 Conceptual Architecture

5.1 Design Principles

We have formulated six design principles based on the aforementioned design requirements. Figure 2 illustrates the relationships between the design requirements and design principles. To select a standardized formulation of design principles in IS literature, we have followed the scheme proposed by Chandra et al. [30] as described in Sect. 3. Boundary conditions always apply to the design principles. For example, it is assumed that the decision-makers agree in general to work with a co-pilot that uses physiological data. To this end, all data protection concerns of the persons involved must be disclosed. Furthermore, the decision-makers must be able to interact with LACII at all times, for example, via a computer.

DP1: Provide the system with diverse physiological sensors and real-time anomaly detection capabilities to afford the system to observe physiological reactions of the decision-maker in order for the system to detect changes in their cognitive load. The decision-maker is observed in real time while working. If LACII recognizes anomalies from patterns in the data [21] that either the decision-maker's cognitive load is getting too high or flow is being experienced, it must be able to react accordingly. This could be that the worker with increased cognitive load gets to take a break, while the worker experiencing flow shall not be interrupted.

DP2: Provide the system with physiological sensors that integrate seamlessly into the familiar working environment of the decision-maker in order to afford decision-makers to work without distraction. To avoid the decision-maker feeling uncomfortable due to being monitored, the measurement instruments should be integrated as discreetly as possible into the familiar work environment. For example,

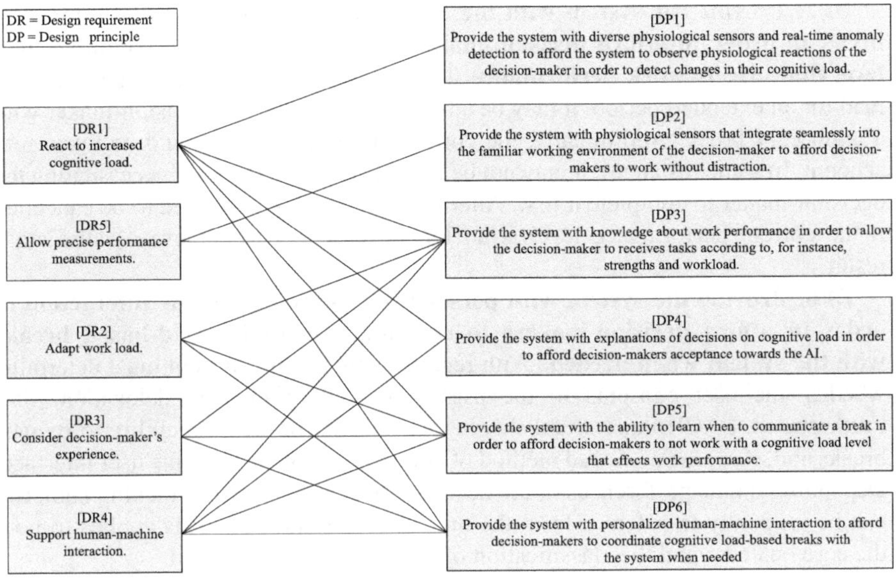

Fig. 2 Visualization of design requirements (DR) and design principles (DP) relationships

measurements can be taken using smart watches, cameras, and more [15]. This allows managers to equip the workplace with sensors and still plan in such a way that workers are not affected. However, if a measurement instrument disturbs the worker, they should be able to customize the workplace according to their own preferences.

DP3: Provide the system with knowledge about work performance in order to allow the decision-maker to receive tasks according to, for instance, strengths and workload. LACII recognizes when the decision-maker performs worse than usual. For example, a worker may be slower than on other days, because they are feeling slightly sick. In this case, LACII controls the work allocation accordingly to relieve the worker. LACII should be able to recognize performance based on objective measurements like time needed per task, but also through the indicators proposed by the NASA-TLX [27] and cognitive load theory [16, 26].

DP4: Provide the system with explanations of decisions on cognitive load in order to afford decision-makers' acceptance towards the AI. During the observation, LACII has to make decisions, which should be fully explained if requested, especially if it directly affects the work of the decision-maker. This is necessary to maintain the integrity of the process and to provide the individual with a transparent explanation of the decision [39]. For example, the explanation could be based on declining performance or unusual physiological sensor data. LACII shall be honest with the worker about what has been found, so that the worker has the opportunity to recognize any self-imposed shortcomings and to understand that a break is recommended to help in the long run. At the same time, LACII shall be sensitive so as not to demotivate the worker.

DP5: Provide the system with the ability to learn when to communicate a break in order to afford decision-makers to not work with a cognitive load level that affects work performance. In the case of an unusually high cognitive load for an extended period, it may be advisable to provide the decision-maker with a temporary break. Additionally, the current and open tasks should be taken into account. In some instances, it may not be feasible to pause the task, necessitating the decision-maker to complete it first. Alternatively, the task may need to be cancelled if the current performance level is no longer sufficient to achieve a satisfactory result.

DP6: Provide the system with personalized human-machine interaction in order to afford decision-makers to coordinate cognitive-load-based breaks with the system when needed. With regard to breaks, the system must determine whether the user consents to the use of technology-induced delegation [35]. Furthermore, LACII must establish whether the user desires the ability to request breaks and, if so, the preferred method of communication through the user interface. Depending on the preferences of the decision-maker, the decision-making authority can be gradually transferred from the human to the machine. This applies to both the core business and the classification of one's own cognitive load.

5.2 Architecture

Figure 3 illustrates the proposed architecture to use LACII. The architecture is divided into four main sections: the decision-maker, LACII, the IIoT production line, and AI-based anomaly detection. Each of these sections is interconnected with the others through the IIoT.

In the IIoT production line (bottom left), there are multiple machines, controllers, and sensors. In most cases, the production machines can be controlled by participants of the CPS. This is necessary to enable the machines to react to changes in the production environment. For instance, the machines can be stopped in the event of an emergency. The symptoms leading to the reactions, such as heat or shaking, are measured with the help of diverse sensors connected to the production machines. These sensors are used to measure as much data as is needed to obtain a comprehensive picture of the behavior and condition of the machines as well as the produced goods. For instance, the temperature of an oven is monitored to predict potential damage or the quality of the final product.

This form of analysis and prediction is enabled by AI-based anomaly detection. The machine sensors and the physiological sensors emit data in real time and send it continuously to the AI in the bottom right of the architecture. The role of the feature extractor is to collect and process the features in the data that are relevant to the further analysis. The analysis is conducted in real time by the machine learning classifier, which uses the previously built machine learning model. The result of the classification determines further actions.

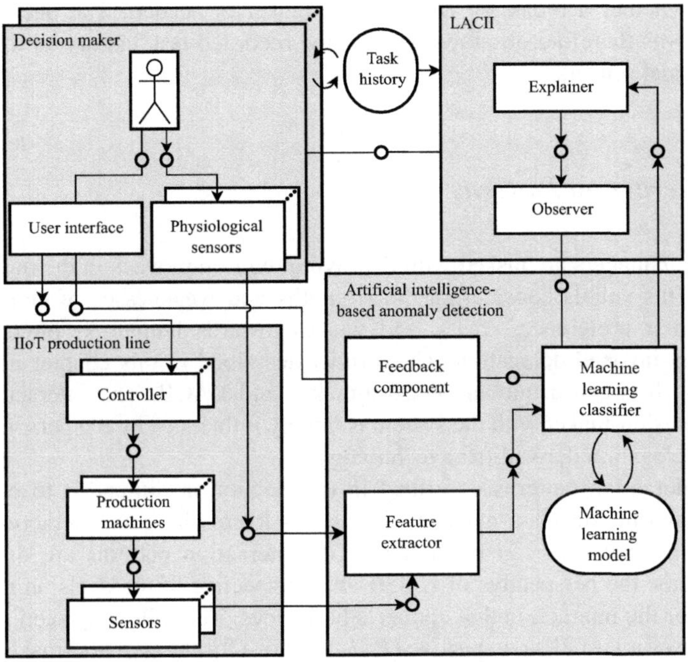

Fig. 3 LACII architecture with cognitive load observation of the decision-maker in an IIoT production

LACII monitors the decision-maker in real time. This is represented in the top right. Upon detecting increased cognitive load, LACII can delegate or recommend a break to the decision-maker via the user interface. This decision is based on the data of the physiological sensors, previous interactions, and also the task history. These factors are collected and evaluated by the machine learning model. For instance, LACII identifies the decision-maker has completed a high amount of tasks. Depending on the rate and quality of the results, task load or cognitive load could be increased. Given the diverse nature of human decision-makers, the form of interaction plays a decisive role. This is why the observer uses an explainer to individualize communication based on preferences. Some decision-makers may wish to understand why a break is recommended. The task history is also used as a knowledge base, since decision-makers have different levels of experience that influences the performance and perceived workload.

The decision-makers (top left) are the main focus point of the architecture. They communicate with every other section of the architecture. The decision-makers can control the IIoT production line through the user interface. It is also used to interact with the feedback component that communicates every decision made by the machine learning classifiers, but also accepts answers to optimize the classifications. The classifications include whether a machine should be controlled, but also whether a break for the decision-maker is needed. The decision-maker themselves is therefore observed through the recorded task history as well as with physiological sensors.

5.3 Interaction Patterns

Company managers are tasked with deploying their employees in the most optimal manner. This entails considering the strengths and weaknesses of employees, as well as their preferences, fears, and wishes towards human-AI interaction and technology-invoked delegation. The artifact described in this chapter necessitates interaction between a human decision-maker and LACII. The interaction of the human decision-maker with the system is largely influenced by the personal attitude towards delegation derived from technology.

The interaction patterns described in this section are intended to provide the manager and the decision-maker with patterns from which an interaction tailored to the respective user can be created. The interaction patterns are divided into three groups: the perspective of LACII, the perspective of the decision-maker, and patterns for the interaction that connects both sides. The following section presents the patterns for LACII's perspective (*PL1–5*), those for the decision-maker (*PDM1–3*), and those for the interaction (*PI1–5*). For our design process, we adapted the levels of automation of Janiesch et al. [36]. Table 1 provides an overview with a brief description of the patterns.

The models of every interaction pattern are designed using the BPMN 2.0. The process models are presented in Figs. 4, 5, 6, 7, 8, 9, 10, 11, 12, 13, 14, 15, 16 and 17. The general structure is as follows. First, the base collaboration diagram (*PI0*) of LACII and the decision-maker is illustrated in Fig. 4. The process begins with the user starting a session. This is indicated by the user sending an (automated) start message to LACII. At this point, both parties are involved in the process, and the decision-maker performs their task. Simultaneously, the user is observed by monitoring the cognitive load. During this period, communication may occur, which could result in a break. Upon the user's termination, due either to a break or completion of the tasks, LACII also ceases monitoring. Both pools contain call activities. This is because within the called activity, the decision power defining the form of messages sent and how to react to messages is defined. The remaining *PIs* are presented later on.

LACII (PL1–5) The interaction patterns of LACII specify the different levels of decision power allocated to LACII that are called by the overlying collaboration diagrams. These are sorted in descending order of decision power.

Therefore, *PL1* represents LACII with very high power. LACII is able to delegate a break as soon as increased cognitive load is found. This is illustrated in the process model in Fig. 5. In general, LACII monitors the cognitive load of the decision-maker for as long as they work on their tasks. When an increase in cognitive load is found,

Table 1 Interaction patterns between LACII and the decision-maker, grouped by the decision power of LACII in descending order

Pattern no.	Description
PL1	LACII detects increased cognitive load and locks the user interface. A break is delegated
PL2	LACII detects increased cognitive load and shows a message announcing that the user interface will be locked soon. A break is delegated
PL3	LACII detects increased cognitive load and shows a message announcing that the user interface should be closed soon. A break is suggested
PL4	LACII detects increased cognitive load and shows an alert in the user interface that suggests a break
PL5	LACII detects increased cognitive load, but does not communicate to the user
PDM1	The decision-maker is able to take a break at any time
PDM2	The decision-maker is able to request a break
PDM3	The decision-maker is not able to request a break
PI1	Following the delegation, the decision-maker gets to choose to either stop now or first complete the current task
PI2	A break is suggested, but the decision-maker gets a veto against LACII
PI3	The decision-maker is only able to request a break when an alert is shown
PI4	The decision-maker is able to request a break even when an alert is not shown, but LACII decides on the exact time
PI5	There are no alerts. The decision-maker is able to request a break at any time, but LACII decides on the exact time

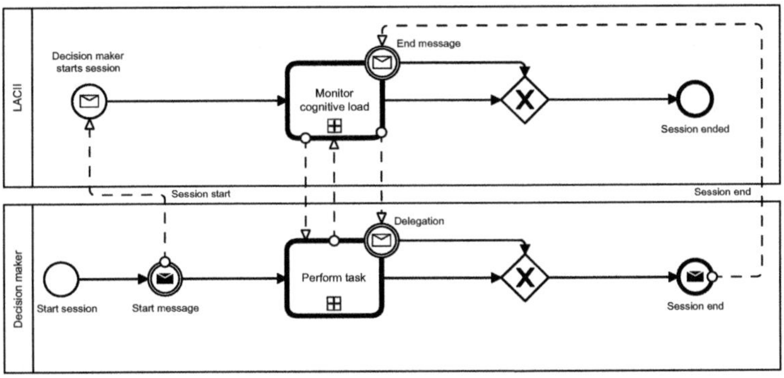

Fig. 4 Base collaboration diagram: pattern interaction 0 (PI0)

the user interface will be locked by the system, attempting to halt the decision-maker's process directly. This delegation is communicated to the user. Nevertheless, the event of detected cognitive load does not immediately stop the observation, as it is possible that the user will not comply with the delegation. The user's potential response to the delegation is discussed later. The focus of this discussion is on the role of LACII.

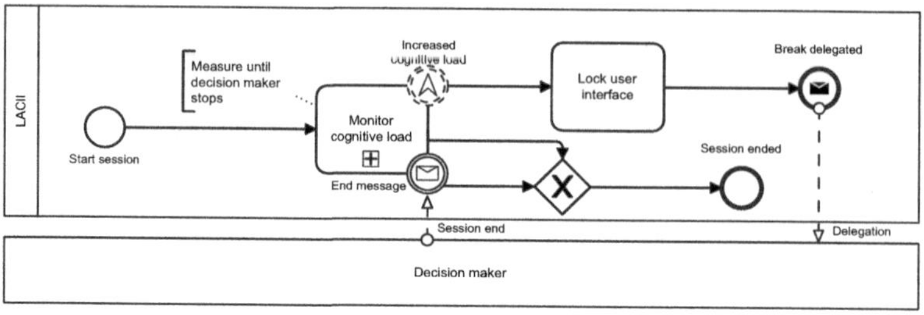

Fig. 5 Pattern LACII 1 (PL1)

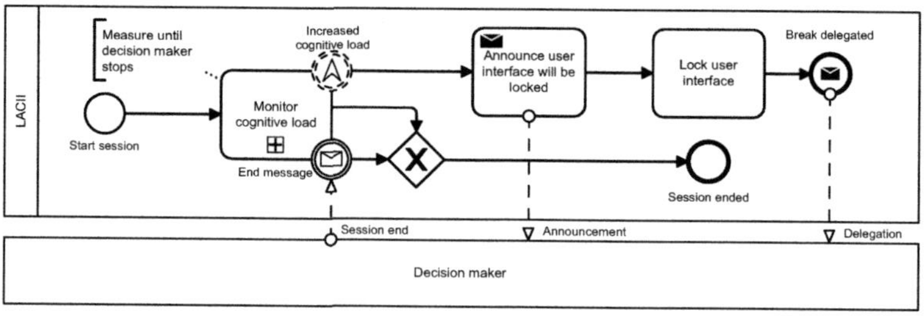

Fig. 6 Pattern LACII 2 (PL2)

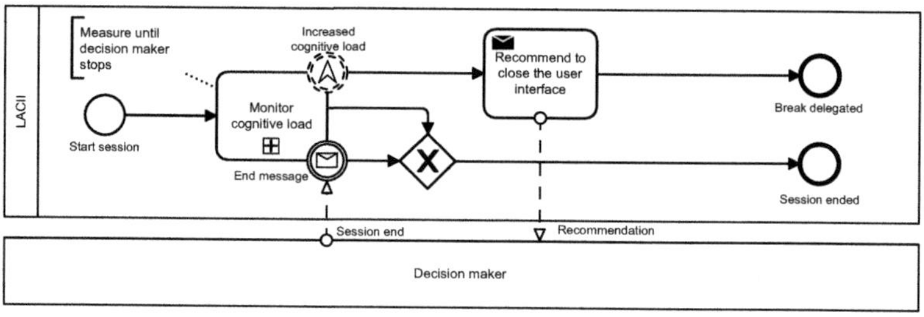

Fig. 7 Pattern LACII 3 (PL3)

PL2 (Fig. 6) employs a comparable approach. As in PL1, the detection of an increase in cognitive load prompts the locking of the user interface and the delegation of a break. However, PL2 diverges from PL1 in that the user is informed of a specified time period, which defines the duration left on the task. It is crucial to note that this time is determined by LACII. Upon expiration of the specified time, the break is initiated. In our study, we do not focus on the precise time that would be optimal. This may vary considerably depending on the nature of the task and the user's preferences. Also, it is conceivable to use a number of worklist items instead of time.

Patterns *PL3* (Fig. 7) and *PL4* (Fig. 8) are comparable and similar, but differ in a crucial manner. In the event of an increased cognitive load, the decision-maker shall be notified. To do that, in *PL3* LACII interrupts the current task of the decision-

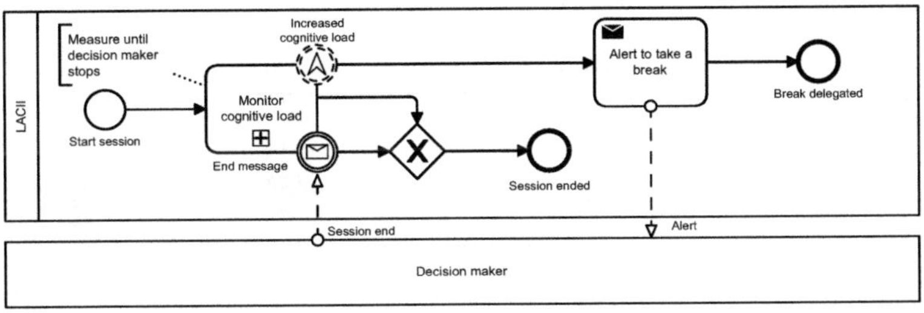

Fig. 8 Pattern LACII 4 (PL4)

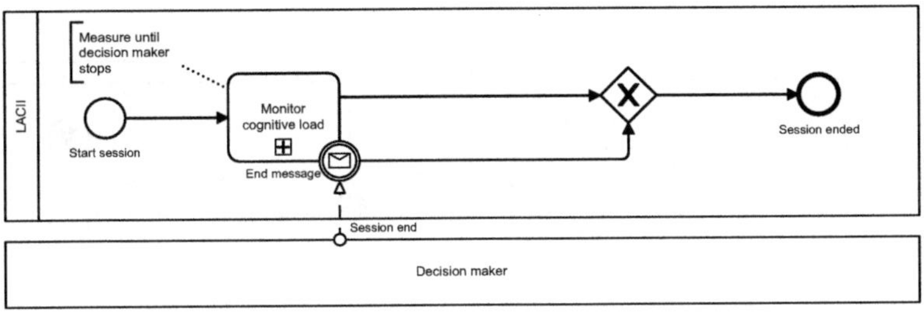

Fig. 9 Pattern LACII 5 (PL5)

maker. A break is then recommended by LACII. The decision-maker can react with continuing or pausing, depending on their own liking and decision power. In contrast, in *PL4* LACII does not interrupt the decision-maker. Instead, a (silent) alert is displayed within the user interface. Upon recognition of the alert by the decision-maker, they are able to respond in an appropriate manner.

PL5 (Fig. 9) represents the lowest possible decision-making authority of LACII. While LACII continues to monitor the cognitive load, it is unable to communicate its findings in any way. Realistically, *PL5* is employed by the manager as a last resort for decision-makers who are unwilling to accept technology-invoked delegation and who require minimal interaction.

Decision-Maker (PDM1–3) The patterns of the decision-maker can be considered analog to the patterns previously discussed. They illustrate different decision power levels starting from high to low.

Once decision-makers begin a session, they proceed to complete their tasks in a sequential manner. The activity is concluded when all tasks have been completed or the usual work shift has ended. LACII's delegation to take a break marks the end of the activity, unless otherwise specified (see *PI1–5*). In Fig. 10 (*PDM1*), the high decision-making power user is permitted to take a break at any time if necessary. This results in the process being terminated with an initiated break that is communicated to LACII.

In *PDM2* (Fig. 11), the decision-maker is not able to take a break at any time. Instead, the user can request a break. LACII decides if and when the break is

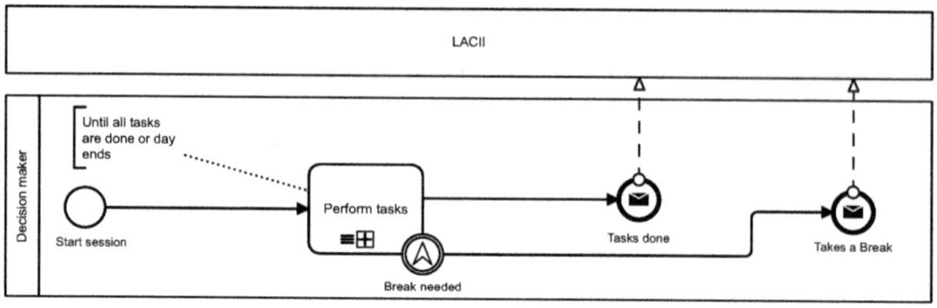

Fig. 10 Pattern decision-maker 1 (PDM1)

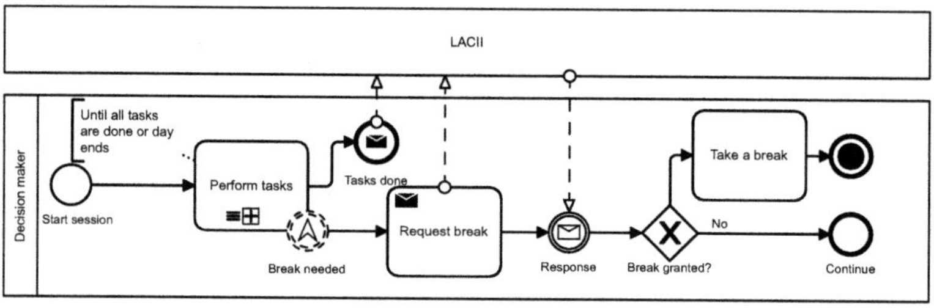

Fig. 11 Pattern decision-maker 2 (PDM2)

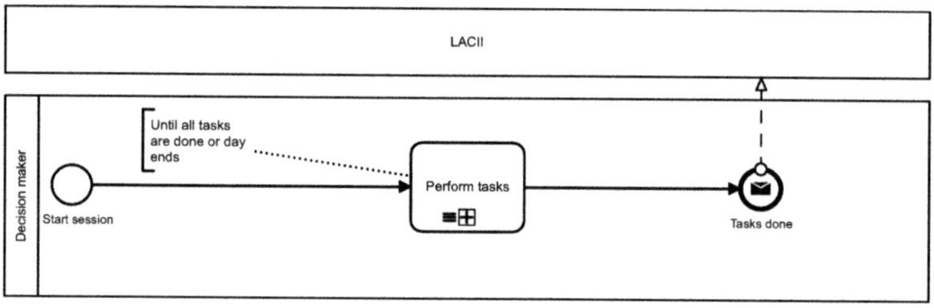

Fig. 12 Pattern decision-maker 3 (PDM3)

granted. Only when the answer is received can the break be initiated. In comparison, the decision-maker in *PDM3* shown in Fig. 12 is not able to request or initiate a cognitive-load-based break at all.

Interaction of LACII and the Decision-Maker (PI1–5) In order to combine the patterns of LACII and the decision-maker, interaction patterns are needed. For example, if *PL1* and *PDM3* are chosen, *PI0* can be used to combine them. The following patterns are suitable for creating more complex interaction combinations.

Some tasks are very complex and it would be inefficient to cancel them during the operation, or perhaps a worker is discontent if a task has to be left unfinished. In this case *PI1* illustrated in Fig. 13 can be used. Here, after delegation, the main task is cancelled to start the break, but the user can choose whether to finish the current task before the break.

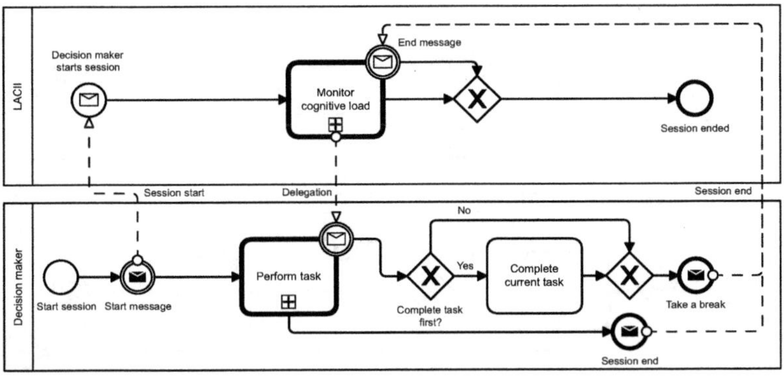

Fig. 13 Pattern interaction 1 (PI1)

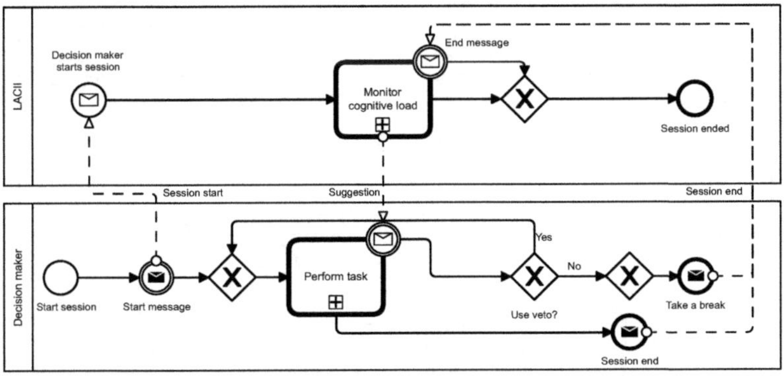

Fig. 14 Pattern interaction 2 (PI2)

In some cases, the user may disagree with LACII's decision. In this case, an interaction pattern could be implemented that allows the human to veto LACII's delegation. In *PI2* (Fig. 14), the delegation to take a break stops the user from performing the task. The decision-maker then has the choice of accepting the delegation or vetoing it and continuing work. In some processes, it may be useful for the manager to allow only a certain number of usable vetoes. For example, to protect decision-makers from being too ambitious and hurting themselves in the long run.

If LACII uses alerts to imply cognitive load to the user, this can be handled in two ways. First, in *PI3* (Fig. 15), the duration of the alert marks a time frame that can be used by the decision-maker to request a break. The alert itself does not stop task performance, but the alert allows the user to stop manually.

PI4 works in a similar way to *PI3* and is shown in Fig. 16. The difference is that in addition to the alert-driven time frame, the user can request a break at any time. This does not stop the task immediately, but LACII is notified and chooses the next best time for a break. This time can be determined as a forecast of the cognitive load based on the current knowledge of the user's cognitive load or by the current tasks

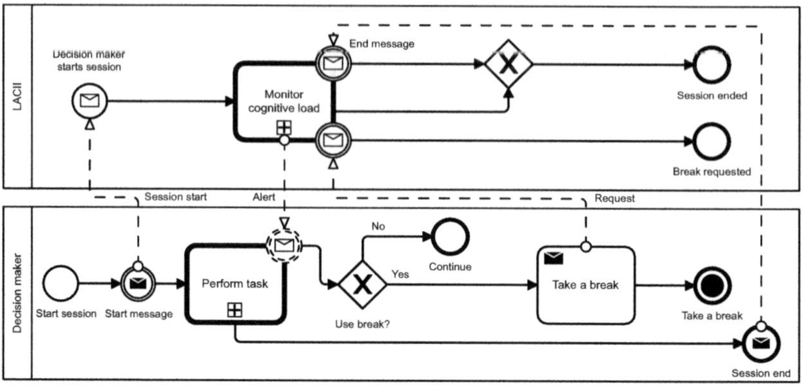

Fig. 15 Pattern interaction 3 (PI3)

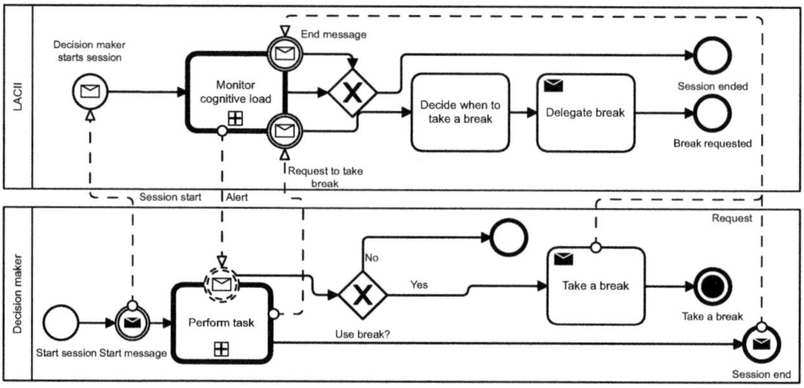

Fig. 16 Pattern interaction 4 (PI4)

in the queue. For example, the system may decide that a certain number of tasks should be done first in order to work efficiently. The users' notification can be used for reinforcement learning to improve the cognitive load detection. *PI5* (Fig. 17), the last interaction pattern presented, works exactly like *PI4* with the exception of alerts. The break can be requested independent of alerts.

6 Illustration

The objective of this section is to demonstrate and evaluate the potential application of LACII in a realistic context. A hypothetical production company and the associated business process are presented using the scenario technique.

Hevner et al. [40] propose a selection of different evaluation methods, with the scenario technique representing a proven option for evaluation in the DSRM [41,

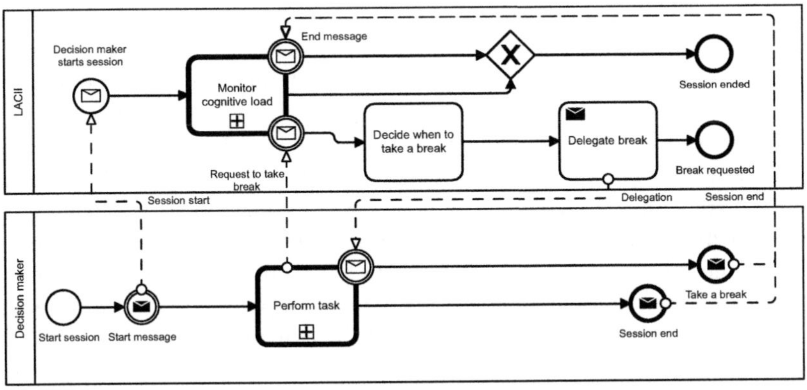

Fig. 17 Pattern interaction 5 (PI5)

42]. The technique is essentially based on informed arguments that demonstrate the utility and how well the artifact contributes to the anticipated problem solution [41]. To begin, we will introduce the scenario company and then define two personas that represent typical users of the artifact.

This production line uses the IIoT with a human-in-the-loop and could benefit directly from LACII. It is therefore essential to consider not only the existing business process, but also how LACII integrates into it. The scenario company produces specialized metal parts, such as car body components, in assembly line production that require a high degree of precision in manufacturing. Each part is subject to a series of production steps that are strictly timed. In the event of a machine malfunction, the entire production process is stopped and waits for the repair.

A part of the production process looks like this: first, the metal parts are welded together. On the one hand, the weld seams must be sufficiently stable. On the other hand, the weld seam must not be too thick so that it does not get in the way during the subsequent production steps. During the welding the temperature is measured to aim for the optimal welt.

In the second step, the metal piece is subjected to a cleaning process using a nozzle that sprays liquid carbon dioxide under high pressure, which ensures the surface is treated evenly and smoothly. It is essential that the entire surface undergoes an identical cleaning process, as this is necessary for the subsequent stages of the process, for example, for the optimal adhesion of glue. Sensors are employed to monitor key parameters, including the ratio of the gas mixture, temperature, and flow rate. An incorrect mixture or alterations in temperature could result in a change in viscosity, which in turn affects the flow rate or leaves unwanted stains. Fluctuations in the flow rate could potentially lead to uneven surface treatment.

To meet the demands of the market, the company is compelled to operate in shifts around the clock. As production cannot be halted, it is essential to coordinate breaks between employees to ensure that no machine is left unattended for too long.

The workers act as decision-makers that receive analyses of the machine data. In the event of an anomaly, it is up to the decision-maker to determine whether the anomaly is significant enough to justify intervention. The decision-maker is confronted with the challenge of balancing the potential consequences of an anomaly, which could include extended downtime, with the risk of unnecessary halting of the process to clean the nozzles in the event of a false-positive anomaly. It is therefore essential to minimize the number of stops to the extent that only those that are truly necessary are enacted. This puts an extended pressure on the shoulder of the decision-makers.

The inclusion of LACII has the goal to increase the process efficiency while increasing the well-being of the decision-maker. To allow the best possible results, sensor technology is added to the decision-maker's workplace. The employee now wears a smart watch that measures the heart rate, an eye-tracking camera is attached to the screen and all inputs such as mouse movements are recorded to measure reaction times.

LACII performs an anomaly detection of these inputs in real time, with the objective of identifying any indications that the decision-maker may be exhibiting increased levels of cognitive load. LACII is also aware of the tasks currently being undertaken by the decision-maker's colleagues and the work they already completed. If LACII determines that the employee will require a break, the system considers that not all employees leave their station simultaneously when determining the precise time, but that the individual does not have to wait an excessive amount of time for the break.

Depending on the preferences of the different employees in the scenario, different configurations of the interaction patterns discussed in Sect. 5.3 can be combined. In the following, we give two scenarios that represent archetypical implementations of LACII.

Figure 18 demonstrates the interaction pattern of a representative employee within the organization. While the decision-maker observes the real-time analysis of the sensor data and awaits the occurrence of anomalies, he is being monitored by LACII. The decision-maker prefers to take flexible breaks, which may be requested at any time (PDM1). LACII, on the other hand, is also endowed with the highest degree of decision-making authority and is able to lock the decision-maker's user interface, thereby delegating a break (PL1). If the decision-maker wishes to reduce LACII's ability to lock the user interface in the future, PI2 could be employed as an alternative. The decision-maker would then be able to veto against the delegation.

A second exemplary employee's workflow could be negatively impacted if they are interrupted by LACII. In this case, alerts could be employed as an alternative solution. While the employee is engaged in tasks related to the user interface and monitoring sensor data to identify any anomalies, silent alerts are generated and displayed in the decision-maker's interface. These alerts serve to remind the employee to take a break when the cognitive load is increased (PL4 in combination

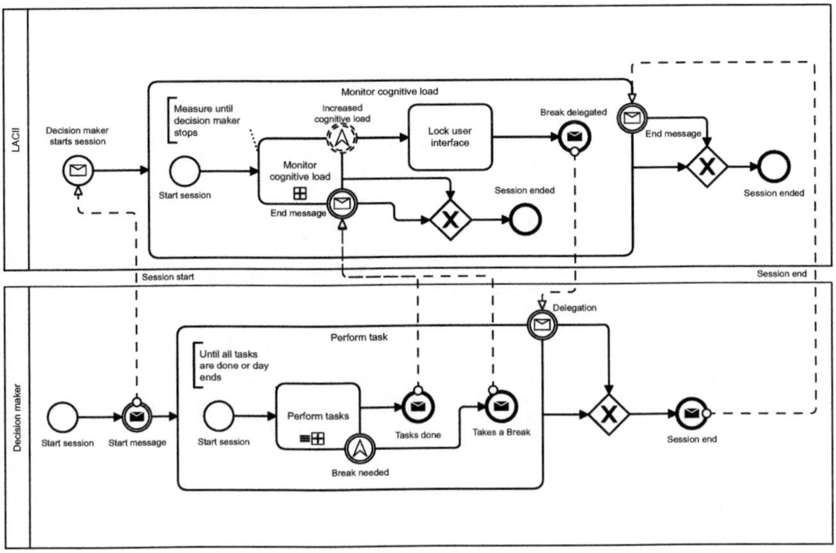

Fig. 18 Exemplary interaction of a decision-maker with LACII (PI0, PDM1, and PL1)

with PI3 or PI4). However, it is possible that the employee may miss some alerts. Therefore, it is essential to provide the employee with the ability to request breaks independently (PDM1 or PDM2).

7 Discussion

In order to achieve the aforementioned balance of business process efficiency and employees' well-being, managers of IIoT companies could decide to implement LACII to use physiological sensors and the already used IIoT together with the interaction patterns discussed in Sect. 5.3. During the implementation, the manager's tasks are to plan how cognitive load-based break times will be determined. This includes how physiological sensors are integrated to emit the corresponding data and how the human-machine interaction is organized. In this section the manager's considerations on determining which interaction pattern should be used will be presented.

The decision-making authority for initiating a cognitive-load-based break can be shared between LACII and the decision-maker. The manager must decide how this division should be organized on a case-by-case basis. For this reason, managers and decision-makers should consult and decide together on the chosen form of interaction. Figure 19 illustrates the trade-offs that organizations must consider when implementing such a system.

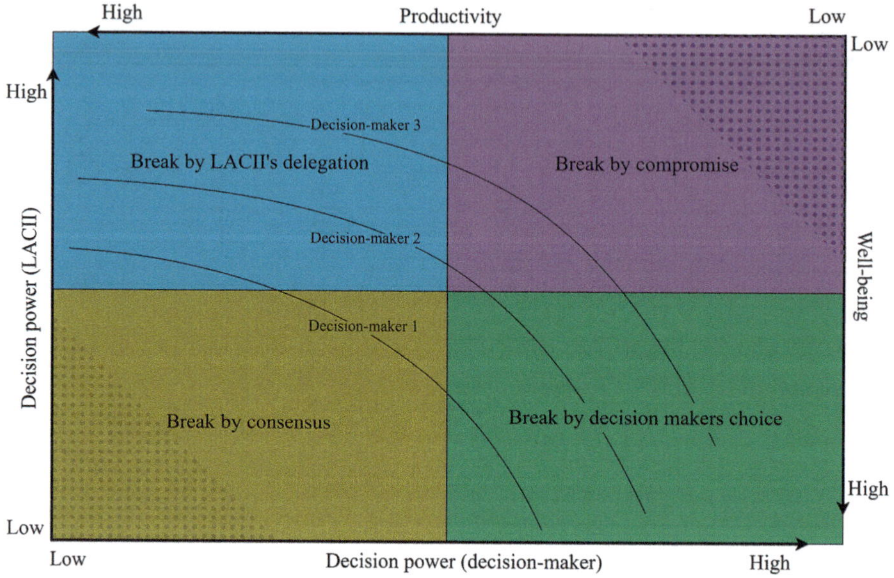

Fig. 19 Decision power has to be balanced between LACII and the human decision-maker. Human decision-makers can be placed depending on their preferences balancing their own well-being and their working output

While at first glance it may appear to be the obvious approach to capture as many data as possible and grant LACII full decision-making authority over the cognitive-load-based breaks [43], to minimize off-times, it is unlikely that a fully paternalistic approach through the machine will be accepted by the decision-maker. Additionally, organizations should not use one predefined interaction pattern for every decision-maker involved. The preferences regarding the interaction of humans and machines depend on a number of variables that should be taken into account. While some may simply be reluctant to relinquish decision-making authority, others condemn the use of their physiological data or may respond with reactance when being delegated by a machine [35]. This is particularly the case when it involves their own perceived workload. For instance, some may perceive the delegation to take a break as a penalty for working inefficiently or with poor results, because they view the technology-invoked delegation as self-threat [35]. Employees may be concerned that through the monitoring of their physiological data their privacy is disregarded and data is misused [38]. Conversely, some employees may appreciate the opportunity to take breaks based on their workload, as it may help them to remember to take regular breaks, benefiting them in the long run.

Figure 19 is divided into two perspectives. Every interaction pattern from Sect. 5.3 can be placed somewhere in the graph. The *first* perspective examines the decision-making power trade-off between LACII and the decision-maker. Four areas in the graph are defined that characterize classes on how the manager could decide to handle the decision power in the system. When choosing the interaction pattern that suits the individual decision-maker, the manager and the affected decision-maker can narrow their decision down by using the four areas. First, the decision power could be in one of the extremes. While one is able to decide everything, the other one gets almost no power. This is the area at the upper left or lower right.

In the field labelled "Break by consensus," both parties have low decision powers on their own. This is because they only signal or ask if breaks should be taken (for example, non-interrupting alarm signals). Consensus is reached, when both indicate the necessity of a break.

"Break by compromise" describes the contrary perspective. Both parties have high decision powers. Both sides must accept it if the other party delegates or takes a break. The interaction is therefore based on mutual concessions. For example, if LACII with high decision-making power delegates a break, the high decision-making power worker is obliged to take the break, even though he did not request it. One potential drawback could be that the decision-maker feels undermined.

The dotted corners represent the unrealistic extremes. With no decision power at all between the parties, no cognitive-load-based breaks will take place. With full decision power on both ends compromises are unrealistic and conflicts are inevitable.

The *second* perspective examines the trade-off between the productivity and the well-being of the decision-makers. In this case, we define productivity as the number of decisions that are made by the decision-maker in a given period of time with data occurring in real time. Each person's well-being is individual and is characterized by success, for example, but also by stress and other factors. If the decision-maker has to work in an environment that does not suit them, their well-being will suffer.

Generally, the aim is to increase both well-being and productivity. This could be done by lowering cognitive load and allowing flow [44]. When deciding on the interaction pattern to use in the system, the manager should consult with the decision-maker. The illustrated decision-maker curves 1–3 in Fig. 19 are exemplary and have different trade-off curves implying that the chosen interaction pattern is based on highly individual preferences. Many studies suggest that users differ in the acceptance of technology and the individual intention to use the technology; see, for example, [45–48]. Therefore, different patterns may fit different user's preferences.

When forced to hand over their own decision power, provide more data than wanted to enable the AI, or obey technology-invoked delegation against their will, the decision-makers' well-being suffers. In addition, we assume that simultaneously very high decision-making power for both parties leads to conflicts that reduce overall productivity.

The manager's objective is to optimize. It is important to ensure that decision-makers are provided with the highest possible level of well-being to retain them in the company in the long term. At the same time, business success depends on maintaining high levels of productivity at all times. While, in theory, it is possible to have high productivity and well-being simultaneously, this may not be possible for every individual. The illustration implies that pattern that promote communication and allow democracy enable the respective decision-makers to work with the optimal combination of well-being and productivity [43]. The extent of desired power therefore depends on the character of the individual.

It is important to note that our work is not without limitations. For instance, our illustration provides a simplified view of a production that does not claim to fully capture the complexity of a real organization. Additionally, the formulated results, such as the design principles, are inherently subjective. One potential approach to address this would be to conduct interviews with experts in the field to discuss the results and refine them in a subsequent DSRM iteration.

8 Conclusion

We started this chapter with the RQ on how a cognitive-load-aware computer-in-the-loop for human-in-the-loop decision-makers should be designed that aims to increase human productivity as well as human well-being. To address this question in a systematic manner, we conducted a design science study using DSRM [12] and selected an approach that emphasizes the interaction between humans and machines. The outcome is our design proposal of LACII as a new participant in a CPS.

With the formalization of design requirements and design principles we developed foundation that we used to build the architectural design and interaction patterns that demonstrate the proposed design. Ultimately, we have addressed the integration of the proposed system within an organizational context, as well as the considerations that a manager using the system with decision-makers would need to take into account.

Our work implies that managers that try to increase the productivity and well-being of their decision-makers should tend towards human-machine interaction pattern that promote communication and allow democracy. This is limited to the decision-maker's preferences in an organization. Managers therefore need to know their decision-maker's preferences in regard to machine interaction, but also their preferences towards technology-induced delegation.

Future research could focus on evaluating systems such as LACII in various experiments. Longitudinal studies offer the opportunity to show how people work with such a system after familiarization and whether they become more satisfied and productive overall. Other influences that could support the desired results can also be investigated.

References

1. Flemisch, F.,Usai, M., Herzberger, N. D., Baltzer, M. C. A., Hernández, D. L., Pacaux-Lemoine, M -P.: Human-machine patterns for system design, cooperation and interaction in socio-cyber-physical systems: Introduction and general overview. In: 2022 IEEE International Conference on Systems, Man, and Cybernetics (SMC), pp. 1278–1283, Prague (2022). https://doi.org/10.1109/SMC53654.2022.9945181

2. Hermann, M., Pentek, T., Otto, B.: Design principles for industrie 4.0 scenarios. In: 49th Hawaii International Conference on System Sciences (HICSS), pp. 3928–3937. Koloa, HI (2016). https://doi.org/10.1109/HICSS.2016.488
3. Janiesch, C., Koschmider, A., Mecella, M., Weber, B., Burattin, A., Di Ciccio, C., Fortino, G., Gal, A., Kannengiesser, U., Leotta, F., Mannhardt, F., Marrella, A., Mendling, J., Oberweis, A., Reichert, M., Rinderle-Ma, S., Serral, E., Song, W., Su, J., Torres, V., Weidlich, M., Weske, M., & Zhang, L.: The Internet of things meets business process management: A manifesto. IEEE Syst Man Cybern Mag **6**(4), 34–44 (2020). https://doi.org/10.1109/MSMC.2020.3003135
4. Turner, C.J., Ma, R., Chen, J., Oyekan, J.: Human in the loop: industry 4.0 technologies and scenarios for worker mediation of automated manufacturing. IEEE Access **9**, 103950–103966 (2021). https://doi.org/10.1109/ACCESS.2021.3099311
5. Dumas, M., La Rosa, M., Mendling, J., Reijers, H.A.: Grundlagen des Geschäftsprozessmanagements. Springer, Berlin (2021). https://doi.org/10.1007/978-3-662-58736-2
6. François, P.A., Borghoff, V., Plattfaut, R., Janiesch, C.: Why companies use RPA: a critical reflection of goals. In: Di Ciccio, C., Dijkman, R., del Río Ortega, A., Rinderle-Ma, S. (eds.) Business Process Management. BPM 2022. Lecture Notes in Computer Science, vol 13420. Springer, Cham (2022). https://doi.org/10.1007/978-3-031-16103-2_26
7. Huang, H., Yang, L., Wang, Y., Xu, X., Lu, Y.: Digital Twin-driven online anomaly detection for an automation system based on edge intelligence. J. Manufact. Syst. **59**, 138–150 (2021). https://doi.org/10.1016/j.jmsy.2021.02.010
8. Stahmann, P., Rieger, B.: Towards design principles for a real-time anomaly detection algorithm benchmark suited to industrie 4.0 streaming data. In: Proceedings of the 55th Hawaii International Conference on System Sciences (2022)
9. Liu, K., Hwang, S., Hsieh, M., Liang, S., Chuang, C.: Systematic layout planning in human-system interface: an evaluation of alarm displays with spatial proximity for accidents diagnosis of advanced boiling water reactor. Int. J. Ind. Ergon. **51**, 30–42 (2016). https://doi.org/10.1016/j.ergon.2014.12.014
10. Paas, F.G.W.C.: Training strategies for attaining transfer of problem-solving skill in statistics: a cognitive-load approach. J. Educ. Psychol. **84**(4), 429–434 (1992). https://doi.org/10.1037/0022-0663.84.4.429
11. Epps, J.: Task load and stress. In: Norman, K.L., Kirakowski, J. (eds.) The Wiley Handbook of Human Computer Interaction Set, pp. 207–223. Wiley Blackwell, Hoboken (2018). https://doi.org/10.1002/9781118976005.ch11
12. Peffers, K., Tuunanen, T., Rothenberger, M.A., Chatterjee, S.: A design science research methodology for information systems research. J. Manag. Inf. Syst. **24**(3), 45–77 (2007). https://doi.org/10.2753/MIS0742-1222240302
13. Sehrawat, D., Gill, N.S: Smart sensors: Analysis of different types of IoT sensors. In: 3rd International Conference on Trends in Electronics and Informatics (ICOEI), pp. 523-528, Tirunelveli (2019). https://doi.org/10.1109/ICOEI.2019.8862778
14. Kurada, S., Bradley, C.: A review of machine vision sensors for tool condition monitoring. Comput. Ind. **34**(1), 55–72 (1997). https://doi.org/10.1016/S0166-3615(96)00075-9
15. Gonzalez-Sanchez, J., Baydogan, M., Chavez-Echeagaray, M.E., Atkinson, R.K., Burleson, W.: Affect measurements: A roadmap through approaches, technologies, and data analysis. In: Jeon, M. (eds.) Emotions and Affect in Human Factors and Human-Computer Interaction, pp. 255–288. Academic, Cambridge (2017). https://doi.org/10.1016/B978-0-12-801851-4.00011-2
16. Klepsch, M., Schmitz, F., Seufert, T.: Development and validation of two instruments measuring intrinsic, extraneous, and germane cognitive load. Front. Psychol. **8**, 1997 (2017). https://doi.org/10.3389/fpsyg.2017.01997
17. Cranford, K.N., Tiettmeyer, J.M., Chuprinko, B.C., Jordan, S., Grove, N.P.: Measuring load on working memory: the use of heart rate as a means of measuring chemistry students' cognitive load. J. Chem. Educ. **91**(5), 641–647 (2014). https://doi.org/10.1021/ed400576n

18. Kurebayashi, I., Maeda, K., Komuro, N., Hirai, K., Sekiya, H., Ichikawa, M.: Mental-state estimation model with time-series environmental data regarding cognitive function. Int. Things **22**, 100730 (2023). https://doi.org/10.1016/j.iot.2023.100730
19. Zu, T., Hutson, J., Loschky, L.C., Rebello, N.S.: Using eye movements to measure intrinsic, extraneous and germane load in a multimedia learning environment. J. Educ. Psychol. **112**(7), 1338–1352 (2020). https://doi.org/10.1037/edu0000441
20. Komuro, N., Hashiguchi, T., Hirai, K., Ichikawa, M.: Development of wireless sensor nodes to monitor working environment and human mental conditions. In: Kim, H., Kim, K.J. (eds.) IT Convergence and Security. Lecture Notes in Electrical Engineering, vol. 712. Springer, Singapore (2021). https://doi.org/10.1007/978-981-15-9354-3_15
21. Chandola, V., Banerjee, A., Kumar, V.: Anomaly detection: a survey. ACM Comput. Surv. **41**(3), 15:1–15:58 (2009). https://doi.org/10.1145/1541880.1541882
22. Galy, E., Caiou, M., Mélan, C.: What is the relationship between mental workload factors and cognitive load types?. Int. J. Psychophysiol. **83**(3), 269–275 (2012). https://doi.org/10.1016/j.ijpsycho.2011.09.023
23. Kirschner, P.A.: Cognitive load theory: implications of cognitive load theory on the design of learning. Learn. Instruct. **12**(1), 1–10 (2002). https://doi.org/10.1016/S0959-4752(01)00014-7
24. Betti, S., Lova, R.M., Acerbi, G., Santarelli, L., Cabiati, M., Del Ry, S., Cavallo, F.: Evaluation of integrated system of wearable physiological sensors for stress monitoring in working environments by using biological markers. IEEE Trans. Biomed. Eng. **65**(8), 1748–1758 (2018). https://doi.org/10.1109/TBME.2017.2764507
25. Brünken, R., Steinbacher, S., Plass, J.L., Leutner, D.: Assessment of cognitive load in multimedia learning using dual-task methodology. Exp Psychol. **49**(2), 109–119 (2002). https://doi.org/10.1027/1618-3169.49.2.109
26. Bijarsari, S.E.: A current view on dual-task paradigms and their limitations to capture cognitive load. Front. Pschol. **23**, 648586 (2021). https://doi.org/10.3389/fpsyg.2021.648586
27. Hart, S.G., Staveland, L.E.: Development of NASA-TLX (task load index): results of empirical and theoretical research. Adv. Psychol. **52**, 139–183. North-Holland (1988). https://doi.org/10.1016/S0166-4115(08)62386-9
28. Hart, S.G.: Nasa-task load index (NASA-TLX); 20 years later. Proc. Human Factors Ergon. Soc. Ann. Meet. **50**(9), 904–908 (2006). https://doi.org/10.1177/154193120605000909
29. Rissler, R., Nadj, M., Li, M.X., Loewe, N., Knierim, M.T., Maedche, A.: To be or not to be in flow at work: physiological classification of flow using machine learning. IEEE Trans. Affect. Comput. **14**(1), 463–474 (2023). https://doi.org/10.1109/TAFFC.2020.3045269
30. Chandra, L., Seidel, S., Gregor, S.: Prescriptive knowledge in IS research: Conceptualizing design principles in terms of materiality, action, and boundary conditions. In: 48th Hawaii International Conference on System Science (HICSS), pp. 4039–4048. Kauai, HI (2015). https://doi.org/10.1109/HICSS.2015.485
31. Sein, M.K., Henfridsson, O., Purao, S., Rossi, M., Lindgren, R.: Action design research. MIS Quart. **35**(1), 37–56 (2011). https://doi.org/10.2307/23043488
32. Hevner, A.R., Chatterjee, S.: Design Research in Information Systems. Springer Science+Business Media, LLC, New York (2010). https://doi.org/10.1007/978-1-4419-5653-8
33. Herm, LV., Steinbach, T., Wanner, J., Janiesch, C.: A nascent design theory for explainable intelligent systems. Electron. Markets **32**, 2185–2205 (2022). https://doi.org/10.1007/s12525-022-00606-3
34. Schuetz, S., Venkatesh, V.: Research perspectives: the rise of human machines: how cognitive computing systems challenge assumptions of user-system interaction. J. Assoc. Inf. Syst. **21**(2), 460–482 (2020). https://doi.org/10.17705/1jais.00608
35. Adam, M., Diebel, C., Goutier, M., Benlian, A.: Navigating autonomy and control in human-AI delegation: user responses to technology- versus user-invoked task allocation. Decis. Support Syst. **180**, 114193 (2024). https://doi.org/10.1016/j.dss.2024.114193
36. Janiesch, C., Fischer, M., Winkelmann, A., Nentwich, V.: Specifying autonomy in the internet of things: the autonomy model and notation. Inf. Syst. E-Bus Manage. **17**, 159–194 (2019). https://doi.org/10.1007/s10257-018-0379-x

37. Elston, D.M.: The Hawthorne effect. J. Am. Acad. Dermatol. (2021). https://doi.org/10.1016/j.jaad.2021.01.085
38. Saifuzzaman, M., Ananna, T.N., Chowdhury, M.J.M., Ferdous, M.S., Chowdhurry, F.: A systematic literature review on wearable health data publishing under differential privacy. Int. J. Inf. Secur. **21**, 847–872 (2022). https://doi.org/10.1007/s10207-021-00576-1
39. Adadi, A., Berrada, M.: Peeking inside the black-box: a survey on explainable artificial intelligence (XAI). IEEE Access **6**, 52138–52160 (2018). https://doi.org/10.1109/ACCESS.2018.2870052
40. Hevner, A.R., March, S.T., Park, J., Ram, S.: Design science in information systems research. MIS Quart. **28**(1), 75–105 (2004). https://doi.org/10.2307/25148625
41. Venable, J., Pries-Heje, J., Baskerville, R.: A comprehensive framework for evaluation in design science research. In: Peffers, K., Rothenberger, M., Kuechler, B. (eds.) Design Science Research in Information Systems. Advances in Theory and Practice. DESRIST 2012. Lecture Notes in Computer Science, vol. 7286. Springer, Berlin (2012). https://doi.org/10.1007/978-3-642-29863-9_31
42. Janiesch, C., Fischer, M., Imgrund, F., Hofmann, A., Winkelmann, A.: An architecture using payment channel networks for blockchain-based Wi-Fi sharing. ACM Trans. Manage. Inf. Syst. **14**1, Article 1, 24 (2023). https://doi.org/10.1145/3529097
43. Sales, S.M.: Supervisory style and productivity: review and theory. Personnel Psychol. **19**(3), 275–286 (1966). https://doi.org/10.1111/j.1744-6570.1966.tb00303.x
44. Fullagar, C.J., Kelloway, E.K.: Flow at work: an experience sampling approach. J. Occupat. Organiz. Psychol. **82**, 595–615 (2009). https://doi.org/10.1348/096317908X357903
45. Venkatesh, V., Morris, M.G., Davis, G.B., Davis, F.D.: User acceptance of information technology: toward a unified view. MIS Quart. **27**(3), 425–478 (2003). https://doi.org/10.2307/30036540
46. Davis, F.D.: A Technology Acceptance Model for Empirically Testing New End-User Information Systems: Theory and Results, pp. 1–291. Massachusetts Institute of Technology, Cambridge (1986)
47. Davis, F.D.: Perceived usefulness, perceived ease of use, and user acceptance of information technology. MIS Quart. **13**(3), 319–340 (1989). https://doi.org/10.2307/249008
48. Sheppard, B.H., Hartwick, J., Warshaw, P.R.: The theory of reasoned action: a meta-analysis of past research with recommendations for modifications and future research. J. Consumer Res. **15**(3), 325–343 (1988). https://doi.org/10.1086/209170

Exploring Process Mining in Human Activity Recognition: Challenges and Future Directions

Dipanwita Thakur, Antonella Guzzo, and Giancarlo Fortino

1 Introduction

Nowadays, academicians and industry-based people are working in the domain of smart healthcare, remote patient monitoring, and assisted living of elder people in which analyzing and monitoring physical daily activities play a crucial role. Process mining offers a promising approach to enhance human activity recognition (HAR) [6, 23], an area focused on the identification and understanding of human behavior through data collected from various sensors. HAR plays a crucial role in fields such as smart healthcare, remote patient monitoring, and assisted living for elderly people, where accurately recognizing daily activities can provide valuable insights into individuals' health, lifestyle, and needs. As such, HAR is increasingly applied in areas ranging from medical adherence monitoring and elderly care to sports training and behavioral analysis in smart homes.

Traditional HAR methods rely heavily on machine learning algorithms, such as decision trees, support vector machines (SVMs), and neural networks [24], which require less manual feature engineering and demonstrate better performance [22, 26], to process sensor data and identify specific activities. While effective, these approaches often focus on classifying isolated activities rather than considering the broader sequences or interdependencies within which these activities occur. This perspective may limit their applicability in complex, real-world scenarios, where human activities are influenced by dynamic factors and often follow certain patterns over time. Process mining can address this gap by providing a way to analyze these sequences of activities as interconnected processes, offering deeper insights into behavior patterns. There has been relatively less emphasis on unsupervised [21] or

D. Thakur (✉) · A. Guzzo · G. Fortino
Department of Informatics, Modeling, Electronics, and Systems, University of Calabria (Unical), Arcavacata, Italy
e-mail: dipanwita.thakur@unical.it; antonella.guzzo@unical.it; giancarlo.fortino@unical.it

G. Fortino, M. Mecella (eds.), *Internet of Things Meets Business Process Management*, Internet of Things, https://doi.org/10.1007/978-3-031-90746-3_6

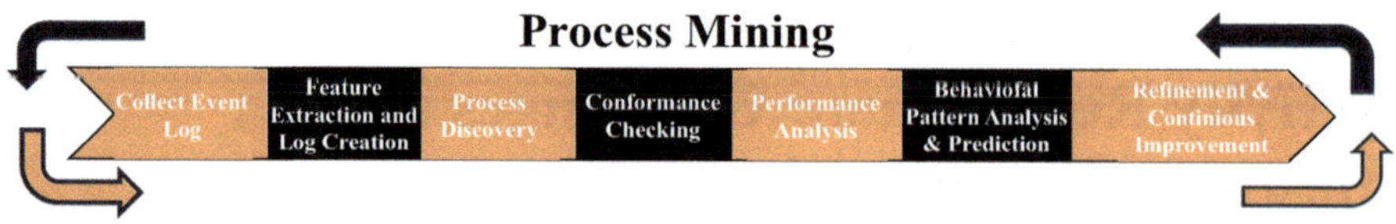

Fig. 1 Process mining methodology

semi-supervised techniques [19] that necessitate fewer annotations [25]. Another notable challenge pertains to the emerging concept of transfer learning [6], which facilitates the deployment of an activity recognition model from one floor to another, taking into account differences in layout, environmental factors, population, and activities.

Process mining emerged as a distinct discipline within the business domain at the turn of the twenty-first century, initially focusing on the automated discovery of business process models. However, over time, the scope of process mining has broadened in various dimensions [27]. It was initially developed to analyze workflows and business processes by extracting knowledge from event logs. Event logs are records of actions or events over time, and they serve as the foundation for process mining, allowing analysts to construct data-driven models of processes. Process mining tools can analyze these event logs to uncover workflows and dependencies within the activities, helping analysts and practitioners build accurate models of how people perform daily tasks [8]. Figure 1 summarizes the main steps of the process mining methodology. In the context of HAR, event logs can be created from sensor data that captures the sequence of human actions. Notably, process mining has found substantial applications in analyzing human behavior [4].

One significant advantage of process mining in HAR is its capacity to reveal real-world processes based on actual event data, rather than relying on assumptions or predefined models. Human activities are often highly variable and influenced by external factors, such as the environment, individual preferences, or even the weather. For example, an individual might drink water more frequently in warm conditions than in colder ones. By applying process mining to HAR, analysts can capture these variations and gain a deeper understanding of how specific external conditions impact activity patterns. This level of detail is often missing in traditional machine learning-based HAR approaches, which tend to focus solely on recognizing discrete activities without considering these contextual influences.

Moreover, process mining addresses a central challenge in HAR: the inherent complexity and variability of human behaviors [7]. Unlike structured workflows in business or industrial processes, human activities are fluid and adaptable and often lack a rigid sequence. People may perform the same activity in different ways, depending on personal habits, physical conditions, or environmental constraints. Process mining allows analysts to model these variations more accurately, capturing common patterns as well as individual-specific deviations. This can be particularly useful in healthcare applications, where understanding unique patient behavior patterns can aid in tailoring interventions, detecting anomalies, and providing personalized care.

Furthermore, process mining facilitates the visualization and communication of human-centered workflows, which is essential for applications in healthcare and assisted living. Process mining tools can generate visual representations of human activities, making it easier for healthcare providers to interpret complex workflows and make informed decisions. For instance, in a smart healthcare environment, a process mining-based HAR model can illustrate the daily routines of elderly patients, including deviations from typical patterns. These deviations could serve as early indicators of health issues, such as reduced mobility, irregular sleep patterns, or changes in physical activity levels, which are often associated with underlying health conditions.

In addition to visualization, process mining can aid in optimizing activity-related processes by identifying inefficiencies or bottlenecks in workflows [16]. By comparing sensor data across multiple individuals, analysts may detect patterns or delays within specific routines that indicate areas for improvement. For example, in an assisted living setting, process mining can reveal common challenges in daily tasks, such as dressing or preparing meals, that may require additional support. This insight enables caregivers to intervene more effectively, potentially enhancing the quality of life for individuals under their care.

Process mining also supports the development of adaptive models that can adjust to changing behavior over time. Human activities are not static; they evolve based on various factors, such as aging, changes in health, or even seasonal shifts in lifestyle. Traditional HAR models often struggle to adapt to these changes, especially when based on rigid classifications. In contrast, process mining provides a framework to continuously monitor and update activity patterns, enabling models that remain relevant and effective as behaviors evolve. This adaptability is crucial for applications in long-term monitoring and care, where the ability to respond to gradual shifts in activity patterns can lead to better outcomes.

Another key benefit of using process mining in HAR is its potential to bridge the gap between high-level workflows and specific task-oriented actions. While HAR often focuses on identifying individual activities, such as walking or sitting, process mining can place these activities within a broader context, showing how they interrelate in the course of a day or week. For example, in a rehabilitation setting, process mining could help track a patient's progress by analyzing the sequence of exercises performed over time, providing insights into adherence and recovery. This level of process-level insight goes beyond simple activity recognition, allowing for a more comprehensive understanding of behaviors and routines.

Additionally, process mining addresses the challenge that many process and system owners lack comprehensive insights into the actual occurrences within their operations. In practice, there is often a disparity between expected or prescribed activities and what actually happens. Process mining endeavors to present an accurate portrayal of reality, which can be instrumental in validating activity models and facilitating system redesign. In HAR applications, this can help analysts

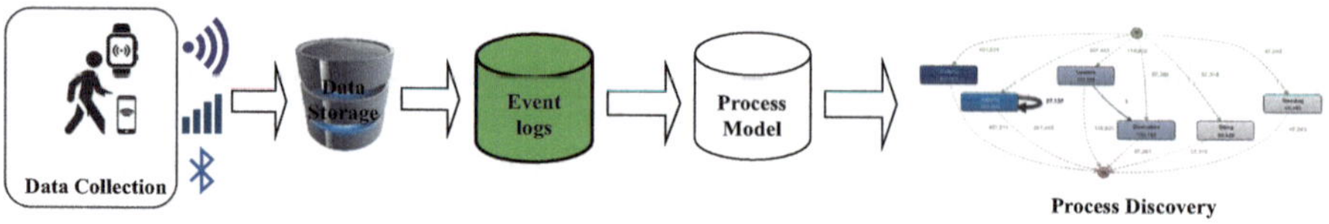

Fig. 2 Process mining for HAR

distinguish between intended activities and actual behaviors, providing a clear view of routines and deviations as shown in Fig. 2.

This chapter aims to examine recent trends and challenges of process mining in HAR. Initially, it covers the fundamental concepts of process mining and its applicability in the HAR domain. Subsequently, it addresses the challenges associated with process mining and its implications for HAR and it proposes solutions to surmount these challenges for future researchers in the realm of process mining within the framework of HAR. Finally, it discusses the unsolved challenges and future directions.

2 Process Mining for HAR in Action

A few crucial steps are involved in applying process mining to HAR to modify conventional methods to the subtleties of activity data, which is frequently obtained from sensor logs or other time-stamped records of human behavior. An overall strategy for using process mining to the identification of human activities is as follows:

1. **Data collection and preparation**
   - Collect event logs: Collecting event logs is a prerequisite for process mining, thus the first step is to gather human activity data. Sensors (such as GPS and accelerometers), smart gadgets, or system logs that timestamped and annotated every action that was observed could be the sources of this data.
   - Data preprocessing: To make sure the data is appropriate for process mining, it is necessary to clean and prepare it. Several methods can be used for data preprocessing such as (1) identification of the event that determines distinct actions or occurrences. Walking, sitting, and running, for instance, could all be noted as distinct events. (2) Timestamp alignment is used to capture the time connection between activities which accurately reflects the sequence in which the actions occurred. (3) Sequencing of events sorts the data into sequences according to the tasks that the same person completed during the given time periods. This could entail breaking up continuous data streams into discrete events that correspond to certain activity or daily routines.

2. **Feature extraction and log creation**
   - Feature extraction and selection: It is necessary to choose pertinent features for every event, including the kind of activity, the start and end timings, the length of the event, and, depending on the use case, any extra elements like location or intensity.
   - Event log creation: Make an event log using the data that has been prepared. Usually, each log entry contains a distinct identifier for every activity sequence, such as a person's daily activity sequence, which is called a case ID, activity, the kind of work that is done, timestamp, the moment the activity began (and maybe finished) and additional attributes, and contextual details that can shed further light on the action, such as the type of sensor or its location.
3. **Process discovery activity**
   - Model human activity sequence: The common activity sequences will be represented by the process model that has been found. A process map could indicate, for instance, that "walking" is typically followed by "sitting," which could later result in "working" or "eating."
   - Process model visualization: The majority of process mining solutions provide visualization features that let you view the activity flow as a process map. This makes it easier to spot regular patterns, transitions, and routines in human behavior.
4. **Conformance checking of human behavior**
   - Define a reference model: When a known or expected pattern of human activities, such as a standard routine, is present, it is advisable to establish a reference model for comparing actual activities against it.
   - Detect deviations and anomalies: Utilize conformance checking to pinpoint discrepancies between observed activities and the reference model. For instance, deviations from the usual sequence of activities, such as "morning exercise" followed by "breakfast," should be brought to attention.
   - Identify anomalous patterns: Conformance checking is effective in unveiling uncommon patterns, such as omitted activities or altered sequences, which could indicate exceptions, errors, or even potential health issues in human activities.
5. **Performance analysis**
   - Timing analysis: Employ process mining tools to analyze the timing of activities. This can assist in identifying bottlenecks or activities that exceed expected durations. For example, it may reveal that "commuting" consistently takes longer than anticipated.
   - Frequency analysis: Assess how frequently specific activity sequences occur. This can aid in understanding routine behavior patterns or identifying frequently repeated activities.
   - Resource utilization: If additional resources such as location or equipment used are logged, analyze how they impact activity sequences. This can be

beneficial for optimizing time spent on specific activities or adjusting routines for improved efficiency.

6. **Behavioral pattern analysis and prediction**

   – Identify common activity patterns: Employ the discovered process model to gain an understanding of typical behavior patterns. For example, common sequences may include "walking" to the "kitchen" → "making coffee" → "drinking coffee."
   – Predict future activities: Utilize the process model and historical data to develop predictive models for forecasting future activity sequences. Process mining can yield insights into probable next activities or likely sequences based on past patterns.

7. **Refinement and continuous improvement**

   – Refine the process model: As additional data is gathered, it is important to update the process model to accurately reflect any changes in activity patterns or behavior.
   – Implement real-time monitoring: Many process mining tools such as celonis offer real-time monitoring capabilities, which can be particularly valuable in industries such as healthcare, where the prompt identification of deviations in activity patterns is crucial.
   – Integration with machine learning: Enhance process mining results by integrating them with machine learning models. For instance, following the identification of a specific routine by the process model, a machine learning model can be trained to categorize or predict specific activities within that routine. Python libraries like pm4py can be utilized for developing custom implementations, enabling the integration of process mining with other machine learning tools.

Through the utilization of process mining for human activity recognition, one can obtain a comprehensive understanding of the sequential structure and timing of activities. This insight facilitates a deeper comprehension of behavior patterns and contributes to the ongoing enhancement of these processes. Several challenges are there in every step of implementation which are discussed in the next section.

3 From Sensor Data to High-Level Activity Traces

3.1 Data Collection and Event Log Representation

Effectively organizing event logs to represent complex human activities in human activity recognition (HAR) systems is challenging due to the diversity of data sources, types of activities, and sensor modalities. A well-structured event log should include essential details about activities, their chronological order, and the surrounding context to support accurate analysis and interpretation [1, 20]. HAR data often consists of a combination of time-series data, images, and text, making it essential to select suitable preprocessing techniques to transform raw data into meaningful event logs.

Activities in event logs can be represented at different levels of granularity [2, 17]. Fine-grained logs capture detailed activities, such as walking or jogging, with high temporal resolution, enhancing model accuracy by allowing them to detect subtle variations in activity patterns. However, this detailed level of logging may introduce noise and complexity, making it harder to identify general trends or patterns. In contrast, coarse-grained logs consolidate actions into broader categories, like physical activity versus sedentary behavior, providing a simpler and more interpretable view of activities. While this approach is easier to analyze, it may overlook important nuances, potentially reducing accuracy in identifying specific activities or transitions.

In real-world scenarios, HAR data is inherently heterogeneous, as it can come from various devices such as smartphones, smartwatches, wearable sensors, cameras, and other vital sign monitoring equipment. Integrating this data into a unified event log is challenging, as data may be stored in different formats (e.g., CSV or JSON), and synchronizing time across devices can be difficult. Variations in sensor sampling rates can affect the granularity and completeness of the event log. Missing data from one sensor may result in an incomplete representation of activities, and diverse data types—such as accelerometer, gyroscope, and GPS data—make aggregation more complex. Additionally, each sensor may interpret activities differently, leading to inconsistent results. Noise-induced variations in readings can further compromise event accuracy, and differences in device or activity usage can introduce inconsistencies. Contextual data, like location and social interactions, can also influence activity interpretation and may not always be captured by sensors. The integration of multiple data sources raises privacy concerns, especially when dealing with sensitive personal information. Ensuring compliance with privacy regulations, such as the General Data Protection Regulation (GDPR) [5], can add another layer of complexity when merging data from multiple devices. The absence of universal standards for event logging and activity classification can hinder device interoperability, and some devices may use proprietary data collection methods, complicating integration. Real-time data fusion from various sources may introduce latency, affecting the immediacy of activity recognition. Furthermore, processing large volumes of heterogeneous data can place a strain on system resources.

Addressing these challenges requires careful planning and the implementation of robust data integration and preprocessing techniques, along with attention to privacy and standardization across devices [9, 10]. Collaboration with domain experts and leveraging machine learning methods can help create more unified and accurate event logs, improving the quality and reliability of HAR analysis. Figure 3 illustrates the event log generation steps in HAR for different activities, such as walking and running.

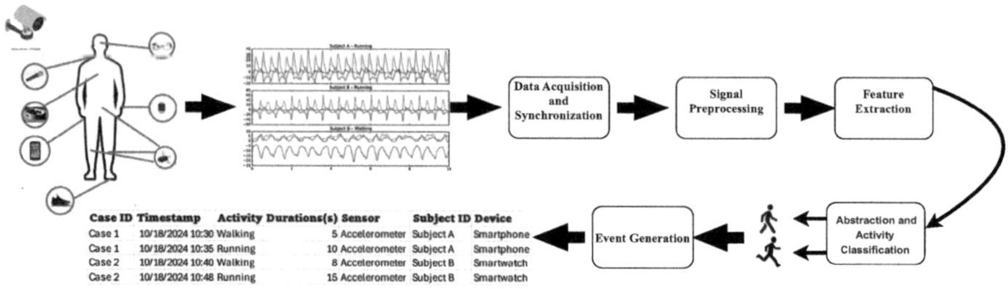

Fig. 3 Event log generation in HAR

3.2 *Handling Noise and Inconsistency in Human Activity Data*

Ensuring the accuracy and dependability of human activity recognition (HAR) systems requires addressing uncertainty and noise in event logs produced by human activity sensors. Here are a few efficient ways to deal with these issues. Data preprocessing techniques such as filtering, windowing, and outlier detection can mitigate the noise and uncertainty in human activity data. Filters (e.g., low-pass filters) are used to smooth out noise in sensor data, especially from accelerometers and gyroscopes. Sliding windows are used to reduce noise and make the data easier to handle, dividing it into digestible portions. Elimination of outliers using statistical techniques could be the result of the removal of noise. Sensor fusion is another commonly used method to handle noise and inconsistency in the data. Multisensor integration and the use of a Kalman filter [3] are two well-known methods for sensor fusion. To increase the activity detection system's resilience, multisensor integration merges data from several sensors (such as accelerometers, gyroscopes, and GPS). Any sensor's noise can be lessened with the use of fusion techniques. Kalman filters or comparable approaches are used for real-time estimates that can help anticipate the true state of the system based on noisy observations. Data augmentation is another method to handle noisy and inconsistent data. Generating synthetic data and noise injection are examples of data augmentation. Adding variability and noise simulation to the training dataset by creating synthetic data can improve the models' ability to generalize to real-world situations. To assist the model to become resilient to noise, deliberate controlled noise injections can be incorporated into training data. Probabilistic models are also used to handle noise and inconsistent data. Hidden Markov models (HMMs) [12] are used to model sequences of observed activities, allowing for the incorporation of uncertainty in state transitions and emissions. Bayesian approaches can also be employed to update the probability of certain activities based on new evidence, allowing a formal way to handle inconsistency. Temporal contextualization is another method to handle noisy and inconsistent data. Utilizing time series analysis techniques that consider temporal correlations in the data can mitigate noise and provide contextual information for activity recognition. Implementation of temporal smoothing techniques, such as moving averages, reduces short-term fluctuations in sensor readings over time.

Feature selection and engineering are the common methods of handling noisy data. It is required to identify and select features that are most informative and less susceptible to noise. Dimensionality reduction techniques (e.g., PCA) can help in this regard [14]. Statistical features (mean, variance, peak frequency) derived from raw sensor data to capture the essence of activities while minimizing the impact of noise are also useful in this regard. Applying cross-validation techniques can ensure the model's consistency across different subsets of the data, reducing the likelihood of overfitting to noisy data. Continuous learning can also be used to mitigate noise and inconsistency issues. Adaptive models can learn and adjust to changes in the environment or sensor characteristics over time, enhancing resilience to noise. Feedback mechanisms allow for continuous feedback and retraining of models based on new data, enabling adaptive models to better handle inconsistency.

It is crucial for process mining techniques to handle missing or incomplete event data in HAR scenarios. It requires a number of tactics to maintain the analysis' relevance and usefulness. Process mining can efficiently tackle these difficulties in several ways. Data imputation is a popular technique to handle missing and incomplete event data. Statistical imputation estimates missing values using statistical techniques based on available data. One can use methods such as regression-based imputation, mean/mode imputation, or median imputation. KNN imputation is also used to estimate missing values by comparing them to the values of the dataset's nearest neighbors. Temporal interpolation is another method to address missing and incomplete event data. By estimating missing values from surrounding data points, linear interpolation can fill in missing values for time-series data. For more intricate data patterns, spline interpolation, which enables a smoother transition between known values, can also be used for the same. Process discovery methods are also used for this purpose. Process mining can still find process models in incomplete data. In spite of missing events, methods such as Alpha, Heuristic, and Fuzzy Miner [13] can create models based on the data at hand and aid in understanding the overall structure. Patterns that are found in complete data segments can be applied to the incomplete portions of the logs in an effort to rebuild the event logs. Multisource data integration is another method to address the missing and incomplete data event issues. To make up for missing data from one source, we can combine data from several sources (such as wearables, mobile devices, and environmental sensors) known as cross-referencing. This can improve the event log's thoroughness. To lessen the impact of missing events from any one source, we can combine data from several sensors using data fusion techniques.

Process mining techniques may effectively manage missing or incomplete event data in HAR situations by a combination of imputation, probabilistic modeling, multisource integration, and robust analysis methods. Even in the face of data limitations, organizations can preserve the integrity of their process analyses and gain insightful knowledge by utilizing these tactics.

4 Activity Modeling with Process Discovery

Traditional process mining algorithms, including Alpha Miner, Heuristics Miner, Inductive Miner, and Fuzzy Miner, exhibit both strengths and limitations when applied to the discovery of process models from HAR data. Table 1 demonstrates the strengths and limitations of traditional process mining algorithms w.r.t HAR. While traditional process mining algorithms can provide valuable insights into human activities, their effectiveness in HAR contexts often necessitates careful preprocessing and adaptation to address the unique challenges of the data. The process mining algorithm should have the mechanism to remove noise and be adaptive in nature to adopt noise to some extent. Specifically, it should incorporate noise removal mechanisms that can identify patterns in the data and apply methods such as adaptive thresholding. Statistical techniques should be used to model predicted behavior and distinguish between noise and actual activity. Additionally, probabilistic models utilizing the current log to infer missing data, along with tools such as HMMs and Bayesian networks, can be valuable. Furthermore, a process discovery strategy should be capable of handling larger datasets by employing imputation techniques to estimate missing values in event logs. Contextual data, including location, time of day, and user status, should be integrated to enhance model accuracy and more precisely represent human behavior. Finally, the system should provide techniques for building hierarchical models that can adapt to data fluctuations by capturing both high-level activities and their sub-activities. In the context of supervised learning, the proposed approach involves training models using labeled data by integrating process mining and supervised learning techniques. This integration aims to enhance the algorithm's capability to identify actions amidst noisy data. Furthermore, by employing ensemble learning to combine multiple models, the resilience of the model to noise and data fluctuations can be improved. The envisioned algorithms are designed to gradually learn from fresh data, thereby adapting to changes in user behavior and continuously enhancing performance. By leveraging the strengths of both traditional process mining methods and contemporary machine learning techniques, the handling of noise and incompleteness can be significantly improved. The establishment of feedback channels is essential to facilitate user-driven adjustments and suggestions, thereby progressively refining the model's accuracy. Additionally, the incorporation of user-driven procedures during the discovery phase allows for valuable participant input, ultimately contributing to the development of more precise models. Combining process mining and machine learning can create a powerful synergy for recognizing and analyzing human activities. This integrated approach leverages the strengths of both methodologies, leading to more accurate predictions, deeper insights, and improved understanding of complex behaviors.

Table 1 Process mining algorithms w.r.t human activity recognition

Method	Strength in HAR	Limitations in HAR
Alpha Miner [28]	– User-friendly, straightforward to implement, and easy to understand	– Exhibits sensitivity to noise and incomplete data, which are prevalent challenges in HAR scenarios
	– Offers a robust theoretical framework for process discovery	– Encounters difficulties with complex or overlapping activities, as it operates under the assumption of a more linear structure
	– Facilitating a clear representation of activities and their interrelationships	– Alpha Miner may fall short in capturing the variability and nuances of human behavior, particularly in dynamic environments where activities can frequently change
	– Effective with simple, well-structured event logs, where activities follow distinct sequences, which is often found in controlled HAR scenarios	– Presumes independence among activities, an assumption that may not be valid in HAR settings where activities can be interdependent and contextually influenced
	– Directly maps event logs to Petri nets, yielding a clear visual representation of the identified activities and their respective workflows	– May exhibit reduced efficiency and produce overly complex models that can be challenging to interpret
Heuristic Miner [29]	– Demonstrates greater resilience to noise and incomplete data when compared to the Alpha Miner, making it particularly valuable in HAR scenarios where data quality can fluctuate	– Relies on various parameters, including thresholds for activity frequency. Selecting inappropriate values may result in suboptimal models or the omission of significant activities
	– Capable of managing complex event logs that include overlapping and interdependent activities, which are prevalent in human behaviors	– Manages complexity more effectively than Alpha Miner; the resulting process models can still become intricate, particularly when dealing with complex or extensive event logs
	– By utilizing frequency-based heuristics, it effectively identifies significant paths within the process	– It does not provide formal guarantees concerning the completeness or soundness of the discovered models, which raises concerns for applications in critical fields
	– The models produced by the Heuristic Miner tend to be more interpretable, as they focus on the most frequent paths and activities	– It primarily focuses on descriptive modeling rather than predictive capabilities, which may limit its applicability in scenarios requiring the forecasting of future activities
	– It accommodates process variability, making it well-suited for dynamic environments where activities may not adhere to strict sequences	– For very large datasets, the algorithm may encounter performance challenges, potentially leading to extended processing times and less responsive outcomes

(continued)

Table 1 (continued)

Method	Strength in HAR	Limitations in HAR
	– Effectively uncovers shorter paths within the process model, which can be instrumental in identifying key activities in HAR without being hindered by less pertinent data	
Inductive Miner [15]	– Offers a reliable methodology for discovering models that are both sound and block-structured	– May diminish when confronted with highly complex event logs containing numerous variations, potentially resulting in less accurate models
	– Models accurately represent valid processes, facilitating their understanding and analysis	– It operates under the assumption of a block-structured process model, which may not be applicable to all HAR scenarios, particularly those characterized by unstructured or highly variable behaviors
	– Demonstrates robustness in the presence of noise and is adept at handling incomplete or imperfect event logs such as HAR, where variations in data quality are common	– Quality of the Inductive Miner's outputs is influenced by the selection of appropriate parameter settings, including thresholds for frequency
	– Excels in discovering hierarchical models, which enable the representation of activities at various levels of abstraction	– It may not provide robust predictive capabilities, which could be a limitation in contexts where forecasting future activities is essential
	– Prioritizes the identification of frequent behavioral patterns, enhancing its effectiveness in recognizing common activities and interactions within human behavior	– For very large datasets, the algorithm's computational complexity can become a significant concern, potentially impacting processing times and the feasibility of its application in real-time scenarios
	– The models produced by the Inductive Miner are not only interpretable but also conducive to effective visualization	
HMM [12]	– Recognized for their efficacy in modeling time-dependent sequences, a critical aspect of HAR, where activities often progress over time	– Operates under the first-order Markov assumption, which posits that the current state is influenced solely by the preceding state, which can result in an inadequate representation of processes characterized by intricate dependencies or prolonged interactions among activities
	– Offers a robust probabilistic representation of activities, which allows for effective handling of uncertainty and noise inherent in sensor data commonly encountered in HAR applications	– Necessitate that the number of hidden states is established a priori, which poses challenges in HAR where the number of potential activity types may be uncertain or evolve over time

	– Capable of inferring hidden states that are not directly observable yet significantly influence the observable activity sequence. In HAR, these hidden states may reflect latent factors such as user intentions or physiological conditions – Facilitate online or real-time activity recognition by processing sequences in real-time as they are observed – Adapted to manage multiple types of activities in HAR, making them particularly effective in complex environments where individuals may transition between several activities	– Designed for sequential data, which limits their effectiveness in recognizing concurrent or interleaved activities in HAR, such as scenarios where individuals simultaneously engage in multiple tasks – Requires extensive labeled training data to accurately estimate transition and emission probabilities – The hidden states within HMMs are frequently not easily interpretable, creating challenges in comprehending the underlying structure of the modeled human activities or processes – May face challenges with high-dimensional sensor data commonly utilized in HAR
Fuzzy Miner [13]	– Simplifies noisy and variable HAR data, making it easier to understand dominant activity patterns – Focuses on significant and frequent activities, reducing the complexity of the model – Provides visual and interpretable models for exploring human activities – By abstracting details, Fuzzy Miner avoids overfitting to complex and noisy data	– Can ignore or downplay rare but important activities – Requires careful threshold selection for significance and correlation – Struggles with parallel or overlapping activities in HAR – Faces challenges when dealing with high-dimensional sensor data – Lacks precision in capturing detailed temporal sequences between activities – May not be ideal for real-time HAR due to its reliance on post hoc analysis and tuning

5 Conformance Checking for HAR

5.1 Activity Patterns and Deviations

Process mining techniques can be quite useful for finding anomalous behaviors, mining activity patterns, and examining human activity logs for abnormalities. Process mining leverages event logs to uncover, monitor, and optimize processes. Hence, detecting anomalies in human activity logs using process mining can be achieved by identifying deviations in different ways such as process discovery and conformance checking, temporal deviations, and sequence and dependency analysis as shown in Table 2. Process discovery generates a model from logs that represents the standard sequence of activities. Conformance checking then aligns real event sequences to this model, highlighting deviations where steps are skipped, repeated unexpectedly, or performed in the wrong order. Process mining can identify deviations in the duration of activities or wait times using timestamps. Unusually short or long task durations may indicate rushed or delayed actions, possibly signaling risky behaviors or anomalies. By evaluating sequence and dependency patterns in activity logs, process mining can detect deviations such as unexpected transitions or dependencies that do not conform to expected patterns. Several process mining metrics are particularly valuable for identifying and evaluating deviations in HAR such as fitness, precision, generalization, and simplicity. This fitness measures how well real activity logs align with the standard process model. A low fitness score indicates that observed sequences deviate from the expected patterns, potentially highlighting abnormal or rare behaviors. In HAR, fitness can help in identifying activities that diverge from the norm. Precision evaluates whether the process model is too permissive (allowing too many variants) compared to the actual logs. A high precision score means the model accurately represents common patterns, while a low precision score may highlight unusual or rare sequences. Generalization ensures that the model captures the essential patterns without overfitting to noise or outliers in the data. Generalization helps determine whether detected deviations are significant or merely artifacts of noisy data. Simplicity gauges whether the model remains interpretable, focusing on the most significant deviations. For HAR applications, simplicity ensures the model does not become too complex and hard to analyze, which is essential for interpretability in critical use cases like health monitoring. Together, these metrics help balance the model to accurately capture typical behavior while highlighting significant deviations, essential for HAR tasks where understanding and interpreting behavior is crucial.

Process mining offers significant advantages in the identification and analysis of abnormal behaviors such as health anomalies or risky activities due to its ability to model both expected and unusual patterns in activity data. Process mining can identify health-related anomalies by monitoring deviations in activity patterns. For instance, in eldercare, deviations such as prolonged inactivity or erratic movement patterns may indicate potential health issues. Metrics such as fitness and time-based analysis can highlight these irregularities for further investigation. In high-risk

Table 2 Birth view of PM techniques to detect deviations in HAR

Concept	Description	Techniques	Applications in HAR
Process discovery	Extracting models of human activities from event logs to understand how individuals perform tasks in real-world scenarios	Alpha Miner, Heuristic Miner, Inductive Miner	Mapping typical human behavior, identifying activity patterns
Conformance checking	Comparing the discovered models of human activities with observed behavior to identify deviations and ensure compliance with expected actions	Model-based checking, alignment methods	Ensuring adherence to protocols, evaluating training effectiveness
Temporal deviations	Analyzing the timing of human activities to detect delays or accelerations that may indicate issues in performance or adherence to expected timeframes.	Temporal analysis, time-series analysis	Monitoring performance against SLAs, identifying efficiency issues
Sequence analysis	Studying the order of activities performed by individuals to reveal typical workflows, variations, or anomalies in behavior	Sequence mining, pattern recognition	Understanding user routines, detecting abnormal behavior patterns
Dependency analysis	Examining the relationships between different human activities to understand how one activity influences the occurrence or timing of another	Dependency graphs, correlation analysis	Identifying inter dependencies in tasks, optimizing workflows

settings like construction or industrial environments, process mining can identify unusual sequences or transitions. For example, skipping critical safety steps or working outside regular hours may indicate risky behaviors, which can be flagged by comparing actual logs to safe operating procedures. Process mining enables the visualization and in-depth exploration of specific behavior patterns, aiding in the identification of the root cause of anomalies. This level of detail supports healthcare or safety analysts in identifying potential interventions or providing support for individuals displaying abnormal behaviors. By integrating historical patterns and real-time monitoring, process mining can offer actionable insights into abnormal behavior, thereby facilitating early intervention or support.

Understanding both frequent and rare patterns in human activity data is critical for behavior analysis and anomaly detection. Process mining provides several techniques like frequent pattern mining, rare pattern mining, sequential pattern mining, and outlier detection to achieve this. Techniques such as the Alpha Miner or Heuristic Miner can extract the most common sequences of activities, aiding in defining "normal" behavior by identifying the most frequent activity paths, sequences, or patterns. By focusing on low-frequency paths or sequences, rare or abnormal patterns can be discovered. Tools like fuzzy mining can dynamically filter less common but potentially important behaviors, aiding in surfacing rare activities without overwhelming the analysis. Algorithms designed for temporal data, such as PrefixSpan or Apriori, can be utilized to find recurring sequences across human activity logs. For instance, in healthcare, this could reveal routines or rare exceptions in daily habits, aiding in the detection of shifts in regular activity patterns. Process mining can leverage statistical or machine learning-based outlier detection algorithms (e.g., Isolation Forest, DBSCAN) to identify uncommon sequences or rare deviations that indicate significant changes in behavior. These mining techniques facilitate the understanding of common activity flows and the identification of rare or unexpected behaviors, making them well-suited for applications in health monitoring, safety, and behavioral analysis. Process mining encompasses a robust array of techniques aimed at identifying deviations, evaluating abnormal behaviors, and uncovering both frequent and rare patterns within human activity logs. Through the use of metrics such as fitness and precision, process models can be refined to effectively discern normal from abnormal behaviors. Furthermore, process mining techniques are uniquely positioned to provide support for HAR in domains requiring a comprehensive comprehension of both common and rare activity patterns, such as health monitoring, security, and behavior analysis [11]. In these contexts, deviations can serve as indicators of potential risks or offer crucial insights into health-related matters.

5.2 Activity Monitoring

When it comes to activity monitoring, conformance checking is a useful tool, particularly when it comes to areas like safety laws or health protocols that demand conformity to preset models. Here are some examples of applications, obstacles, and modifications for real-time monitoring and feedback in industries such as fitness or healthcare related to conformity checking. Conformance checking entails the comparison of actual activities with predefined models to ensure adherence. In the context of activity monitoring, these models may encompass health protocols, fitness routines, or safety procedures. Commence by establishing the process model that encapsulates ideal or expected behaviors, such as a standard operating procedure (SOP) [18] in healthcare or a correct sequence of movements in a fitness routine. This model may be derived from historical data on normal behaviors or sequences defined by experts. Human activities, whether in real time or logged, captured through sensors, wearables, or video analysis, are transcribed into event logs. These logs may comprise timestamps, activity types, and sequence order. Subsequently, conformance checking aligns each event log with the predefined model, evaluating whether steps are executed correctly, in the correct order, and within acceptable timeframes. This process can identify deviations, such as omitted steps in a fitness routine or variations from a health protocol. This method ensures that observed human activities conform to the anticipated patterns, facilitating the monitoring of compliance with health and safety standards.

Real-time conformance checking for activity monitoring presents several challenges, particularly in dynamic domains such as healthcare or fitness. Real-time monitoring necessitates swift analysis and immediate feedback, which can be hindered by the intricate nature of conformance-checking algorithms. Processing delays may diminish the efficacy of immediate alerts. In real-time scenarios, human activity data collected via sensors is frequently susceptible to inaccuracies. Noise or missing data can distort the logs, resulting in false positives or overlooked deviations. Human activities exhibit a wide spectrum of variation, and deviations may not consistently signify noncompliance. For instance, alternative exercise routines may still align with fitness objectives, and patients may have customized health protocol adjustments. Stringent models may encounter challenges in accommodating this variability. Real-time monitoring necessitates contextual understanding, such as an individual's health status or environmental factors. Without this contextual insight, certain deviations may be inaccurately flagged, while others may go unnoticed. Addressing these challenges requires the development of robust, adaptable, and efficient conformance-checking methods tailored to real-time applications.

In order to accommodate flexible and highly variable human behaviors, conformance checking can be adapted in several ways such as flexible models with variants and tolerances, machine learning integration, context-sensitive conformance checking, and anomaly detection for outliers. Process models can be adjusted to allow for different acceptable variants of an activity. For example, fitness routines can have acceptable ranges of movement types, speeds, or rest periods. This tolerance makes models less rigid and more adaptable to real-world behavior. Machine learning can enhance conformance checking by identifying common deviations that still meet overall objectives. By learning from previous interactions, models can adapt to individual variations, such as a personalized exercise pace for fitness routines or custom thresholds in health monitoring. Including contextual data (e.g., a patient's age, medical conditions, or an athlete's training level) can make conformance checking more relevant. Adding contextual layers enables models to be conditionally adaptive and evaluate behavior based on individual requirements rather than a one-size-fits-all standard. Instead of rigid alignment to a single model, anomaly detection techniques can identify outliers in activity logs, which can serve as a flexible way to flag unusual patterns that deviate from normal behavior, while still allowing for inherent variability in human activities. These adaptations make conformance checking suitable for environments with significant variability, enhancing its applicability to HAR.

Real-time conformance monitoring systems can be developed to trigger alerts when activities deviate from expected models. This necessitates the integration of process mining with real-time data processing and alerting mechanisms. The system must continuously process data streams from sensors or devices. Data preprocessing techniques help ensure accuracy and filter out noise, while efficient data pipelines facilitate near-instantaneous processing. The system should continuously perform conformance checks using lightweight and fast algorithms suitable for real-time applications. By monitoring deviations as they occur, the system can promptly detect and flag noncompliant behaviors. When deviations are detected, alert mechanisms can be triggered based on defined thresholds or tolerance levels. For example, in healthcare, an alert could be sent if a patient's movements deviate significantly from recovery protocols, or in fitness if incorrect posture is detected repeatedly. Feedback should be provided to users (e.g., patients, athletes, workers) to correct deviations. Real-time notifications, visual cues, or audio feedback can guide users back to compliant behaviors, enhancing adherence and safety. Such real-time monitoring systems benefit from edge computing, where conformance checks are performed closer to data sources, reducing latency and allowing for immediate feedback. This system design makes real-time alerts both feasible and impactful in high-stakes applications like healthcare and safety.

In health and safety environments in particular, conformance checking is invaluable for guaranteeing conformity to prescribed activity models. Real-time monitoring can greatly improve compliance and safety in HAR applications by utilizing real-time data, modifying process models for flexibility, and putting quick conformance-checking algorithms and alerting systems in place. This method facilitates prompt interventions in high-stakes areas such as fitness and healthcare while also guaranteeing protocol adherence.

5.3 *Sequential and Temporal Dependencies in Human Activities*

Process mining is a valuable tool for evaluating human actions, particularly when it comes to temporal limitations and sequential dependencies. Here is a detailed look at how time-aware models can improve human activity recognition and how process mining can be modified to better capture these aspects. In the realm of human activities, specific sequences with interdependent steps are often observed, each possessing unique timing constraints. To adapt process mining for capturing sequential dependencies and temporal constraints, several strategies can be considered. The enrichment of event logs with start and end times for each activity allows for the direct capture of temporal constraints. This enhancement enables mining algorithms to account for both sequence and timing, resulting in more accurate models depicting the flow and duration of human activities. Standard process discovery algorithms can be adapted by incorporating temporal conditions. For example, Heuristic Miners can encompass frequency thresholds and minimum/maximum time constraints for dependencies, thus representing prevalent time-bound relationships in activity flows. Temporal dependency graphs prove valuable for representing and visualizing sequences of events with time constraints. These graphs can delineate the order of activities and the permissible or typical time intervals between them, thereby elucidating and structuring the dependencies. In temporal pattern mining, the amalgamation of process mining with sequential pattern mining algorithms, such as PrefixSpan and GSP, facilitates the identification of sequential dependencies while accommodating time windows and interval constraints between activities. This approach is particularly pertinent in HAR. Hybrid models that integrate process models with temporal logic offer an effective means of capturing the time-bound nature of human activities. These models enable the encoding of specific timing rules, such as "Activity A must be followed by Activity B within 5 minutes," by leveraging temporal logics like linear temporal logic (LTL) or timed Petri nets. Notably, these constraints are particularly advantageous for representing allowable time intervals and sequence order in complex, time-sensitive activities, such as those prevalent in healthcare settings. Furthermore, the incorporation of timed transitions in models, such as timed automata or timed Petri nets, facilitates the comprehensive representation of both the sequence and specific timing conditions for transitions between activities. This capability allows for the encoding of time-bound dependencies and the development of precise activity flows that incorporate temporal thresholds. Additionally, the integration of hidden Markov models (HMMs) with temporal constraints in hybrid models enables the inclusion of probabilistic transition timing, thereby accommodating the inherent variability in human activities. By capturing the likelihood of transition timings, HMMs make the model adaptable to human behavior variations while upholding essential temporal dependencies. Given these capabilities, hybrid models stand as a well-suited solution for time-sensitive activity monitoring. They facilitate conformance checking not only of the sequence but also of the duration of activities, thereby supporting compliance with time-bound standards. In order to effectively model the real-time temporal dynamics of human activities, it is imperative for time-aware process mining techniques to accommodate ongoing, evolving sequences with temporal dependencies. In real-time scenarios, activities can be continuously monitored using a sliding time window. This method facilitates the capture of recent patterns without necessitating a complete log, thereby enabling real-time analysis and feedback based on the temporal dynamics within each window. By implementing predictive models that leverage past timing information, the system can predict expected durations for upcoming activities. Time-aware predictive process models (e.g., utilizing machine learning or reinforcement learning) aid in anticipating the remaining time for current tasks and identifying activities at risk of exceeding their normal duration. The incorporation of temporal weights, or time decay functions, to recent events allows for the prioritization of more recent data over older information. This approach enables the model to adapt to short-term changes in activity flows, particularly beneficial for HAR where real-time monitoring necessitates dynamic adjustments to current patterns. Integrating conformance checking with temporal boundaries enables the triggering of alerts in the event of deviations in timing. For instance, in eldercare monitoring, prolonged inactivity or delayed movements can automatically signal potential health issues, thereby enhancing response times and outcomes. These time-aware techniques empower real-time monitoring systems to accurately reflect the current state of human activities while maintaining alignment with expected models. The impact of variations in activity duration on the accuracy of process mining models, particularly in the context of HAR, presents distinctive challenges. High variability in activity durations can lead to reduced precision if the model assumes fixed durations. The misalignment of process models can occur when there is significant variation in durations, as typical transition times are no longer consistent. To mitigate this issue, adaptive duration thresholds can be integrated. These thresholds adjust based on historical data for each activity type, aiding the model in distinguishing between acceptable variability and genuine deviations. In cases where activity durations exhibit significant natural variation, probabilistic timing models such as Gaussian mixture models or Bayesian networks can depict expected duration ranges. This allows the model to accommodate normal variability while identifying true anomalies. Allowing for multilevel granularity, where activity durations can be analyzed at varying levels (e.g., seconds, minutes, or hours), assists in creating models that are resilient to short-term timing differences while still detecting substantial deviations. This approach is particularly valuable in activity recognition settings where minor timing variations do not necessarily signify a change in behavior. By incorporating these adaptations, process mining models can uphold accuracy and relevance, even in fields like HAR, where timing variability is prevalent.

In conclusion, to effectively capture sequential and temporal dependencies in human activities, process mining techniques can be adapted with hybrid models, time-aware methods, and probabilistic approaches that account for natural variation in human behavior. Through the utilization of temporal logic, adaptive thresholds, and predictive analysis, process mining becomes highly effective for real-time, time-sensitive monitoring. This capability allows HAR models to provide precise, dynamic insights into human activities, facilitating the detection of significant deviations and the delivery of timely alerts in critical applications such as healthcare, fitness, and safety monitoring.

6 Multidimensional Process Mining

6.1 Contextual Factors in Human Activity Recognition

When modeling and analyzing complicated activity patterns influenced by multiple external factors, contextual considerations can greatly improve the analysis' accuracy and relevance when incorporated into process mining for HAR. In order to improve activity recognition, we can incorporate contextual information such as location, time of day, and environmental conditions into event logs in process mining. This means adding attributes such as location data (e.g., GPS coordinates), time (e.g., morning, afternoon, evening), and environmental data (e.g., temperature, noise level) to each activity. We can use process mining techniques like conformance checking to align activity sequences with context-specific models. For example, a model for activities during "rush hours" may differ from a model used for "nighttime" activities, to account for likely variations in behavior. Additionally, we can integrate advanced machine learning methods, such as contextual clustering and dynamic time warping, with process mining to identify context-specific features. This helps improve recognition accuracy by dynamically adapting to varying contextual influences on activity. Process mining enables the identification of variations by segmenting data based on context and uncovering patterns within each segment. For instance, by grouping event logs according to time or location, distinct walking patterns may emerge, such as faster walking speeds during commuting hours and more leisurely walking during off-peak hours. Process mining tools can compare context-dependent process variants, illustrating how activities deviate from typical patterns in different contexts. For example, process models for varying weather conditions (sunny vs. rainy) may unveil unique patterns in outdoor activities. Tailoring algorithms such as Inductive Miner or Fuzzy Miner allow the creation of models specific to different contextual clusters. These models capture variations in activity flows for each context, shedding light on how and why activities change across different contexts. Multidimensional process mining involves characterizing events across multiple dimensions, such as activity type, location, and time. This approach enables analysts to dissect data across dimensions, leading to the generation of distinct process views. By concurrently examining multiple contextual dimensions, this method can reveal exclusive patterns, such as identifying specific activity sequences occurring only under particular weather conditions and times. The construction of an activity cube, an extension of OLAP for process mining, facilitates the analysis of dimensions such as location, time, and environmental factors across multiple layers. This method allows for the extraction of insights, such as pinpointing activity "hot spots" at specific times and locations. Time-aware and space-aware process mining models can dynamically adapt to context. For instance, temporal process models can accommodate recurring daily patterns (e.g., morning vs. evening routines), while spatial models can integrate location-specific behaviors (e.g., activities at the gym vs. at home). Visualization techniques can overlay contextual factors on process maps to show where activities differ based on context. For example, heatmaps can be used to indicate activity density in different locations or times, while color-coding can represent different contextual conditions (e.g., using different colors for morning vs. evening activities). Plots that show sequences of events can be enhanced with contextual markers. For instance, a timeline of activities can include icons or color codes indicating the time of day, location, or weather, helping users visually interpret how context influences activity transitions. Interactive dashboards can allow users to explore context by selecting different dimensions (location, time, environmental factors) and observing how process models adjust in response. This real-time, interactive approach helps uncover patterns hidden within contextual layers. Visualization tools like split screens or layered models can compare process variants across contexts. This can reveal, for instance, how activity patterns shift from weekdays to weekends, aiding the identification of context-specific adaptations in activity sequences.

These approaches provide a structured way to enhance the interpretability and accuracy of process models in HAR, making them more responsive to the dynamic influences of contextual factors. The key is to integrate contextual data into both model construction and visualization to effectively capture the real-world complexity of human activities.

6.2 Real-Time Process Mining for HAR

The use of real-time process mining to HAR in dynamic contexts has distinct problems and opportunities. This is because the data-driven, real-time method necessitates quick processing of high-frequency sensor input and constant adaption. Several challenges are involved in applying real-time process mining to HAR, such as data variability and complexity, real-time data processing, handling unpredictable events and sensor limitations, and fusion. Human activities can vary widely across individuals, contexts, and environments, creating noise and complexity in data streams. Dynamic environments introduce additional variability, such as changes in lighting, temperature, or people moving between rooms, which complicates the detection of consistent patterns. Processing sensor data in real time requires managing large volumes of continuous data streams. Maintaining low latency is crucial to providing timely feedback and recognition. In dynamic environments, unexpected activities or anomalies can disrupt established patterns, necessitating flexible and adaptive process models that can respond to novel events. Real-time HAR often relies on data from multiple sensor types, such as motion sensors, cameras, and wearable devices, each with its limitations and noise characteristics. Effective sensor fusion techniques are necessary to synthesize this data in real time. To achieve real-time HAR, it is crucial to develop algorithms that can update process models incrementally as new data arrives, rather than relying on traditional batch processing. This requires optimizing model updating methods to minimize computational costs. Efficiency can be improved by designing algorithms tailored for stream processing engines (e.g., Apache Flink, Kafka) and event-driven frameworks, which enable real-time action upon recognition of specific patterns in continuous data streams. Edge computing and distributed processing play a critical role in minimizing latency and optimizing resource usage. Therefore, it is essential to deploy efficient, lightweight algorithms that can run on constrained hardware, such as wearable devices. Integrating machine learning with process mining, for example, using neural networks for activity classification and process mining for pattern detection, could significantly enhance the robustness and adaptability of real-time HAR systems. The process of conformance checking entails establishing normative behavior patterns, such as regular activities or workflows, against which observed behavior is assessed. This necessitates creating a baseline of "normal" activities and devising methods to detect deviations or anomalies in real time. Efficient real-time pattern recognition techniques play a crucial role in promptly identifying deviations. For instance, temporal logic and symbolic sequence matching can swiftly identify nonconforming patterns in data streams. Continuous monitoring must incorporate real-time feedback mechanisms, which may alert users or trigger automated responses when activities deviate from expected patterns. This is particularly critical in domains such as healthcare, where deviations may signify health issues (e.g., a fall or unusual inactivity). Conformance checking should be context-sensitive to accommodate normal variations in behavior. This necessitates adaptive models capable of integrating environmental cues and discerning the disparity between significant deviations and natural variations in activity. The development of real-time process mining models for HAR necessitates a thoughtful consideration of the trade-offs between accuracy, responsiveness, and computational cost. In accuracy versus responsiveness, achieving high accuracy often necessitates the use of complex models, potentially leading to reduced responsiveness due to longer processing times. Conversely, simplified or heuristic models can enhance responsiveness but may sacrifice detail, thereby increasing the likelihood of misclassification. Highly accurate complex models can place significant computational demands, particularly in resource-constrained environments such as wearables and IoT devices. It is imperative to strike a balance between computational cost and model complexity to ensure efficient real-time processing without overburdening system resources. Conformance checking at a high frequency can yield more immediate responses but at the cost of increased computational load. Conversely, lower-frequency checking conserves resources but may overlook rapid changes in activity patterns, affecting responsiveness. Upholding data quality through denoising and filtering is vital for accuracy but introduces additional preprocessing time and costs. Thoughtful consideration of efficient preprocessing algorithms or selective preprocessing on key sensors is essential in optimizing this trade-off.

The realm of real-time process mining for HAR necessitates a nuanced approach to these trade-offs to construct systems capable of delivering timely and accurate insights, especially within dynamic environments.

7 Evaluation and Benchmarking

Process mining is concerned with finding, monitoring, and optimizing processes based on event logs; in contrast, HAR focuses on identifying and categorizing physical or digital activities. This makes comparing and evaluating process mining techniques in HAR a complex task. To these aims we present the most useful public dataset that can be employed as benchmark in Table 3. Many HAR datasets contain sensor data (e.g., accelerometers, gyroscopes) collected through mobile devices or wearable sensors. Common datasets include the UCI HAR dataset, a widely used benchmark dataset with sensor data from smartphones, covering activities like walking, sitting, and standing. PAMAP2 contains sensor data from various body locations for activities such as cycling, running, and walking. Opportunity dataset provides multimodal sensor data (wearable, environmental) capturing activities in a kitchen setting. WISDM focuses on activity recognition data using smartphones and includes activities such as walking and jogging. Datasets used in process mining (e.g., the BPIC datasets) typically include event logs from different application domains (healthcare, finance) but may not contain physical activity data. Creating or sourcing datasets that contain both continuous sensor data (for HAR) and discrete event logs (for process mining) would be ideal. This may require fusing sensor data from HAR datasets with event logs derived from timestamped sensor data. For benchmarking, custom datasets might be required where both HAR data (continuous time series) and process mining logs (discrete events) are included, such as synthetic datasets combining HAR logs with simulated process logs. Both HAR and process mining necessitate accuracy, precision, and recall. However, the interpretation of these metrics may vary, as HAR traditionally focuses on classifying individual activities, while process mining may assess the completeness or fidelity of discovered processes. Temporal alignment measures the alignment between recognized activities and process events over time. Trace fidelity metric could evaluate how well the sequence of identified activities matches a process model, particularly useful for sequence-sensitive HAR applications. Activity transition accuracy measures the accuracy of detecting transitions between activities as they map to process events, which is crucial in multistep process tasks. Activity-to-event matching accuracy measures the accuracy in mapping activity instances from HAR data to events in the process log. Metrics that combine HAR's classification metrics and process mining's sequence accuracy or model fitness can effectively address the gap between the two fields. Several challenges are involved in developing large-scale datasets for HAR and process mining such as data collection complexity, event log generation, privacy and ethics, and dataset annotation. The process of gathering sensor data across diverse human activities for HAR, while also capturing corresponding events in a process log format, presents a significant challenge. It necessitates the consistent labeling of data, which can be labor-intensive, especially in scenarios involving multiple activities or users. The translation of continuous sensor data (e.g., accelerometer) into discrete event logs poses a challenge. Determining a threshold for activity recognition that reliably translates to an

Table 3 Popular HAR datasets and related applications

Dataset	Description	Key feature	Application	Link
MOPED	– Consists of real-world event logs from a manufacturing process. – Includes detailed information about human-machine interactions, task sequences, and timing. – Helps to understand how humans engage with industrial machinery and carry out tasks in a manufacturing environment.	– Event logs capturing human-machine interaction – Timestamped events for process modeling – Can be used to track and analyze task workflows	Process mining of industrial operations, tracking human activities in a manufacturing context	MOPED dataset
PAMAP2	– The PAMAP2 dataset is widely used for human activity recognition. – Contains data from wearable sensors worn by individuals. – Includes various activities such as walking, running, sitting... – The dataset features time-series data from accelerometers and heart rate monitors and gyroscopes	– Sensor data including accelerometer, gyroscope, and magnetometer readings – Activities such as walking, running, cycling, sitting, standing, and household activities – Timestamped sensor readings for HAR	HAR, fitness monitoring, activity tracking	PAMAM2
Opportunity	– Most comprehensive HAR datasets, designed for activity recognition in pervasive environments. – Consists of recordings of daily human activities using a large number of body-worn and ambient sensors.	– More than 240 sensors capture data from accelerometers, gyroscopes, and magnetometers – Complex human activities such as opening/closing doors, drinking, and cleaning – Designed for recognizing both basic and complex activities	HAR, context-aware systems, process mining of daily human activities	OPPORTUNITY

(continued)

Table 3 (continued)

Dataset	Description	Key feature	Application	Link
WISDM	– Most popular smartphone-based HAR datasets. – It contains accelerometer data from smartphones, recorded while participants performed daily activities such as walking, jogging, and climbing stairs.	– Data collected from smartphone sensors (mainly accelerometers) – Includes activities such as walking, jogging, standing, sitting, and stair climbing – Timestamped data for HAR analysis	HAR using mobile and wearable devices, process mining of mobile user activities	WISDM
UCI HAR	– Contains accelerometer and gyroscope data from smartphones. – Collected for six different activities: walking, walking upstairs, walking downstairs, sitting, standing, and lying down. – It is a benchmark dataset for HAR.	– Data from the accelerometer and gyroscope of smartphones – Six activities, labeled for supervised learning and recognition – Well-suited for training machine learning models for HAR	Activity recognition, mobile health monitoring	UCI HAR
BPI challenge datasets	– Widely used in the process mining community. – Consists of event logs from real-world business processes, such as incident management or healthcare processes. – While not directly HAR-related, they can be used to model and mine human decision-making processes and workflows.	– Event logs from real-world business processes – Timestamped events and human actions in different contexts (e.g., finance, healthcare) – Useful for modeling workflows and human interactions in business processes	Process mining, analyzing human activities in organizational settings	BPI
Cornell activity	–Contains RGB-D videos of people performing household activities.	–RGB-D sensor data (depth and color images) for activity recognition	HAR, visual recognition of human-object interactions, activity modeling	CAD-120

	– Designed for activity recognition and involves both object interactions and human movements.	– 120 sequences of 10 high-level activities (e.g., cooking, cleaning) –Includes both human poses and object interactions		
Collection of social log	– Contain real-world event logs from municipal services in the Netherlands. – Capture human actions and workflows within public service processes. – Ideal for process mining and modeling human decision-making in administrative processes.	–Event logs of public administrative processes –Focus on human workflows in decision-making processes –Time-based event sequences	Process mining, analyzing administrative decision-making processes	CoSeLoG
CASAS Smart Home	– Include sensor data from smart home environments, where sensor readings are captured as people perform daily activities. – Can be used for both HAR and process mining in the context of smart homes.	– Smart home sensor data for HAR – Daily activities such as cooking, cleaning, and exercising – Event logs with timestamps, allowing process mining techniques to analyze human workflows in smart homes	HAR, smart home automation, process mining in intelligent environments	CASAS

event is difficult, as is ensuring consistency across users. Both HAR and process mining data collection often involve sensitive personal data. Privacy concerns, particularly for large datasets that include both sensor and log data, can impede the availability of public datasets. Comprehensive annotation is indispensable to ensure that activity recognition accurately maps to the discrete events in process mining. The maintenance of consistency and accuracy in labels within large datasets is challenging. The collaboration between research groups in process mining and HAR could facilitate the development of synthetic or semisynthetic datasets. Additionally, the implementation of privacy-preserving techniques, such as data anonymization or federated learning, could help mitigate privacy concerns. Developing a framework tailored to handle both HAR and process mining evaluations would yield significant benefits. This framework could encompass data transformation capabilities to convert sensor data into event logs, thus enabling standardized processing across HAR-to-process-mining benchmarks. The definition of specific HAR tasks, such as recognizing a sequence of activities in a workout session or a workflow in a manufacturing setting, can standardize benchmarking tasks and facilitate a fair comparison of process mining techniques. The establishment of baseline algorithms for common HAR-process mining tasks (e.g., event log discovery, and activity transition modeling) would streamline a unified comparison. Baselines could encompass simple event discovery methods or standard HAR classifiers. The availability of open-source toolkits, potentially with dedicated modules for HAR-process mining hybrid tasks, would promote reproducibility. Libraries such as ProM for process mining, coupled with HAR-specific libraries, could form the foundation of a modular benchmarking environment. A standardized methodology for comparing process mining techniques in HAR should prioritize traceability, accuracy in activity-event mapping, and robustness across diverse datasets. The establishment of benchmarks in this hybrid field will necessitate community-driven efforts and datasets that bridge the gap between HAR and process mining domains.

8 Future Directions

Process mining can be profitable used in HAR to enhance activity detection accuracy, comprehend intricate human behavior patterns, and optimize processes involving human interactions. Below are some recent applications and possible future works:

1. *Hybrid models for improving activity detection:* Process mining is being increasingly combined with machine learning techniques to create hybrid models for better detection and classification of activities. By using process mining to analyze sequences and patterns of activities, these models can recognize complex activities more accurately and with less reliance on large datasets. Hybrid models are used in environments where activities are structured but vary slightly across instances, such as in healthcare monitoring (e.g., identifying

different patterns of patient mobility) or smart home systems (e.g., detecting complex daily routines).

2. *Process discovery for activity pattern recognition:* Recent advancements in process discovery algorithms have enabled the extraction of common activity patterns from sensor data. This allows for the modeling of not only individual activities but also sequences of activities over time. This technology has various applications, including in elderly care facilities to identify deviations that could indicate issues such as a fall or a health problem. Additionally, process mining is utilized in fitness monitoring to evaluate the completeness and accuracy of exercise routines.
3. *Compatibility checking for compliance and anomaly detection:* Compatibility checking involves comparing observed activities with expected behavior models to identify deviations in real time. This is crucial for ensuring compliance in various areas. For example, in workplace safety, process mining checks can detect noncompliant actions, such as skipping safety steps on machinery. In security-sensitive environments, it can identify unauthorized or unusual activities, such as employees entering restricted areas or accessing confidential information. These are helpful for HAR in Industrial 4.0.
4. *Predictive activity monitoring and proactive intervention advances:* With the use of predictive capabilities in process mining, HAR systems can now predict future activities or potential issues based on past behavior. This is especially helpful in dynamic environments where human actions affect ongoing processes. In the healthcare field, predictive activity monitoring can anticipate potential health risks for patients based on their movement patterns. For example, if a patient's movement indicates fatigue or difficulty walking, the system can alert caregivers for proactive intervention.
5. *Real-time process mining for adaptive systems:* Real-time process mining improvements enable HAR systems to adapt to changing human activities and provide real-time feedback. This is particularly valuable in adaptive systems that dynamically respond to human behavior. In smart home automation, real-time process mining can adjust environmental controls (such as lighting and temperature) based on the recognized activities of inhabitants. In industrial settings, it can adjust machinery parameters or work schedules in real time based on the detected activities of workers.
6. *Integration with IoT and edge computing for scalable HAR solutions:* With the evolution of IoT and edge computing, process mining has become more feasible at the edge, allowing for faster data processing closer to where it is generated. This reduces latency and makes HAR more scalable and responsive. In smart cities, HAR systems combined with process mining analyze pedestrian movements, optimize traffic signals, and improve public safety. On factory floors, edge-based process mining helps monitor worker movements, optimizing task assignments and enhancing productivity in real time.
7. *Event abstraction and contextualization in HAR:* Process mining has made progress in abstracting and contextualizing event data, which aids in understanding high-level activities from low-level sensor data. This capability

improves HAR by providing more meaningful insights into human behavior. In retail environments, event abstraction helps in identifying patterns of customer movement and engagement, which can be used to make informed decisions about store layout. In healthcare, contextualized activity data from patients can result in personalized care recommendations based on their specific movement patterns.

8. *Automated workflow recommendations and process optimization:* Through the implementation of automated recommendations, process mining in HAR can provide suggestions for workflow improvements, thereby enhancing productivity and user experience. This is particularly relevant for workflows involving repetitive or manual human tasks. Within office environments, HAR can propose methods for optimizing desk layouts and recommend task reassignments to balance workloads based on observed activity patterns. In the logistics sector, this leads for optimizing pick-and-pack workflows by analyzing worker movement and task completion times.
9. *Leveraging process mining for activity recognition in collaborative work environments:* Recent progress in the field involves the utilization of process mining to scrutinize team activities, offering valuable insights into collaborative behaviors and streamlining team workflows. In industrial environments emphasizing collaboration, the utilization of process mining to analyze HAR data facilitates an understanding of team members' interactions with each other and with equipment. This comprehension can result in enhanced task allocation, improved coordination, and an overall increase in team efficiency.
10. *Advanced visualization for complex activity analysis:* Significant progress has been made in developing enhanced visualization techniques to effectively represent complex activity flows and interactions in an easily understandable manner. This advancement facilitates stakeholders in comprehending and analyzing activity patterns and their influences on processes. In healthcare settings, the use of process mining visualizations for activity data assists caregivers in understanding patient routines and identifying patterns that may require intervention. Similarly, in the retail sector, these visualizations offer valuable insights into customer journey patterns, enabling the optimization of layout for enhanced customer engagement.

In summary, the combined utilization of process mining and HAR represents a potent alliance capable of revolutionizing various industries by offering actionable insights into human behavior and activity patterns, ultimately enhancing efficiency, safety, and compliance.

References

1. Aalst, W.M.P.V.D., Adriansyah, A.A., Dongen, B.F.V.: Replaying history on process models for conformance checking and performance analysis. WIREs Data Mining Knowl. Discovery **2**, 182–192 (2012). https://doi.org/10.1002/widm.1045

2. Augusto, A., Conforti, R., Dumas, M., Rosa, M.L., Maggi, F.M., Marrella, A., Mecella, M., Soo, A.: Automated discovery of process models from event logs: review and benchmark. IEEE Trans. Knowl. Data Eng. **31**(4), 686–705 (2019)
3. Basar, T.: A New Approach to Linear Filtering and Prediction Problems, pp. 167–179 (2001)
4. Bertrand, Y., Van den Abbeele, B., Veneruso, S., Leotta, F., Mecella, M., Serral, E.: A survey on the application of process discovery techniques to smart spaces data. Eng. Appl. Artif. Intell. **126**, 106748 (2023)
5. de Hert, P., Papakonstantinou, V.: The new general data protection regulation: still a sound system for the protection of individuals? Comput. Law Security Rev. **32**(2), 179–194 (2016)
6. Dhekane, S.G., Ploetz, T.: Transfer learning in human activity recognition: A survey. ArXiv **abs/2401.10185** (2024)
7. Di Federico, G., Burattin, A., Montali, M.: Human behavior as a process model: Which language to use? In: Proceedings of the 1st Italian Forum on Business Process Management, pp. 18–25. CEUR Workshop Proceedings, CEUR-WS (2021); 1st Italian Forum on Business Process Management, ITBPM 2021; Conference date: 06-09-2021 Through 10-09-2021
8. Dumas, M., Rosa, M.L., Mendling, J., Reijers, H.A.: Fundamentals of Business Process Management. Springer, Berlin (2018)
9. Fortino, G., Guzzo, A., Ianni, M., Leotta, F., Mecella, M.: Exploiting marked temporal point processes for predicting activities of daily living. In: IEEE International Conference on Human-Machine Systems, ICHMS 2020, Rome, September 7–9, 2020. pp. 1–6. IEEE (2020)
10. Fortino, G., Guzzo, A., Ianni, M., Leotta, F., Mecella, M.: Predicting activities of daily living via temporal point processes: approaches and experimental results. Comput. Electr. Eng. **96**(Part), 107567 (2021)
11. Fortino, G., Greco, C., Guzzo, A., Ianni, M.: Neural network based temporal point processes for attack detection in industrial control systems. In: 2022 IEEE International Conference on Cyber Security and Resilience (CSR), pp. 221–226 (2022). https://doi.org/10.1109/CSR54599.2022.9850333
12. Ghahramani, Z.: An introduction to hidden markov models and bayesian networks. Int. J. Pattern Recogn. Artif. Intell. **15**(1), 9–42 (2001)
13. Günther, C.W., van der Aalst, W.M.P.: Fuzzy mining – adaptive process simplification based on multi-perspective metrics. In: Alonso, G., Dadam, P., Rosemann, M. (eds.) Business Process Management, pp. 328–343. Springer, Berlin (2007)
14. Jolliffe, I.T., Cadima, J.: Principal component analysis: a review and recent developments. Philos. Trans. R. Soc. A Math. Phys. Eng. Sci. **374**(2065), 20150202 (2016)
15. Leemans, T., Fahland, D., van der Aalst, W.M.P.: Discovering block-structured process models from event logs. In: Proceedings of the International Conference on Business Process Management (BPM) (2013)
16. Mans, R.S., Schonenberg, M.H., Song, M., van der Aalst, W.M.P., Bakker, P.J.M.: Application of process mining in healthcare – a case study in a dutch hospital. In: Fred, A., Filipe, J., Gamboa, H. (eds.) Biomedical Engineering Systems and Technologies, pp. 425–438. Springer, Berlin (2009)
17. Marin-Castro, H.M., Tello-Leal, E.: Event log preprocessing for process mining: a review. Appl. Sci. **11**(22) (2021)
18. Merriam-Webster: Standard operating procedure. In: Merriam-Webster.com Dictionary (2023). webster.com/dictionary/standard%20operating%20procedure
19. Paola Patricia, A.C., Rosberg, P.C., Butt-Aziz, S., Marlon Alberto, P.M., Roberto-Cesar, M.O., Miguel, U.T., Naz, S.: Semi-supervised ensemble learning for human activity recognition in casas kyoto dataset. Heliyon **10**(8), e29398 (2024)
20. Sani, M.F.: Preprocessing event data in process mining. In: CAiSE (Doctoral Consortium), pp. 1–10 (2020)
21. Thakur, D., Biswas, S.: Smartphone-based human activity pattern identification using unsupervised learning. In: Saraswat, M., Roy, S., Chowdhury, C., Gandomi, A.H. (eds.) Proceedings of International Conference on Data Science and Applications, pp. 549–559. Springer Singapore, Singapore (2022)

22. Thakur, D., Fortino, G.: A Comprehensive Review of Deep Learning for Activity Recognition, pp. 67–95. Springer Nature Switzerland, Cham (2024)
23. Thakur, D., Pal, A.: Human Activity Recognition: Trends and Challenges, pp. 161–182. Springer Nature Switzerland, Cham (2024)
24. Thakur, D., Pal, A.: Subsampled randomized hadamard transformation-based ensemble extreme learning machine for human activity recognition. ACM Trans. Comput. Healthcare **5**(1), 1–23 (2024)
25. Thakur, D., Roy, S., Biswas, S., Ho, E.S.L., Chattopadhyay, S., Shetty, S.: A novel smartphone-based human activity recognition approach using convolutional autoencoder long short-term memory network. In: 2023 IEEE 24th International Conference on Information Reuse and Integration for Data Science (IRI), pp. 146–153 (2023)
26. Thakur, D., Guzzo, A., Fortino, G.: Intelligent adaptive real-time monitoring and recognition system for human activities. IEEE Trans. Ind. Inf. **20**, 1–11 (2024)
27. van der Aalst, W.M.P.: Process Mining: A 360 Degree Overview, pp. 3–34. Springer International Publishing, Cham (2022)
28. Van der Aalst, W.M.P., Belchior, A., Dumas, M.: The alpha algorithm: A formal basis. In: Proceedings of the International Conference on Business Process Management (BPM) (2003)
29. Van der Aalst, W.M.P., Dumas, M., Hofstede, A.H.M.: Mining business processes: Data science in action. In: Proceedings of the International Conference on Business Process Management (BPM) (2003)

Event-Log Granularity for IoT Process Discovery

Massimiliano de Leoni, Faizan Ahmed Khan, Benoît Depaire, Greg van Houdt, and Niels Martin

1 Introduction

The Internet of Things (IoT) is the inter-networking of physical objects, which can range from sensors and actuators to software systems and devices [25]. The interest for IoT systems is gaining momentum. In general, IoT systems collect and exchange data via local networks or the Internet. Analyzing these data opens up many opportunities to improve IoT systems from many viewpoints, e.g., considering human satisfaction. The data harvested from the sensors can be used to discover the process executed by the devices that interact with the IoT system. The discovery can potentially provide valuable input to guide an improvement of the IoT processes.

Process mining aims to discover, monitor, and improve processes by extracting knowledge from event logs that record, for each case (i.e., for each process instance), the execution of process' activities, the execution timestamp, and the executing entity/person [47]. As process mining does not assume anything related to the nature of these processes, it can also be applied to IoT processes, as also indicated in the manifesto outlining challenges and opportunities related to IoT process management [25]. However, the same manifesto points out that "*a challenge is to bridge the gap between clouds of sensor data and event logs for process mining*" [25]. As a matter of fact, "*sensor data must be aggregated and interpreted to detect activities that can be used as input for process mining algorithms that support decision making*" [25]. Otherwise, the risk is that the results are given in terms of sensor measurements, which are too low-level to provide interpretable insights.

M. de Leoni (✉) · F. A. Khan
University of Padua, Padova, Italy
e-mail: massimiliano.deleoni@unipd.it; faizanahmed.khan@phd.unipd.it

B. Depaire · G. van Houdt · N. Martin
Hasselt University, Hasselt, Belgium
e-mail: benoit.depaire@uhasselt.be; greg.vanhoudt@uhasselt.be; niels.martin@uhasselt.be

G. Fortino, M. Mecella (eds.), *Internet of Things Meets Business Process Management*, Internet of Things, https://doi.org/10.1007/978-3-031-90746-3_7

Indeed, several approaches that apply process mining in IoT scenarios overlook the importance of finding the right event granularity of the event logs exported from IoT systems (see, e.g., [8, 9, 16, 24, 42]). Setting up an inadequate granularity can lead to extremely complex low-level models, which provide little insights for analysts. The initial temptation is to tackle the intrinsic variability by "sweeping the actual problem under the carpet" and focusing on the main paths, thus potentially filtering out plenty of relevant behavior.

Note that this chapter assumes that a low-level event log has already been extracted. Several research works already discuss successful techniques to extract IoT event logs for process mining, such as those introduced by Liming Chen et al. [30], van Eck et al. [48], Yadav et al. [52]. This is generally not a trivial problem, as it sometimes boils down to converting time-series sensor data to event logs and thus tuning the frequency of sensor data sampling, which might be adaptive.

Section 2 reports on an analysis of the literature on setting the right granularity in event-log abstraction. Section 3 illustrates one technique for event-log abstraction by de Leoni and Dündar [10], which has also been successfully applied in IoT (cf. de Leoni and Pellattiero [14]). This technique was also chosen because it requires very little user configuration to be considered as unsupervised; provides an efficient, workable implementation; and features an explanation of how low-level events have been aggregated. While an experience of the technique's application in domotics has been reported in de Leoni and Pellattiero [14], we felt that an additional experience in a different IoT scenario was worthwhile being discussed: Sect. 4 illustrates the application of the event-log abstraction to mine a model for a smart-manufacturing process, while Sect. 5 concludes the work, highlighting avenues for future work.

2 A Survey of Event-Log Abstraction Techniques

A large body of research work exists for the problem of abstracting an event log to a suitable level of granularity. This section reports on the major research works on this topics, building on and extending the state-of-the-art analyses by van Zelst et al. [51] and by Van Houdt et al. [50].

Each work has been analyzed with respect to three relevant criteria: *(i)* whether or not it requires additional information (i.e., it is supervised or unsupervised), *(ii)* it is equipped with a public implementation, and *(iii)* it provides an explanation about how the abstraction has been carried out (e.g., which low-level events have been clustered together). Table 1 provides the result of this analysis, where $+$ or $+/-$ is assigned for each work to each criterion depending whether the criterion is met in part or in full.

Several observations follow from Table 1. Firstly, several techniques are supervised, in the sense that an IoT process analyst needs to provide some input (models, patterns, rules, etc.), which is leveraged to facilitate the discovery of high-level activities. Oftentimes, these patterns are hard to provide by analysts because the

Table 1 Summary of the existing techniques, with respect to three important dimensions. The gray color is used to highlight the approaches that have positive scores on all criteria. Symbols + or +/− are assigned to a criterion for a given work, if the work meets the criterion in full or in part

Reference	Unsupervision	Implementation availability	Abstraction explanability
Bose and Van der Aalst [7]	+		+
Günther et al. [23]	+	+	+
Ferreira et al. [21]			+
Baier et al. [3]		+	+
Folino et al. [22]	+	+	
Mannhardt et al. [33], Tax et al. [39]	+	+/−	+
Tax et al. [41]		+	+
Begicheva and Lomazov [5]		+	+
Leonardi et al. [27]		+	+
Amirah et al. [2]	+		
Baier et al. [4]			+
Bernard and Andritsos [6]			+
Fazzinga et al. [18]			+
Fazzinga et al. [19]		+	+
Tax et al. [40]		+	+
Mannhardt et al. [34]		+	+
Sánchez-Charles et al. [38]	+		
Acheli et al. [1]	+	+/−	+
Rehse and Fettke [37]	+		+
de Leoni et al. [11, 14]	+	+	+
Tello et al. [43]		+	
Li et al. [28]	+		+
Fazzinga et al. [20]			
Li et al. [29]	+	+/−	+
Di Federico and Burattin [15]	+	+	+/−
Rebmann et al. [36]	+	+	

analysts do not have the technical competencies to draw those models, or they do not have the practical insights to determine these patterns. Furthermore, asking for the patterns' provision might be risky at times: the provision of wrong patterns might produce event logs recording the supposed execution of wrong high-level activities. In fact, none of the existing supervised approaches investigates their robustness for increasingly less accurate patterns used as input. Secondly, several techniques do not provide a public implementation that is easily available for testing and use, or the implementation does not scale for larger real-life event logs. Unfortunately, some implementations are no longer available or requires outdated libraries, compilers, or operating systems, due to a significant amount of time passed since the work was put forward; others require complex installations or infrastructures, which were beyond the scope of this analysis. Table 1 points out that, fortunately, several works provide some form of explanations of how the abstraction has been carried out. This is indeed crucial: analysts need to be given an explanation of how event logs have been abstracted so that they can verify the plausibility and, if needed, vary the algorithm's parameters to make the abstraction more reasonable. This requires a transparent mapping of how sequences of low-level activities (and, consequently, events) are mapped to high-level activities/events.

Space limitation prevents us from providing a detailed description of every approach in the table. However, most of approaches fall into one of two categories: a first category groups the approaches that build on sequence pattern mining, and a second category leverages on different techniques based on clustering. The main research works in the two categories are discussed in Sects. 2.1 and 2.2.

2.1 Research Works Based on Frequent-Pattern Mining

The research works that fall into this category are characterized by a first phase in which patterns of sequences of events are discovered with event-log traces. These patterns represent common sequences in which events appear in traces. In a second phase, each occurrence of these sequence patterns is abstracted into one high-level activity associated with the patterns. However, this is not trivial in general, as certain sequences might not exactly map to any pattern: therefore, it is necessary to map sequences to the closest patterns. Also, patterns are not always of the same length: it is thus unclear how many low-level events need to be grouped together into a pattern. The remainder focuses on various research works that address these challenges.

LPM mining [39] aims to describe frequent behavior in an event log in terms of local process-like patterns. The LPM algorithm mainly works with frequency as a quality metric, i.e., the more frequent a pattern occurs in the event log, the higher the assigned quality score will be. The LPM algorithm has also been extended to include utility functions and constraints for pattern selection [40]. Research is also being done on improving the detection of more meaningful local process models, especially in the case of event logs with many activity classes [35]. After discovering local patterns, they can be fed to the abstraction technique proposed by Mannhardt et al. [33], which generates the actual high-level event log. It does so by calculating optimal alignments, returning the most likely mapping between events and the high-level patterns. The combination of both techniques has been tested in the works of [32] and [49]. A derivative of LPM is the Combination-based Behavioral Pattern Mining (COBPAM) algorithm by Acheli et al. [1]. COBPAM is inspired by the LPM algorithm as it attempts to tackle its main weaknesses, namely, *(i)* that LPM discovery is limited in size, *(ii)* the LPM algorithm follows a generate-and-test approach where each generated pattern is immediately tested and removed if not applicable, *(iii)* LPM discovery neglects certain patterns such as two infrequent patterns that become frequent when joined by a choice operator, and finally *(iv)* LPM mining relies on alignments which are computationally heavy. In short, the COBPAM algorithm guarantees the discovery of compact and maximal patterns while leveraging pruning strategies to gain efficiency. The RefMod miner [37] also produces process-like patterns representing subprocesses, called reference models. A reference model is a conceptual model built on frequently appearing process model building blocks such as isolated activity sets. An input log is vertically divided, and the activities are hierarchically clustered based on their spatial proximity to the log. A reference model is described in the form of a process model.

Note that event-log traces are treated so far as a simple sequence of activity names. However, events are more complex and contains attributes: air ventilation, temperature, light intensity, and other IoT-sensor information can be attached to the events. Potentially, these additional attributes and their values can be beneficial to define different patterns. CvAMoS is based on this intuition and aims to discover the so-called motifs, recurring low-level event sequences that incorporate this additional information [15]. Although the evaluation of CvAMoS does not fully highlight the advantages of using the additional event's attributes, we acknowledge that this additional information can be beneficial to create more precise patterns, if compared with other frequent-pattern-based approaches that only focus on the low-level activity names.

2.2 Research Works Based on Clustering

The approaches based on clustering are based on the idea that traces are split into sub-traces, which are grouped in clusters. Each cluster aims to identify one high-level activity, and event sequences that are associated with one given cluster are replaced by the high-level activity identified for that cluster.

The seminar work on clustering-based event-log abstraction is certainly by Günther et al. [23], which is in fact the first advocating the importance of event-log abstraction. A graph is built from the low-level event log where a direct arc between a pair of low-level activities A and B exists if, at least in one trace, one low-level event for activity A is followed by a low-level event for B. This is the so-called directly-follow graph [46], and arcs are also annotated by the frequency of the occurrence of pairs of low-level events. The annotation with frequency allows the technique to feature the possibility to filter out the infrequent paths. Low-level activities are then clustered, looking at the strongly connected components: every low-level event associated with activities in one strongly connected component is assigned to one high-level activity that is determined by component. Unfortunately, this technique is limited by assigning each low-level activities to exactly one strongly connected component, and hence, every low-level event of the same low-level activity is associated to the same high-level activity. In reality, the same low-level event can be associated to different high-level activities, depending on which other low-level events are in proximity.

The limitation that low-level activities need to be associated with only one high-level activity is also present in the work by Li et al. [28]. It specifically proposes a framework to mine a hierarchy of event-log abstractions: low-level events are associated to high-level activities, which in turn can be associated to higher-level activities, thus creating such a hierarchy. In process discovery, this enables discovering models at different levels of abstraction. Unfortunately, this also shares the limitation that one low-level event can only be associated to one high-level activity.

The work by de Leoni et al. [13, 14] also leverages clustering techniques and is based on the concept of sessions: a session is considered as a subsequence of an event-log trace for which time gap with the preceding and following event exceeds a user-defined threshold. Sessions are then encoded into vectors, which consider the frequency of each low-level activity into the sessions; the vectors are later clustered, where each cluster identifies one high-level activity.

2.3 Concluding Remarks

Table 1 points out that only two research works and their associated extensions meet the three criteria (see the gray-highlighted rows). Acheli et al. [1] and Li et al. [29] score a value $+/-$ for the implementation-availability criterion: an implementation is indeed available, but the experiments by van Houdt et al. [50] illustrate that the implementations do not scale, never terminating on the Flemish supercomputer. A value $+/-$ was also given to the score for the implementation-availability criterion for the research work that is reported in Mannhardt et al. [33] and extended in Tax et al. [39]: the reference implementation required a lot of configuration to be run and was thus considered as sufficiently simple to be executed. For the column related to explainability of the low-level-to-high-level mapping, the approach by Di Federico et al. [15] scores $+/-$ because the implementation makes it no longer possible to trace back the sequence of low-level events behind the abstracted activity, although the theoretical framework would not have this limitation.

However, the technique by Günther et al. [23] does not allow low-level activities, and consequently low-level events, to be associated to different high-level activities depending on the other low-level events in proximity. We thus made the decision to focus the rest of the chapter on the technique by de Leoni et al. [13, 14].

3 A Technique Based on Clustering

This section introduces the technique by de Leoni et al. [13] as the ideal candidate for a technique based on clustering.[1] This technique aims to abstract a low-level

[1] A fully-fledged implementation is available in Python at https://github.com/Faizanunipd/IOT-based-event-log abstraction.

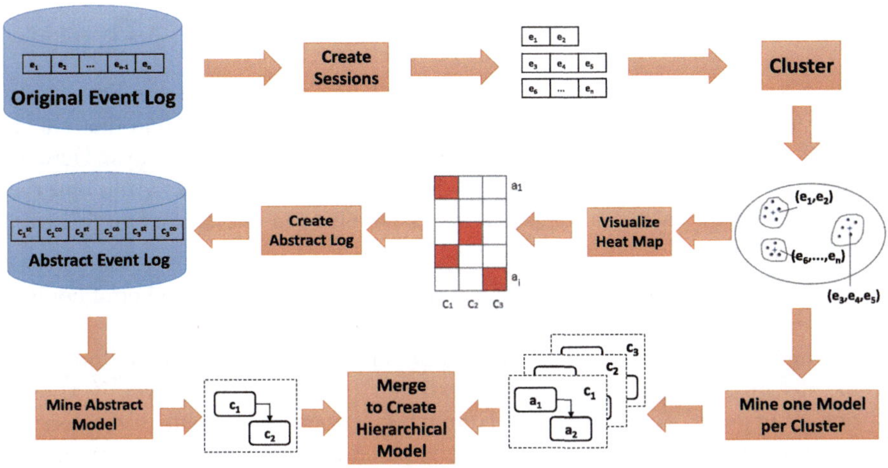

Fig. 1 The abstraction-technique framework for process discovery, extended from [12]. For compactness, we assume that each cluster c_i is given name c_i

event log, by creating sequences of low-level events, and subsequently cluster them, using sequence-to-vector encoding and clustering techniques. Each sequence is then abstracted according to the cluster to which it belongs.

The starting point is thus **event log**, which is composed by a set $\mathcal{E}$ of events. Events are arranged in traces $\sigma = \langle e_1, \ldots, e_n \rangle \in \mathcal{E}^*$ that group events that refer to executions of activities within the same instance of process (e.g., every event that refers to human activities in a domotics house within a certain day). Each event $e \in \mathcal{E}$ carries additional information. Hereafter, we denote the activity $\lambda_A(e_i)$ to which e refers and the timestamp e when the event occurred. Figure 1 summarizes the idea behind the technique: the framework is composed of seven steps, which are discussed below in detail.

3.1 *Create Sessions*

All the traces of the event log are split into sessions; each session of a log trace is the smallest subsequences of the trace such that, for each session, the difference between the timestamp of the last event of the session and that of the first event of the session that follows is larger than a user-defined threshold. For instance, let us consider a trace $\sigma = \langle a_1, b_3, c_4, a_{10}, d_{13} \rangle$. The letter indicates the activity name associated with the event, and the subscript is the timestamp of the event's occurrence (e.g., d occurred at time 13). Assume that the time interval $\Delta = 5$. One can easily see that the time difference between the second occurrence of a and the first of c is greater than Δ: $\lambda_T(a_{10}) - \lambda_T(c_4) = 6 > \Delta = 5$). This yields two sessions: $Ⓢ_\Delta(\sigma) = \langle s_1, s_2 \rangle$ where $s_1 = \langle a_1, b_3, c_4 \rangle$ and $s_2 = \langle a_{10}, d_{13} \rangle$. Note that their concatenation results in σ.

3.2 Cluster

Each session s is converted into a vector through an encoding function ENCODE(s) that abstracts the behavior observed in the session. All these vectors that encode sessions are clustered. The encoding can be obtained in many ways. Here we consider ENCODE(s) $= (v_{x_1}, \ldots, v_{x_n})$, where $x_1, \ldots, x_n$ are the activities observed in the log and, for each $i \in \{1, \ldots, n\}$, v_{x_i} is obtained in one of the two following alternatives:

Frequency-based encoding. The value of dimension v_{x_k} is the number of completion events for activity x_k in s.

Duration-based encoding. The value of dimension v_{x_k} is the average duration of the instances of activity x_k in the session s (a zero value is given if x_k does not occur in s). If the start and completion events are present in the log, the duration of activity instances can be computed exactly. Otherwise, they need to be estimated as in de Leoni and Dündar [12].

Note that the output of the cluster phase can be abstracted as a set $C \in 2^{\mathcal{E}^*}$ of clusters where each cluster is a set of sessions (i.e., sequences).

3.3 Visualize Heatmap

Clusters are visualized on heatmap to provide analysts with some insights on how sessions were clustered. Figure 2 illustrates an example of heatmap, stemming from the IoT case study in smart manufacturing that is discussed in Sect. 4. The rows refer to the ten different clusters, from 0 to 9; the column refers to the low-level activities. For each cluster c (i.e., row), the color referring to a certain low-level activity a (i.e., column) is related to the value of that component for activity a (either the number or the average duration of a's occurrences) for the centroid of c. For comparability of the dimension values at centroid, the values are normalized. The normalization of a given centroid $(y_1, \ldots, y_n)$ is achieved by dividing by the sum of the centroid's values: $(\frac{y_1}{sum}, \ldots, \frac{y_n}{sum})$ where $sum = y_1 + \ldots + y_n$. The heatmap visualizes the colors in a palette that ranges from yellow (value equal to 0) to dark red (value equal to 1).

3.4 Create Abstract Event Log

The third step has focused on visualizing the centroids of the clusters on a heatmap, thus providing information about the most predominant activities for low-level events. This information is used by process analysts to assign cluster names NAME(c) to each cluster $c \in C$. The fourth step creates the abstract event log. For each log trace σ, we compute the sessions' sequence $Ⓢ_\Delta(\sigma) = \langle s_{\sigma_1}, \ldots, s_{\sigma_m} \rangle$ and

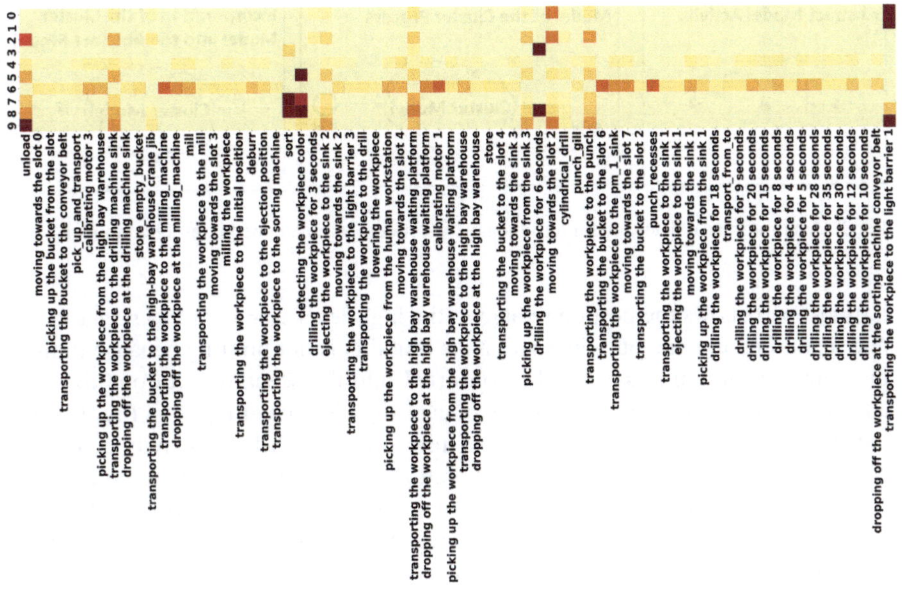

Fig. 2 Heatmap of the clusters obtained via DBSCAN for the case study on smart manufacturing discussed in Sect. 4, when the duration-based encoding was employed

then the abstract trace $\sigma_{\text{ABST}} = \langle f^{st}_{\sigma_1}, f^{co}_{\sigma_1} \ldots, f^{st}_{\sigma_n}, f^{co}_{\sigma_n} \rangle$ where, for every $1 \leq i \leq m$, $f^{st}_{\sigma_i}$ and $f^{co}_{\sigma_i}$ are the abstract events for session s_{σ_i}: the activity names of $\lambda_A(f^{st}_{\sigma_i})$ and $\lambda_A(f^{co}_{\sigma_i})$ are equal to the name $\text{NAME}(c_i)$ of the cluster c_i to which s_{σ_i}.

3.5 *Mine Abstract Model and Mine One Model per Cluster*

Here, we can employ any discovery algorithms available in literature [45]. The step to mine one model per cluster is only necessary when one wants to create a hierarchical model. In particular, it aims to mine the subprocess related to each activity of the abstract model. Each cluster $c \in C$ consists of a set of sessions, which is in fact an event log to be used as input to mine a model of the behavior of the sessions within the cluster. Here we also assume to employ a miner that returns strongly connected Petri nets.

3.6 *Merge to Create Hierarchical Model*

Figure 3 illustrates how one cluster model can be incorporate into the abstract model to create the hierarchical model, where the BPMN notation is employed. The

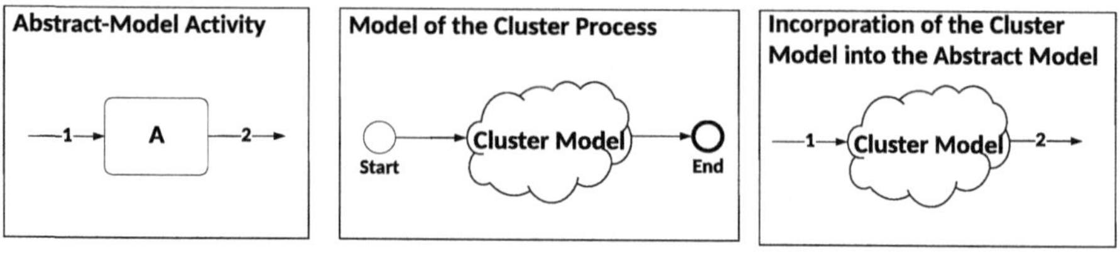

Fig. 3 Merging one cluster model into the abstract model to create the hierarchical model

leftmost part shows the portion related to the high-level activity a in a scenario with one incoming and one outgoing arc. Note that this is not limiting because multiple incoming or outgoing arcs can be converted into the scenario with one incoming and outgoing arc through the right use of gateways. The central part of Fig. 3 depicts the structure of the model of the cluster referring to high-level activity a, with the start and end BPMN events. The incorporation of the cluster model into the abstract model is illustrated in the rightmost part of the figure and is straightforward: activity a is simply replaced with the entire cluster model. This procedure is repeated for each cluster, thus finally producing the hierarchical model.

4 An IoT Case Study in Smart Manufacturing

This section focuses on a case study related to the IoT-enriched process that is executed in a smart factory [31]. The factory comprises two production lines that operate autonomously, yet are interconnected for the purpose of exchanging workpieces. Each production area is equipped with six identical machines. In order to facilitate research associated with Business Process Management (BPM), a service-based architecture is employed to govern the production lines through a process-oriented approach, as composed by Workflow Management Systems. The data generated by the machines is transmitted to Apache Kafka2. The events refer to low-level activities performed by the different machines of the factory to produce the items of the final products. This ultimately enables the extraction of an event log consisting of 301 traces consisting of 9471 events referring to 21 low-level activities.

4.1 Evaluation Methodology

The evaluation of the event-log abstraction focuses on the quality of the abstracted event log to mine a better process model. In particular, we aim to compare the quality of the hierarchical model obtained via the technique in Sect. 3 and the quality of the so-called flat model, which is obtained by naively applying discovery algorithms on the low-level event log.

In line with the process mining literature [46], we used the metrics of fitness and precision to evaluate the quality of the discovered process models. Both metrics are computed with respect to the event log used as input and returns a value between 0 and 1 with 1 being the best values. Fitness aims to quantify how well the process model is capable to represent the behavior observed in an event log. Fitness shares commonalities with recall, which is used in machine and deep learning, in that it focuses on verifying the extent to which a model can identify behavior legitimately allowed by the process. Precision aims to measure the amount of behavior that has not been observed in the input event log but that the model allows for.

Every event log, either low-level or abstracted, has been divided into 70%, which was used for process discovery, and 30% that was used to compute fitness. Conversely, precision was computed on the whole event log. All models were discovered via Inductive Miner [26], with a noise threshold of 0.2 (default value), which produce Petri nets [44], for which fitness and precision were computed. However, every discovered Petri net can easily be converted into the BPMN notation [17], which is certainly easier to interpret, and consequently more known among process analysts and practitioners. Therefore, we decided to use BPMN to visualize process models in the remainder of this chapter.

4.2 Parameter Tuning for the Abstraction Technique and Abstract Model Discovery

The clustering-based abstraction technique requires one to make the following choices:

1. The value of the session threshold
2. The type of session encoding between frequency- and duration-based (cf. Sects. 3.1 and 3.2)
3. The choice of the clustering algorithms

In the remainder, we assume the application of two clustering algorithms: DBScan and K-means. Without further information at hand, a heuristic to set the session threshold is based on the computation of every timestamp difference between each pairs of consecutive events in each trace (see Fig. 4): the threshold is then chosen according to the median value. The rationale behind is that, if events are significantly spaced out in time, they are part of consecutive sessions: the choice of significant value is here determined as the median value. An alternative heuristic could have been two or three times of the standard deviation plus the mean value (Fig. 4).

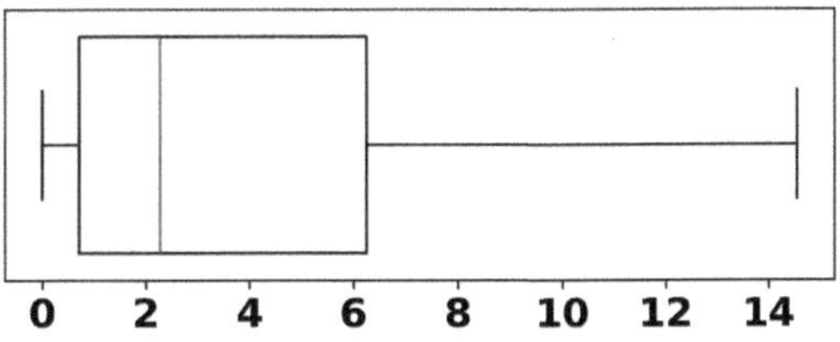

Fig. 4 Box plot displaying the distribution of the timestamp differences between consecutive events. Differences are shown in minutes

Table 2 Performance of the abstraction techniques in terms of fitness and precision of the models discovered by Inductive Miner on the abstract event log

Encoding	Clustering	Fitness	Precision	Harmonic mean
Frequency	K-Means	0.94	0.68	0.79
Duration	K-Means	0.75	0.64	0.69
Frequency	DBScan	0.96	0.69	0.80
Duration	DBScan	0.96	0.69	0.80

No simple indication can be provided for the choice of the type of session's encoding, or for the best clustering algorithms, because this largely depends on the specific case study. The only solution is thus to try all combinations, which are however still limited, e.g., to four (two encoding types multiplied by two clustering algorithms, K-means and DBScan).

The DBScan and K-means algorithms need to be configured in crucial parameters:

K-Means. It requires specifying the number of clusters to generate. To help make a clustering decision, we employed the widely used *elbow method*. It calculates the error curve for a number of different potential cluster numbers and then presented this curve graphically. As the number of cluster is set larger and larger, the error becomes smaller and smaller, until the error tends to stabilize around a certain value. By visually inspecting the error curve, one can more easily estimate the clusters' number such that a further increase of the set number of cluster is negligible.

DBScan. It is largely based on two parameters named *epsilon* and *MinPoints*. Epsilon sets the radius of a region to be considered in every iteration, by that, the cluster size. On the other hand, MinPoints is the minimum number of points required to declare a high-density region. The estimation of the appropriate epsilon value was made through the calculation of the distance to the nearest neighbor for all samples. Based on these distances, a value of epsilon that covers most of the samples was selected. For every sample, the number of neighbors within this epsilon radius was computed, and the results were tabulated based on the number of neighbors. The resulting distribution pointed at an initial peak, which means most of the samples were not surrounded by too many of their neighbors within this distance, indicating the density of points around most samples was not too high. Then, the number gradually decreased and reached a local minimum. This point is key in that it proposes an optimum MinPoints value, marking the threshold between the low- and high-density areas.

As mentioned in Sect. 4.1, an abstract event log was constructed for the best parameter configuration for K-means and DBScan, using the frequency- or duration-based encoding, and these event logs were used as input for Inductive Miner [26] with noise threshold of 0.2 (default value) to discover abstract process models.

Table 2 illustrates the quality of the discovered models for each event log. The quality was measured in terms of fitness, precision, and their harmonic means. The

Table 3 Overview of clusters based on our abstraction method

Cluster number	Cluster prominent activities	Cluster name
0	Transporting the workpiece to the light barrier 1, moving towards slot 2	Transporting the workpiece to the light barrier_0
1	Transporting the workpiece to the light barrier 1	Transporting the workpiece to the light barrier_1
2	Unload, moving towards slot 2, transporting the workpiece to the punch	Unload workpiece_2
3	Drilling the workpiece for 6 seconds	Drilling the workpiece for 6 seconds_3
4	Moving towards slot 4	Moving towards slot_4
5	Detecting the workpiece color	Detecting the workpiece color_5
6	Transporting the workpiece to the milling_machine, calibrating motor 1, transporting the bucket to slot 6, punch_recesses	Transporting the bucket to slot_6
7	Sort, detecting the workpiece color	Sort bucket_7
8	Sort, detecting the workpiece color, drilling the workpiece for 6 seconds	Detecting the workpiece color_8
9	Unload, transporting the workpiece to the light barrier 1	Unload workpiece_9

harmonic means is preferrable over the arithmetic means because it penalizes further if the fitness and precision values are unbalanced. The results indicate that DBSCAN clusters tend to produce superior abstracted models in terms of the harmonic mean of fitness and precision. Additionally, the choice of the encoding scheme does not seem to affect the quality of the discovered model when DBSCAN is used for clustering. Conversely, the frequency-based and duration-based encodings influence the quality when K-means is used, but either cases enable generating models of lower quality. In fact, these findings confirm similar ones observed for the IoT case study on domotics, reported in de Leoni and Pellattiero [14].

We ultimately chose to proceed the event log that was abstracted via DBSCAN and duration-based encoding. The clustering produces the heatmap depicted in Fig. 2. While no dominant low-level activity can be observed for cluster 4 (no orange-red cell stands out), the other clusters highlights the 2–3 dominant low-level activities. Table 3 also details the associations between cluster numbers, names, and dominant activities, where each cluster's name corresponds to the concatenation of its most prominent activity followed by the cluster number. The flat process model obtained via Inductive Miner using the non-abstracted event log is shown in Fig. 5, while the process model resulting from the abstracted log, which is used to assess the impact of abstraction on process discovery, is visualized using the BPMN notation in Fig. 6.

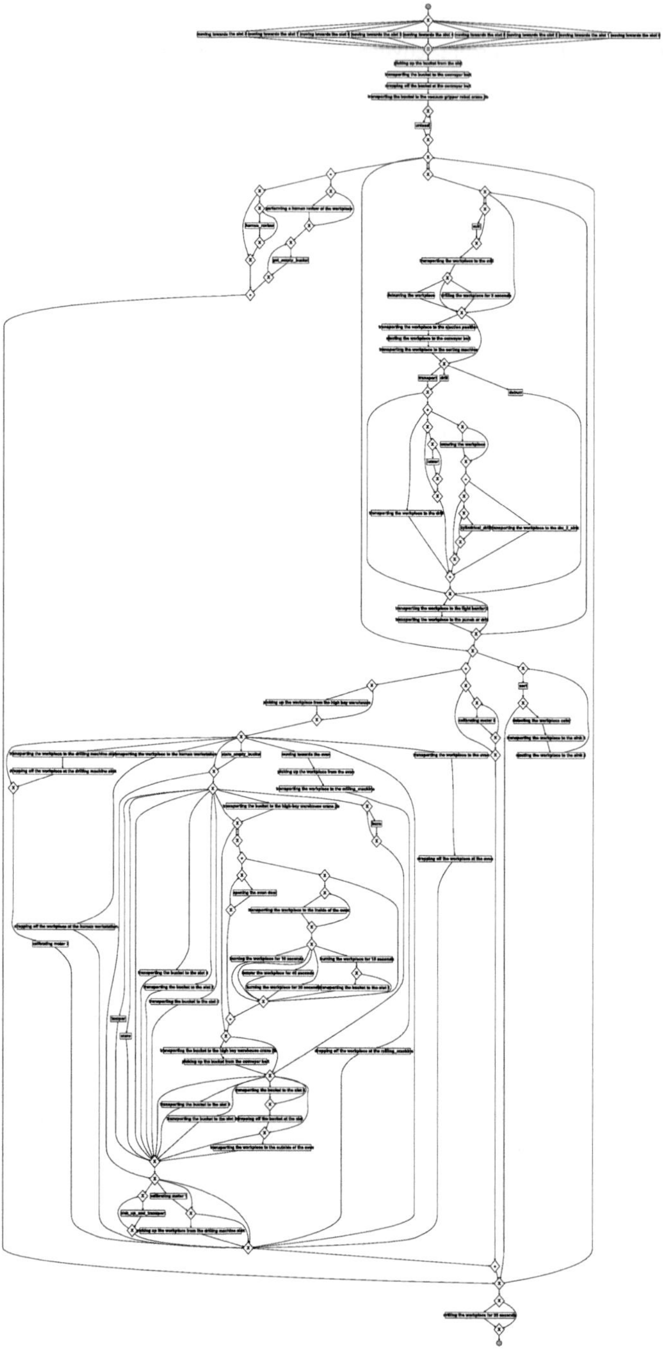

Fig. 5 Flat model discovered on the smart-factory transaction data via Inductive Miner

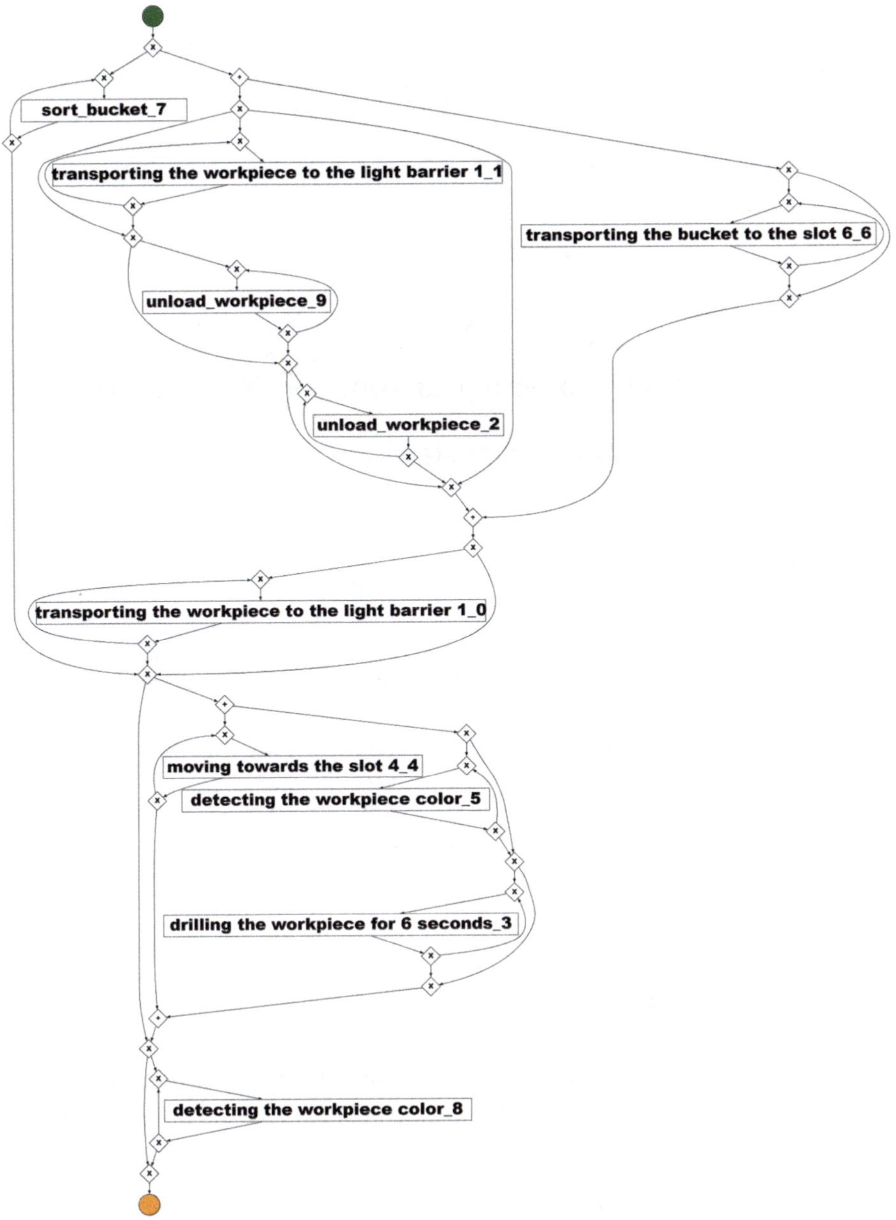

Fig. 6 The abstract model that was produced by Inductive Miner with a noise threshold of 0.2, where the event-log abstraction technique used the duration-based session encoding and DBSCAN for clustering

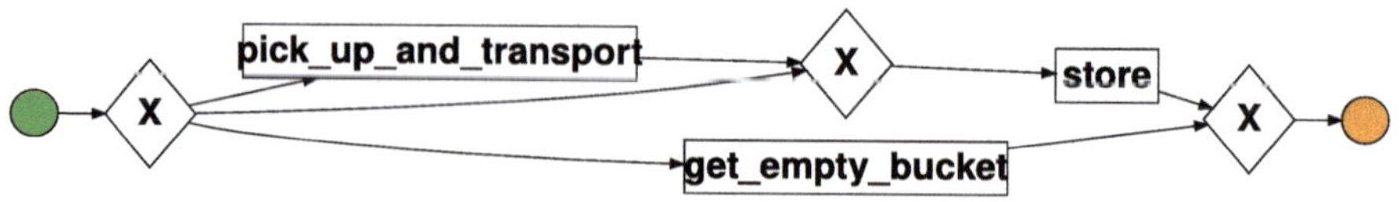

Fig. 7 BPMN model for cluster number 3 named drilling the workpiece for 6 seconds

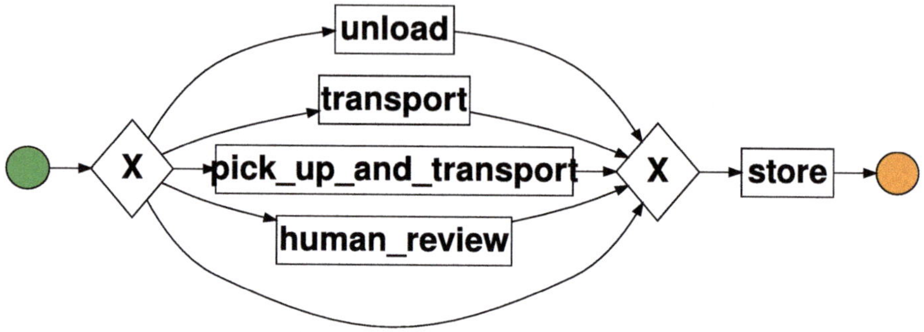

Fig. 8 BPMN model for cluster number 8 named detecting the workpiece color

Table 4 Performance metrics for hierarchical models obtained by the techniques using two session encodings and clustering algorithms. The metrics are also compared with the flat model, namely, obtained without abstraction

Model	Clustering	Encoding	Fitness	Precision	Harmonic mean	Simplicity
Flat			0.94	0.21	0.34	0.60
Hierarchical	DBScan	Frequency	0.99	0.59	0.74	0.59
Hierarchical	DBScan	Duration	0.99	0.59	0.74	0.59
Hierarchical	K-Means	Frequency	0.97	0.63	0.76	0.61
Hierarchical	K-Means	Duration	0.93	0.60	0.73	0.61

4.3 *Hierarchical Model*

The last optional step of the methodology focuses on mining one process model for each cluster, which then replaces the corresponding high-level activity inside the model (cf. Sect. 3.6). This is in fact a hierarchical model. Figures 7 and 8 present models for the behavior observed for cluster 3 (drilling the workpiece for 6 seconds) and 8 (detecting the workpiece color), respectively. As mentioned, we replaced each one of the high-level activities in the model shown in Fig. 6 with a whole process model that represents observed in the session cluster. Therefore, the hierarchical model uses the same.

Table 4 compares the quality of the flat model, which is obtained without abstraction, and the quality of the hierarchical model, which has been obtained as discussed in the previous paragraph. While the fitness of the hierarchical models is similar to the flat model, their precision is three times greater. This constitutes a major improvement: the facilitation of setting the right level of event granularity enables building an abstract model that also allows to better capture of the behavior and order relations that exist between the low-level activities.

5 Conclusion

The discovery, analysis, and improvement of processes executed via devices of an IoT system requires setting the right event granularity. Typical events of IoT systems refer to logging sensor's measurements and very technical information of the sensors and actuators of the IoT system. As a consequence, these events may not be directly linked to the actual activities of the IoT processes: this ultimately means that, e.g., the discovered models are defined over low-level activities, and hence, they convey limited information to the process owners and analysts. Furthermore, IoT systems might record the activities related to events in several orders, given a certain degree of uncertainty of the frequency to harvest IoT-system logging information. This clearly causes problems when one wants to discover the IoT process model, because the model tries to accommodate all the different orderings, thus finally resulting on a underfitting model (i.e., with low precision) that represents an excessive number of alternatives in the allowed ordering of process' activities

Event-log abstraction focuses on aggregating low-level events, namely, events related to low-level activities, to higher concepts that are closer to the activities that subjects understand: in IoT systems, event-log aggregations enable linking process' activities to the IoT processes that IoT process actors, designers, and analyst understand, while abstracting out the intrinsic arbitrariness of the order of the low-level IoT events that are likely aggregated.

This chapter discusses the vast repertoire of techniques for event-log abstraction, which can be grouped in two categories: those based on clustering or based on behavioral patterns. The first category has the advantage to require very little input information, especially in comparison with the second category where behavioral patterns need to be defined by experts or discovered from the data, with all the risks related to providing/discovering unrealistic patterns. Previous research has shown that, without the right amount of information, clustering techniques might often provide better event-log abstraction. In the second part of the chapter, experience is provided on applying a clustering-based technique to two case studies, related to smart manufacturing and home automation, respectively. The experience shows that the use of the used event-log abstraction technique enables to discover an abstract process model, namely, a model that refers to the high-level (i.e., abstracted) activities that can convey much more insightful information for IoT process analysts and owners, if compared with the information that one could obtained from a naive, flat model, namely, discovered on low-level events. Furthermore, the technique is also able to use the abstract event log to build the backbone to discover a hierarchical process model, where each high-level activity a in the process model is replaced with an entire subprocess that models the patterns observed for the sequence of low-level activities that were mapped to high-level activity a. In the hierarchical process model, the activities again refers to the low-level events, but the backbone built on the high-level activities allows discovery algorithms to mine process models of higher quality.

This chapter has only illustrated the benefits of abstracting event logs to address the problem of discovering a process model. However, the event-log abstraction benefits go beyond that. Performance-related questions (such as bottleneck detections) can also be better answered via event-log abstraction, because they provide insights that are easier to interpret because they ignore the low-level technicalities of actions and measurements in IoT systems, providing a helicopter view of the major issues related to the performance of the high-level activities, which are likely those relevant for IoT-system analysts and stakeholders. Predictive and prescriptive analytics can also take advantages to be employed on high-level activities: usually a smaller set of high-level activities are used for the analysis, instead of a large, chaotic set of low-level events and related activities. See that, indeed, more than 40 low-level activities are reduced to 10 high-level activities. This clearly speeds up the training of the models behind predictive and prescriptive analytics, which are typically machine or deep learning models. By reducing the dimensionality of the input data, these models would also be less prone to over- and under-fitting the input datasets. These are two extra application scenarios that add to the model discovery for which event-log abstraction techniques are beneficial: however, one can think of many other application scenarios.

In spite of the satisfactory results obtained by event-log abstraction, a lot of unanswered research questions still exist for this problem:

1. Event logs often contain noise, missing events, or anomalies that complicate abstraction. What methods are most effective for abstraction when logs are incomplete or noisy?
2. There is no consensus on how to evaluate the quality of event-log abstraction methods. In fact, the metrics of fitness and precision mentioned in the case study of Sect. 4 are only indirectly measuring the quality of the abstraction, because they focus on the quality of the models discovered via abstraction. What universal metrics should be used to assess the effectiveness of abstraction techniques?
3. Event-log abstraction should facilitate downstream tasks such as process discovery, conformance checking, and predictive analytics. Is the abstraction method universally valid? Or would it depend on the type of use that is made of the abstract event logs?
4. Related to the point above, the abstraction may hide critical behaviors, making it impossible to detect certain issues. Should this not be considered when the abstraction method is designed?
5. Real-world event logs can be massive, with high-dimensional data and complex interdependencies. How do existing abstraction methods scale efficiently with log size? What techniques can manage dimensionality reduction (e.g., directly

removing less relevant low-level events), without this affecting the quality of the abstraction?

6. Almost all abstraction techniques reported in Sect. 2 assume that each low-level sequence that is abstracted to one high-level activity does not interleave with other low-level sequences. In other words, there are no parallel high-level activities. This is a strong assumption because multiple activities might be performed in parallel in a single execution in the IoT system. Consider the situation in which two parts are made at the same time in smart manufactory. How can existing techniques be adjusted, or new be developed to detect parallel high-level activities?

References

1. Acheli, M., Grigori, D., Weidlich, M.: Efficient discovery of compact maximal behavioral patterns from event logs. In: Giorgini, P., Weber, B. (eds.) Advanced Information Systems Engineering. Lecture Notes in Computer Science, vol. 11483, pp. 579–594. Springer International Publishing, Cham (2019). ISBN 978-3-030212-89-6, 978-3030212-90-2.
2. Amirah, A., Andy, B., Johnson, O.A.: Towards unsupervised detection of process models in healthcare. Stud. Health Technol. Inf. **247**, 381–385 (2018). ISSN 0926-9630
3. Baier, T., Mendling, J., Weske, M.: Bridging abstraction layers in process mining. Inf. Syst. **46**, 123–139 (2014). ISSN 0306-4379
4. Baier, T., Ciccio, C.D., Mendling, J., Weske, M.: Matching events and activities by integrating behavioral aspects and label analysis. Softw. Syst. Model. **17**(2), 573–598 (2018)
5. Begicheva, A.A., Lomazov, I.A.: Discovering high-level process models from event logs. Model. Analy. Inf. Syst. **24**(2): 125–140 (2017). ISSN 2313-5417, 1818-1015
6. Bernard, G., Andritsos, P.: CJM-ab: Abstracting customer journey maps using process mining. In: Mendling, J., Mouratidis, H. (eds.) Information Systems in the Big Data Era. Lecture Notes in Business Information Processing, vol. 317, pp. 49–56. Springer International Publishing, Cham (2018). ISBN 978-3-319929-00-2, 978-3-319929-01-9.
7. Bose, R.P.J.C., Van der Aalst, W.M.P.: Abstractions in process mining: A taxonomy of patterns. In: International Conference on Business Process Management, pp. 159–175. Springer, Berlin (2009)
8. Cameranesi, M., Diamantini, C., Potena, D.: Discovering process models of activities of daily living from sensors. In: Business Process Management Workshops. LNBIP, vol. 308. Springer, Berlin (2018)
9. Coltellese, S., Maggi, F.M., Marrella, A., Massarelli, L., Querzoni, L.: Triage of IoT attacks through process mining. In: On the Move to Meaningful Internet Systems: OTM 2019 Conferences. Springer International Publishing, Cham (2019)
10. de Leoni, M., Dündar, S.: From low-level events to activities—A session-based approach (extended version). arXiv.org, abs/1903.03993 (2019). http://arxiv.org/abs/1903.03993
11. de Leoni, M., Dundar, S.: From Low-Level Events to Activities—A Session-Based Approach (Extended Version). arXiv:1903.03993 [cs] (2019)
12. de Leoni, M., Dündar, S.: Event-log abstraction using batch session identification and clustering. In: Proceedings of the 35th Annual ACM Symposium on Applied Computing, SAC '20, pp. 36–44. Association for Computing Machinery, New York (2020). ISBN 978-14-50368-66-7. https://doi.org/10.1145/3341105.3373861.
13. de Leoni, M., Dündar, S.: Event-log abstraction using batch session identification and clustering. In: Proceedings of the 35th Annual ACM Symposium on Applied Computing, pp. 36–44 (2020)

14. de Leoni, M., Pellattiero, L.: The benefits of sensor-measurement aggregation in discovering IoT process models: A smart-house case study. In: Marrella, A., Weber, B. (eds.) Business Process Management Workshops, pp. 403–415. Springer International Publishing, Cham (2022). ISBN 978-3-030943-43-1
15. Di Federico, G., Burattin, A.: CvAMo–event abstraction using contextual information. Future Int. **15**(3), 113 (2023). ISSN 1999-5903
16. Dimaggio, M., Leotta, F., Mecella, M., Sora, D.: Process-based habit mining: Experiments and techniques. In: 2016 IEEE 13th International Conference on Ubiquitous Intelligence & Computing, pp. 145–152 (2016)
17. Dumas, M., La Rosa, M., Mendling, J., Reijers, H.A.: Fundamentals of Business Process Management. Springer, Berlin (2013). ISBN 978-3-642331-42-8. https://doi.org/10.1007/978-3-642-33143-5
18. Fazzinga, B., Flesca, S., Furfaro, F., Masciari, E., Pontieri, L.: Efficiently interpreting traces of low level events in business process logs. Inf. Syst. **73**, 1–24 (2018). ISSN 0306-4379
19. Fazzinga, B., Flesca, S., Furfaro, F., Pontieri, L.: Process discovery from low-level event logs. In: Krogstie, J., Reijers, H.A. (eds.) Advanced Information Systems Engineering. Lecture Notes in Computer Science, vol. 10816, pp. 257–273. Springer International Publishing, Cham (2018). ISBN 978-33-19915-62-3, 978-3-319915-63-0
20. Fazzinga, B., Flesca, S., Furfaro, F., Pontieri, L.: Process mining meets argumentation: explainable interpretations of low-level event logs via abstract argumentation. Inf. Syst. **107**, 101987 (2022). ISSN 0306-4379
21. Ferreira, D.R., Szimanski, F., Ralha, C.G.: Mining the low-level behaviour of agents in high-level business processes. Int. J. Busin. Proc. Integrat. Manag. **6**(2), 146 (2013). ISSN 1741-8763, 1741-8771
22. Folino, F., Guarascio, M., Pontieri, L.: Mining multi-variant process models from low-level logs. In: Abramowicz, W. (ed.) Business Information Systems. Lecture Notes in Business Information Processing, pp. 165–177. Springer International Publishing, Cham (2015). ISBN 978-3-319190-27-3
23. Günther, C.W., Rozinat, A., Van Der Aalst, W.M.P.: Activity mining by global trace segmentation. In: International Conference on Business Process Management, pp. 128–139. Springer, Berlin (2009)
24. Hemmer, A., Badonnel, R., Chrisment, I.: A process mining approach for supporting iot predictive security. In: Proceedings of OMS IEEE/IFIP Network Operations and Management Symposium (OMS 2020) (2020)
25. Janiesch, C., Koschmider, A., Mecella, M., Weber, B., Burattin, A., Ciccio, C.D., Fortino, G., Gal, A., Kannengiesser, U., Leotta, F., Mannhardt, F., Marrella, A., Mendling, J., Oberweis, A., Reichert, M., Rinderle-Ma, S., Serral, E., Song, W., Su, J., Torres, V., Weidlich, M., Weske, M., Zhang, L.: The internet of things meets business process management: a manifesto. IEEE Syst. Man Cybern. Mag. **6**(4), 34–44 (2020)
26. Leemans, S.J.J., Fahland, D., van der Aalst, W.M.P.: Discovering block-structured process models from event logs—A constructive approach. In: Proceedings of the 34th International Conference on Application and Theory of Petri Nets and Concurrency (Petri Net 2013). LNCS, vol. 7927, pp. 311–329. Springer, Berlin (2013)
27. Leonardi, G., Striani, M., Quaglini, S., Cavallini, A., Montani, S.: Towards semantic process mining through knowledge-based trace abstraction. In: Ceravolo, P., Van Keulen, M., Stoffel, K. (eds.) Data-Driven Process Discovery and Analysis. Lecture Notes in Business Information Processing, vol. 340, pp. 45–64. Springer International Publishing, Cham (2019). ISBN 978-3-030116-37-8, 978-3-030116-38-5.
28. Li, C.-Y., van Zelst, S.J., van der Aalst, W.M.P.: An activity instance based hierarchical framework for event abstraction. In: 2021 3rd International Conference on Process Mining (ICPM), pp. 160–167 (2021). https://doi.org/10.1109/ICPM53251.2021.9576868
29. Li, C.-Y., Van Zelst, S.J., Van Der Aalst, W.M.P.: A framework for automated abstraction class detection for event abstraction. In: Abraham, A., Pllana, S., Casalino, G., Ma, K., Bajaj, A. (eds.) Intelligent Systems Design and Applications. Lecture Notes in Networks and Systems,

vol. 715, pp. 126–136. Springer Nature Switzerland, Cham (2023). ISBN 978-3-031355-06-6, 978-3-031355-07-3.
30. Chen, L., Hoey, J., Nugent, C.D., Cook, D.J., Yu, Z.: Sensor-based activity recognition. IEEE Trans. Syst. Man Cybern. Part C (Appl. Rev.) **42**(6), 790–808 (2012). ISSN 1094-6977, 1558-2442. https://doi.org/10.1109/TSMCC.2012.2198883
31. Malburg, L., Grüger, J., Bergmann, R.: An IoT-Enriched Event Log for Process Mining in Smart Factories. arXiv preprint arXiv:2209.02702 (2022)
32. Mannhardt, F., Tax, N.: Unsupervised event abstraction using pattern abstraction and local process models. In: Gulden, J., Nurcan, S., Reinhartz-Berger, I., Guédria, W., Bera, P., Guerreiro, S., Fellmann, M., Weidlich, M. (eds.) CEUR Workshop Proceedings, vol. 1859, pp. 55–63. CEUR-WS.org (2017)
33. Mannhardt, F., de Leoni, M., Reijers, H.A., van der Aalst, W.M.P., Toussaint, P.J.: From low-level events to activities—A pattern-based approach. In: La Rosa, M., Loos, P., Pastor, O. (eds.) Business Process Management. Lecture Notes in Computer Science, vol. 9850, pp. 125–141. Springer International Publishing, Cham (2016). ISBN 978-3-319453-47-7, 978-3-319453-48-4
34. Mannhardt, F., de Leoni, M., Reijers, H.A., van der Aalst, W.M.P., Toussaint, P.J.: Guided process discovery–a pattern-based approach. Inf. Syst. **76**, 1–18 (2018)
35. Peeva, V., Mannel, L.L., Van Der Aalst, W.M.P.: From place nets to local process models. In: Bernardinello, L., Petrucci, L. (eds.) Application and Theory of Petri Nets and Concurrency. PETRI NETS 2022. Lecture Notes in Computer Science, vol. 13288. Springer, Cham (2022)
36. Rebmann, A., Pfeiffer, P., Fettke, P., Van Der Aa, H.: Multi-perspective identification of event groups for event abstraction. In: Montali, M., Senderovich, A., Weidlich, M. (eds.) Process Mining Workshops. Lecture Notes in Business Information Processing, vol. 468, pp. 31–43. Springer Nature Switzerland, Cham (2023.) ISBN 978-3-031278-14-3, 978-3-031278-15-0
37. Rehse, J.-R., Fettke, P.: Clustering business process activities for identifying reference model components. In: Daniel, F., Sheng, Q.Z., Motahari, H. (eds.) Business Process Management Workshops. Lecture Notes in Business Information Processing, vol. 342, pp. 5–17. Springer International Publishing, Cham (2019). ISBN 978-3-030116-40-8, 978-3-030116-41-5.
38. Sánchez-Charles, D., Carmona, J., Muntés-Mulero, V., Solé, M.: Reducing event variability in logs by clustering of word embeddings. In: International Conference on Business Process Management, pp. 191–203. Springer, Berlin (2017)
39. Tax, N., Sidorova, N., Haakma, R., van der Aalst, W.M.P.: Mining local process models. J. Innovat. Digital Ecosyst. **3**(2), 183–196 (2016). ISSN 2352-6645
40. Tax, N., Dalmas, B., Sidorova, N., van der Aalst, W.M.P., Norre, S.: Interest-driven discovery of local process models. Inf. Syst. **77**, 105–117 (2018). ISSN 0306-4379
41. Tax, N., Sidorova, N., Haakma, R., van der Aalst, W.M.P.: Event abstraction for process mining using supervised learning techniques. In: Bi, Y., Kapoor, S., Bhatia, R. (eds.) Proceedings of SAI Intelligent Systems Conference (IntelliSys) 2016. Lecture Notes in Networks and Systems, vol. 15, pp. 251–269. Springer International Publishing, Cham (2018). ISBN 978-3-319569-93-2, 978-3-319569-94-9
42. Tax, N., Sidorova, N., Aalst, W.M.P.: Discovering more precise process models from event logs by filtering out chaotic activities. J. Intell. Inf. Syst. **52**(1), 107–139 (2019)
43. Tello, G., Gianini, G., Mizouni, R., Damiani, E.. Machine learning-based framework for log-lifting in business process mining applications. In: Hildebrandt, T., Van Dongen, B.F., Röglinger, M., Mendling, J. (eds.) Business Process Management. Lecture Notes in Computer Science, vol. 11675, pp. 232–249. Springer International Publishing, Cham (2019). ISBN 978-3-030266-18-9, 978-3-030266-19-6.
44. van der Aalst, W.M.P.: The application of petri nets to workflow management. J. Circ. Syst. Comput. **08**(01), 21–66 (1998). https://doi.org/10.1142S0218126698000043
45. van der Aalst, W.M.P.: Process Mining - Data Science in Action. Springer, Berlin (2016). ISBN 978-3-662498-50-7
46. van der Aalst, W.M.P.: Process Mining: Data Science in Action. Springer, Berlin (2016). ISBN 978-3-662498-50-7, 978-3-662498-51-4

47. van der Aalst, W.M.P., Adriansyah, A., de Medeiros, A.K.A., Arcieri, F., Baier, T., Blickle, T., Bose, J.C., van den Brand, P., Brandtjen, R., Buijs, J., Burattin, A., Carmona, J., Castellanos, M., Claes, J., Cook, J., Costantini, N., Curbera, F., Damiani, E., de Leoni, M., Delias, P., van Dongen, B.F., Dumas, M., Dustdar, S., Fahland, D., Ferreira, D.R., Gaaloul, W., van Geffen, F., Goel, S., Günther, C., Guzzo, A., Harmon, P., ter Hofstede, A., Hoogland, J., Ingvaldsen, J.E., Kato, K., Kuhn, R., Kumar, A., La Rosa, M., Maggi, F., Malerba, D., Mans, R.S., Manuel, A., McCreesh, M., Mello, P., Mendling, J., Montali, M., Motahari-Nezhad, H.R., zur Muehlen, M., Munoz-Gama, J., Pontieri, L., Ribeiro, J., Rozinat, A., Pérez, H.S., Pérez, R.S., Sepúlveda, M., Sinur, J., Soffer, P., Song, M., Sperduti, A., Stilo, G., Stoel, C., Swenson, K., Talamo, M., Tan, W., Turner, C., Vanthienen, J., Varvaressos, G., Verbeek, E., Verdonk, M., Vigo, R., Wang, J., Weber, B., Weidlich, M., Weijters, T., Wen, L., Westergaard, M., Wynn, M.: Process Mining Manifesto, pp. 169–194. Springer, Berlin (2012). ISBN 978-3-642281-08-2
48. van Eck, M.L., Sidorova, N., van der Aalst, W.M.P.: Enabling process mining on sensor data from smart products. In: 2016 IEEE Tenth International Conference on Research Challenges in Information Science (RCIS), pp. 1–12. IEEE (2016)
49. Van Houdt, G., Depaire, B., Martin, N.: Unsupervised event abstraction in a process mining context: A benchmark study. In: Leemans, S., Leopold, H. (eds.) Process Mining Workshops. Lecture Notes in Business Information Processing, vol. 406, pp. 82–93. Springer International Publishing, Cham (2021). ISBN 978-3-030726-92-8, 978-3-030726-93-5
50. Van Houdt, G., de Leoni, M., Martin, N., Depaire, B.: An empirical evaluation of unsupervised event log abstraction techniques in process mining. Inf. Syst. **121**, 102320 (2024). ISSN 0306-4379. https://doi.org/10.1016/j.is.2023.102320. https://www.sciencedirect.com/science/article/pii/S0306437923001564
51. van Zelst, S.J., Mannhardt, F., de Leoni, M., Koschmider, A.: Event abstraction in process mining: literature review and taxonomy. Granular Comput. **6**, 719–736 (2020). ISSN 2364-4966, 2364-4974
52. Yadav, S.K., Tiwari, K., Pandey, H.M., Akbar, S.A.: A review of multimodal human activity recognition with special emphasis on classification, applications, challenges and future directions. Knowl.-Based Syst. **223**, 106970 (2021). ISSN 0950-7051. https://doi.org/10.1016/j.knosys.2021.106970

Process Mining on Sensor Location Event Data

Dominik Janssen, Agnes Koschmider, and Felix Mannhardt

1 Introduction

The increasing amount of Internet of Things (IoT) sensor event data in domains like smart cities or smart homes demands efficient analysis techniques that give valuable and new insights from data. Process mining can provide valuable insights into the underlying behaviour in terms of discovering processes from raw data recorded by systems. It allows to identify bottlenecks or anomalies in these processes. Despite the many advantages of process mining on raw sensor event data, several challenges remain to be addressed. The main challenge is that IoT environments create flexible processes (i.e. processes that do not follow a clear structure or the process case/instance notion is unclear) [1]. This means that process mining without efficient data preparation and data aggregation steps results in hardly understandable spaghetti processes with little utility to the user.

This paper suggests an approach for process discovery on sensor location event data in single occupation settings with non-overlapping activities. Unlabelled sensor event data is used as input, which undergoes unsupervised learning (clustering) to aggregate events. We assume that the data consists of a stream of raw location sensor data (triggered, not-triggered) and that no additional labelling, such as the activity of interest, has been applied. Process mining is then used to discover process models from the aggregated event data. A feedback loop (Fig. 1) with a sensitivity analysis measures how sensitive the discovered process model is to parameter changes when processing and aggregating sensor location event data in

D. Janssen (✉) · A. Koschmider
Group Process Analytics, Bayreuth University, Bayreuth, Germany
e-mail: dominik.janssen@uni-bayreuth.de; agnes.koschmider@uni-bayreuth.de

F. Mannhardt
Group Process Analytics, Eindhoven University of Technology, Eindhoven, The Netherlands
e-mail: f.mannhardt@tue.nl

G. Fortino, M. Mecella (eds.), *Internet of Things Meets Business Process Management*, Internet of Things, https://doi.org/10.1007/978-3-031-90746-3_8

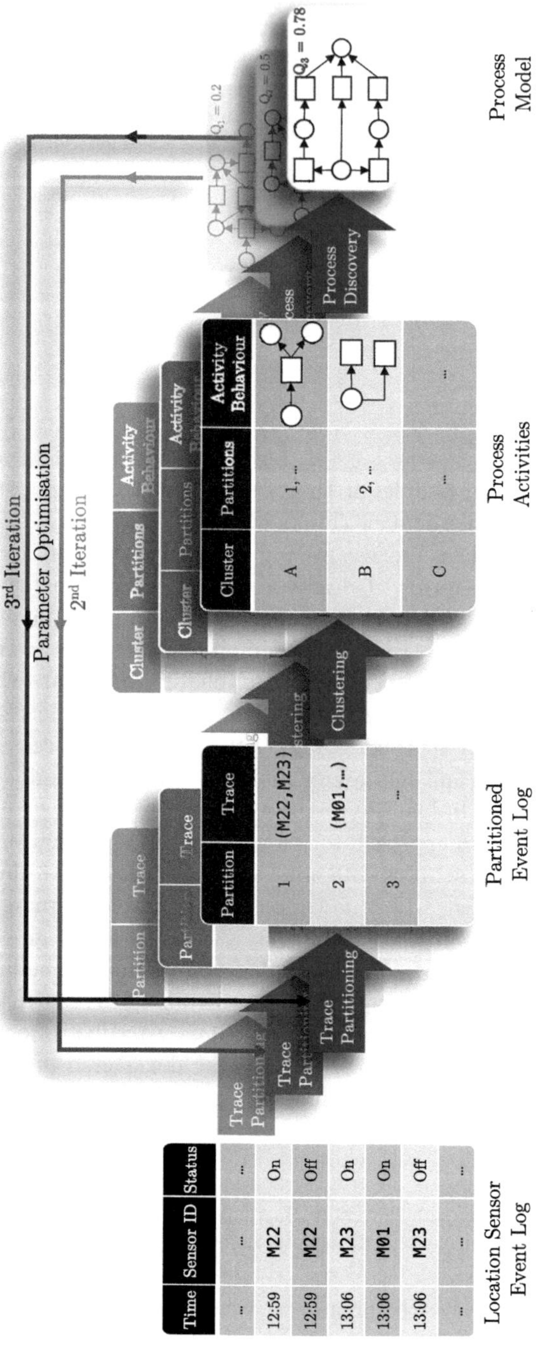

Fig. 1 Our approach: from sensor location event data to discovered processes

an automated manner. Thus, the primary and novel idea is to use the discovered process model as essential quality feedback for activity abstractions and noise filtering. Based on the discovered process models, valuable insights can be obtained regarding nonconforming, delayed or unnecessary activities. In this way, our approach promises to facilitate more widespread use of process mining in use cases relying on sensor location event data.

In our previous work, we suggested a framework for process discovery from sensor event data and a technique allowing us to translate sensor event data into higher-level, discrete event data [2]. We extend this work with the newly inserted concept of giving activities pseudo-labels first and a feedback loop, laying the foundation for sensitivity analysis regarding feedback for clustering. The iterative evaluation of the process model quality gives insights if the number of clusters is appropriate, which we had not evaluated before. This is the first paper iteratively exploiting the road from sensor event data to process discovery and back.

The application of our approach has been evaluated on two sets of smart home sensor data. Our approach is not restricted to a particular domain. Instead, it can be applied to any IoT environment that tracks sensor event data, including time and location, of an individual's movements. The paper investigates three main ideas beyond the existing work on process discovery directly on IoT data [3–5].

Auto-Adjustment: Our approach measures how sensitive the discovered process model is to parameter changes. This allows us to adapt to different scenarios automatically and find an optimal combination of (hyper)parameters. Each step of the framework (e.g. clustering, including noise filtering) can be individually configured for the specific purpose of the application domain.

Unsupervised Learning: We use unsupervised learning to reduce the effort to label activities and introduce a new concept of pseudo-labelling for improved activity recognition.

Evaluation Feedback: The analysis of process model quality gives feedback on the efficiency of noise filtering and clustering. Evaluating filtering and clustering quality is challenging when considering it in isolation. This is an essential step towards improving the clustering techniques that are used for activity recognition in our approach.

2 Motivation

The discovery of insights regarding process models on raw sensor data is currently highly requested. Domains that track a high volume of sensor data have a high demand for efficient process analytics techniques. Although process mining is suitable for giving valuable insights into data, the challenge remains to bring structure into data sourced from unstructured data like sensor location event data. Figure 2 illustrates the challenge using an example. Figure 2a shows a discovered graph from sensor location event data without pre-processing. Each rectangle represents a

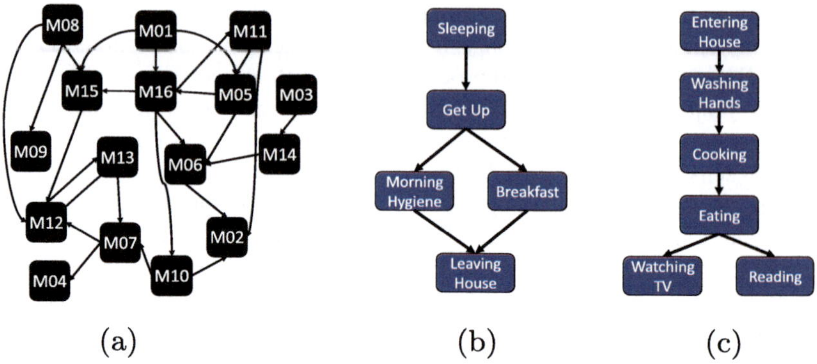

Fig. 2 Directly-follows graph of different aggregation levels. (**a**) Unstructured model: sensor location event data. (**b**) Process example: daily morning routine. (**c**) Process example: daily evening routine

motion sensor, and the directed arcs represent directly-follows relations between the sensors. The discovered process is a spaghetti process since unstructured processes are usually hard to put in a fixed order for processing activities.

The challenge remains to bring structure in process discovery into this kind of data to obtain a structured process model as shown in Fig. 2b, c. Four main objectives have to be addressed:

1. *Data Cleaning:* Noise and outliers must be filtered efficiently since they directly impact the process mining result. If simultaneous sensor activations cannot be explained by the movement of a single person, we exclude them from further evaluation (see Sect. 6).
2. *Feedback Loop:* The number of clusters is (typically) unknown a priori. For this purpose, our approach includes a feedback loop: the algorithm evaluates the resulting model and returns a quality measure that serves as a basis for the next iteration's parameter selection, like the number of clusters.
3. *Case-ID Notion:* Discovering the structure or any routine in sensor location event data requires incorporating process mining notions from the beginning of the data analytics pipeline. Our approach considers the notion of a case ID and traces from the beginning when pre-processing sensor location event data.
4. *Modular Architecture:* Each architecture module is self-contained and can be upgraded with new techniques. Additionally, an automatic approach will mainly relieve domain experts from data labelling. Our approach reduces the effort of data labelling and, thus, contributes towards a (semi)automatic approach.

Figure 3a gives an example of the sensor location event data used as input for our approach. Besides timestamps, the sensor's ID and current status are included. The purpose is to detect process routines like *eating*, *showering* or *reading* from this kind of data as shown in Fig. 3b. The only assumption in our approach is that the sensor activations reflect the behaviour of a single person.

Timestamp	SensorID	Status
...	...	...
2021-01-01 12:59:57	M22	ON
2021-01-01 13:06:21	M22	OFF
2021-01-01 13:06:22	M23	ON
2021-01-01 13:06:23	M01	ON
2021-01-01 13:06:24	M23	OFF
2021-01-01 13:06:24	M01	OFF
2021-01-01 13:06:24	M07	ON
2021-01-01 13:06:30	M07	OFF
...	...	...

(a)

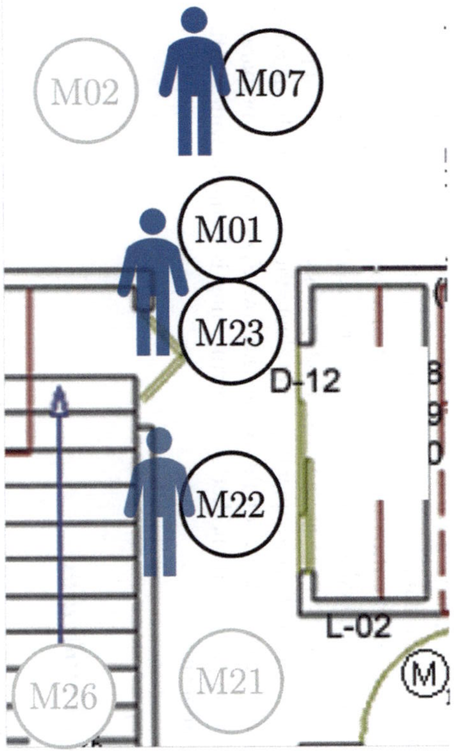

(b)

Fig. 3 Input Data for our approach. (**a**) Excerpt of sensor location event data. (**b**) Movement path of an entity

3 Process Mining on Raw Sensor Event Data

Figure 1 shows our approach for process mining on sensor location event data. Sensor location event data is used as input. Next, the sensor location data is partitioned so that one partition can be mapped onto exactly one process activity. Then, clustering is applied to identify similar traces. One cluster represents the behaviour of a specific process activity. Finally, process mining is applied to discover a process model. Subsequently, we evaluate the quality of the discovered process model. In case of a low-quality value (i.e. low F_1-score), a hyperparameter optimisation is applied, and another trace partitioning is initiated with a different parameter combination.

3.1 Trace Partitioning

This step splits the sensor location event log into traces and filter noise. The challenge is partitioning the sensor location event data so that one activity can be mapped onto one trace. Our approach allows three different types of partitions:

Number of activations: The sensor event data is split into traces with a fixed number of activated sensors. Assume that a log with n sensor location event data entries exists. If the number of allowed activations per trace is i, the sensor event log is then split up into $\frac{n}{i}$ traces.

Activation time: This approach splits the sensor location event log into traces, assuming every activity requires the same time. The user defines a time threshold t_{max}. Events are added to the trace as long as the time difference between the first event in a trace t_1 and the added event with the timestamp t_i is smaller than a threshold ($t_i - t_1 \leq t_{max}$). If the threshold is exceeded, a new trace starts.

Area: Activities are strictly limited to one room or area. Every time an entity leaves an area, the current activity ends, and a new one begins, regardless of whether the number of activations or the time threshold is exceeded. This assumption excludes multi-room activities.

An example of the partitioning is given in Table 1. In the *Split by Activation* column, there is a limit of **five activations**. As a result, the first partition 1_{Act} is completed at 13:06:33, and a sequence of sensor activations characterises it with a length of five: `M22`, `M23`, `M01`, `M07` and `M06`. The *Split by Time* column has a time limit of **5 minutes** for each partition. Once this limit is reached, the partition is considered complete. Consequently, the first partition 1_{time} ends at 13:06:21, as the partition length reaches 6 minutes and 24 seconds. Limiting traces to a fixed length (time or activation) is a simplification and, due to the different lengths of human activities, will not necessarily lead to an ideal splitting.

Table 1 Example of different partitioning methods

			Split by Activation		Split by Time	
Timestamp	SensorID	Status	Counter	Partition	Duration	Partition
...	...	...	...	...	...	...
12:59:57	M22	ON	1	1_{Act}	0:00	1_{Time}
13:06:21	M22	OFF		1_{Act}	6:24	1_{Time}
13:06:22	M23	ON	2	1_{Act}	0:00	2_{Time}
13:06:23	M01	ON	3	1_{Act}	0:01	2_{Time}
13:06:24	M23	OFF		1_{Act}	0:02	2_{Time}
13:06:24	M01	OFF		1_{Act}	0:02	2_{Time}
13:06:24	M07	ON	4	1_{Act}	0:02	2_{Time}
13:06:30	M07	OFF		1_{Act}	0:08	2_{Time}
13:06:30	M06	ON	5	1_{Act}	0:08	2_{Time}
13:06:33	M06	OFF		1_{Act}	0:11	2_{Time}
13:06:33	M05	ON	1	2_{Act}	0:11	2_{Time}
13:06:35	M06	ON	2	2_{Act}	0:13	2_{Time}
13:06:45	M06	OFF		2_{Act}	0:23	2_{Time}
14:30:20	M05	OFF		2_{Act}	83:58	2_{Time}
...	...	...	...	...	...	...

3.2 Clustering

To prepare the traces for clustering, the partitioned traces have to be transformed into *trace vectors*. The length of the *trace vector* τ_j is the number of different activated sensors (S_j) with $|\tau_j| = \#S$. Our approach supports three different trace vector representations. A trace of activated sensors S_j and their respective activation time t_i (S_1: t_1 $\rightarrow$ S_7: t_2 $\rightarrow$ S_6: t_3 $\rightarrow$ S_5: t_4 $\rightarrow$ S_6: t_5) can be transformed according to quantity, time and its concatenation:

Quantity: Counts how often a sensor has been activated:
$\tau_1^Q = (\#S_1, \#S_2, \#S_3, \#S_4, \#S_5, \#S_6, \#S_7) = (1, 0, 0, 0, 1, 2, 1)$
Time: Counts how long each individual sensor has been activated:
$\tau_1^T = (t(S_1), t(S_2), t(S_3), t(S_4), t(S_5), t(S_6), t(S_7)) = (t_1, 0, 0, 0, t_4, t_3 + t_5, t_2)$
Quantity & Time: Concatenates the time and quantity vector:
$\tau_1^{QT} = (1, 0, 0, 0, 1, 2, 1, t_1, 0, 0, 0, t_4, t_3 + t_5, t_2)$

Based on the trace vector representations, clustering is initiated. At this point, each trace vector describes exactly one activity. The clustering aims to map similar traces to the same cluster. Each cluster represents an activity (e.g. *sitting*, *eating*). Our approach supports three clustering algorithms: k-means, k-medoids and self-organising map (SOM). We implemented a custom distance function for the k-medoids clustering, so we do not have to rely on any vector transformation. Instead of calculating the Euclidean distance between vectors, the customised k-medoids approach uses the distance matrix of the smart home and the actual distances of sensors in the smart home to cluster similar traces. To order the process activities (i.e. *eating* has been performed after *sitting*), a graph representation in terms of a directly-follows graphs (DFG) is used for each cluster. Finally, clusters have to be labelled. To reduce manual labelling, we introduce a new concept of pseudo-labelling. Initially, we assign pseudo-labels to each cluster. Assume that n clusters have been identified. Then, each cluster is labelled with a placeholder resulting in clusters $1, 2, \ldots, n$. Only after the hyperparameter optimisation has terminated will the domain expert replace the pseudo-labels with concrete process activity labels for the process model with the highest process model quality. This has three advantages: First, we do not label the clusters in isolation but rather in the context of preceding and succeeding clusters or process activities. Second, labelling is done only once for a discovered process of high quality. Third, in the case of duplicate pseudo-labels, an inspection of the clusters based on discovered processes might reveal further insights like two clusters are not identical but instead refer to different activities. This reduces the effort of manual labelling.

3.3 Process Discovery

This step aims to discover process models composed of high-level activities as shown in Fig. 2b, c. The input for the process discovery step are activities with pseudo-labels (`A`, `B`, `C`, ...) instead of real activity names.

Clusters with similar sensor activation patterns can be aggregated into a process. An example of cluster aggregations according to different time dimensions is shown in Table 1. The aggregation can be done according to the following time dimensions:

Week partition: Usually, the behaviour of an entity (e.g. inhabitants or workers) differs daily (weekends vs. weekdays). We therefore consider three ways to partition the week:

- *None:* No partition according to days or weekends, respectively. Every day is treated the same.
- *Weekday:* Difference between each day of the week (Mon, Tue, ...). We are assuming that behaviour is different every day.
- *Workday/Weekend:* Difference between workdays (Mon, Tue, ...) and weekends (Sat, Sun). We assume that behaviour during the week stays the same independent of the weekday and only differs on weekends.

Day Partition: With the partitioning of the day, one day is further divided into smaller parts reflecting repetitive routines, e.g. morning routines or rest routines in the evening. The number of smaller units impacts the number of discovered processes. This is evident since a morning routine results in a different process than a daily routine.

A possible partition in smart factories would be between day and night shifts. Finally, process mining is applied. Since IoT environments rely on plenty of infrequent behaviours, as discussed above, we selected the Heuristic Miner [6] for process discovery from the PM4PY [7] implementation. However, our modular architecture also allows other process discovery algorithms to be used.

3.4 Parameter Optimisation

The discovered process model is used as essential quality feedback for activity abstractions in this step. To evaluate the process model quality, we use the harmonic mean of the token-based precision and token-based fitness value [8], also known as the F_1-score. The purpose of this stage of the framework is to optimise the F_1-score. With the introduction of *routines* (see Figs. 2c, b), we aim to discover one process model for every segment of the day/week instead of one single process model for the entire day. Our analysis shows that comparing multiple F_1-scores for every segment is not feasible since the number of events differs by day/week. Therefore, we use a weighted F_1-score for a weighted average.

Our approach's final step is parameter tuning to enhance the process model quality based on user-defined values. The parameter pre-selection would be the responsibility of a domain expert. To support the generability of our approach to other domains like smart factories, we decided against a fixed set of parameters. Instead, we assume that the efficient parameters are found independently for every scenario. The specific range of parameter values depends on the application domain and can be narrowed down as needed to speed up the search process.

4 Evaluation

4.1 Experimental Setup

For the evaluation, we selected two single-occupancy data sets provided by the CASAS smart home project [9]. We chose the test beds **Kyoto-05** and **Aruba-19**. Each data set has more than 150,000 sensor events. Both smart homes are equipped with multiple motion sensors per room, and a room layout exists for both. Periods were excluded from the evaluation where the activity of one single person could not explain the sensor activations. To validate our results, we executed a hyperparameter search with 150 iterations. By taking advantage of vectorised operations in NumPy [10] and pandas [11], each iteration took less than 1 minute to compute. The framework implementation is accessible online[1].

4.2 Results

To evaluate the hyperparameter tuning, we examined the influence on the F_1-score in terms of which parameter values yield average results consistently for both data sets. The number of clusters or the number of partitions in a day does not appear to influence the resulting F_1-score strongly: $corr(F_1, \text{\#Clusters})$ and $corr(F_1, \text{\#Partitions})$ for both data sets in the interval of $[-0.02; 0.21]$. The duration of the trace or the number of activations by trace affects the F_1-score. Both are positively correlated with the resulting F_1-score (no significance for Aruba-19 and the #Activation), as shown in Table 2.

	Aruba-19	Kyoto-05
$corr(F_1, \text{\#Activation})$	0.0897	0.4996
$corr(F_1, \text{Duration})$	0.5042	0.3374

The box plots in Fig. 4 show how the F_1-score is affected by the different hyperparameters. The top half shows the results for the Aruba-19 data set, while the bottom half refers to the Kyoto-05 data set. Limiting activities to a single area (e.g. kitchen) shows a consistently inferior F_1-score compared to evaluation runs with multi-room activities permitted (Fig. 4b, f). Figure 4c, g show the correlation between the data set and a clustering algorithm. For instance, k-means clustering delivers solid results for the Kyoto-05 data set. However, for the Aruba-19 data set, the simple k-means algorithm outperforms the more complex self-organising map (SOM) algorithm. For both data sets, the activation time (Sect. 3.1) leads to a high F_1-score. A custom distance metric between traces does not always seem beneficial, as shown in Fig. 4d. Instead, the optimal vectorisation depends on the data set. Figure 5 shows the function value (F_1-score) depending on the number of iterations.

Table 2 shows the five best iterations. It seems that neither a partition by weekday nor the differentiation between weekend and workday results in different significant results. Also, a custom distance between traces does not yield better (F_1-score) results. The best iteration of the Kyoto-05 data set discovered four daily routines. Two are shown in Fig. 6a, c. The *Routine—Day Partition* has a value of 4, and then the day is divided as shown in Fig. 6b. We used the interactive heuristic miner [12] for visualisation purposes (95% of the activities, dependency threshold $= 0.95$).

[1] https://github.com/Process-Analytics-Group/BPM2023-PM_on_Sensor_Data

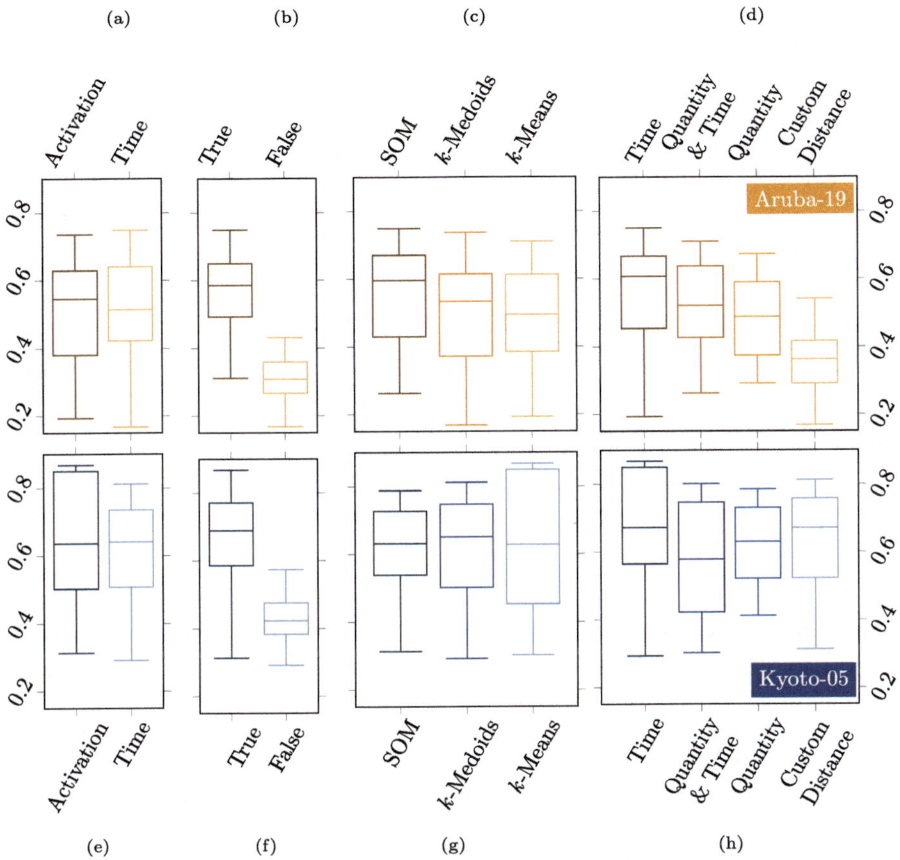

Fig. 4 F_1-score for different parameter values in the hyperparameter optimisation. (**a**) Trace partition. (**b**) Multi room. (**c**) Clustering method. (d) Vectorisation type. (**e**) Trace partition. (**f**) Multi room. (**g**) Clustering method. (**h**) Vectorisation type

The evening routine (Fig. 6a) has sequentially executed activities 6, 0 and 2. Activity 6 involves sensor activations in the kitchen and living room, hinting at the activity combination *cooking* and *eating*. Activity 0 starts with sensors in the living room and proceeds with sensors in the downstairs hallway and the stairs. The routine ends in the upper hallway, representing the activity *going to bedroom*. Activity 2 is shown as a directly-follows graph in Fig. 7 (created with Disco[2], displaying 100% of activities.). The behaviour here indicates that the observed entity is *sleeping* or *resting in bed*. The night routine (Fig. 6c) consists mostly of activities 2, 10 and 0. Activities 2 and 10 primarily contain motion sensors above the bed, indicating that both activities 2 and 10 are *sleeping* (Fig. 7). During the night routine, activity 2 is

[2] https://fluxicon.com/disco/

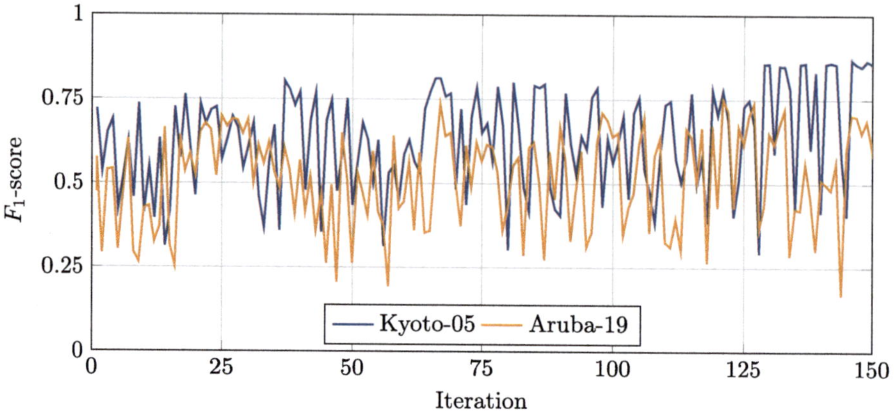

Fig. 5 Function value of the hyperparameter optimisation

Table 2 Results of the best 5 out of 150 iterations

	F_1	Precision	Fitness	Clustering	Trace length	Cluster	Day partition	Partition method
Aruba-19	0.7494	0.6540	0.8926	SOM	28 min	10	5	Time
	0.7380	0.6529	0.8709	SOM	28 min	10	5	Time
	0.7357	0.6358	0.8742	*k*-medoids	45 act.	16	4	Act.
	0.7187	0.6309	0.9216	SOM	28 min	8	2	Time
	0.7157	0.6158	0.8792	SOM	28 min	10	5	Time
Kyoto-05	0.8679	0.8158	0.9279	*k*-means	75 act.	13	4	Time
	0.8656	0.8134	0.9258			13	4	
	0.8599	0.8078	0.9205			13	4	
	0.8587	0.7975	0.9325			13	4	
	0.8579	0.7966	0.9307			13	4	

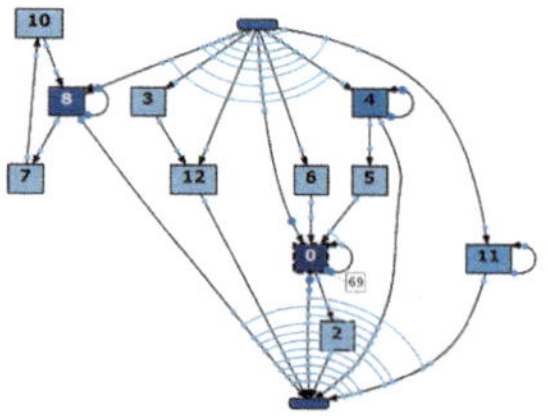

(a)

Morning	06:01 – 12:00
Afternoon	12:01 – 18:00
Evening	18:01 – 00:00
Night	00:01 – 06:00

(b)

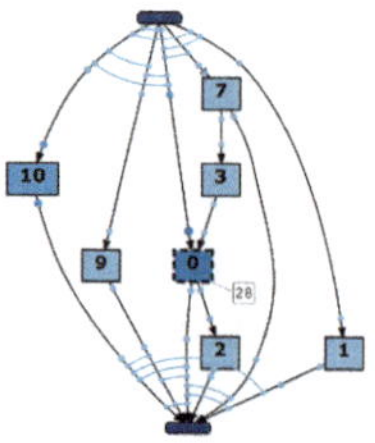

(c)

Fig. 6 Daily routines—Kyoto-05. (**a**) Routine—evening (bindings: 0.10, all-connected heuristic). (**b**) Day partition. (**c**) Routine—night (bindings: 0.0, all-connected heuristic)

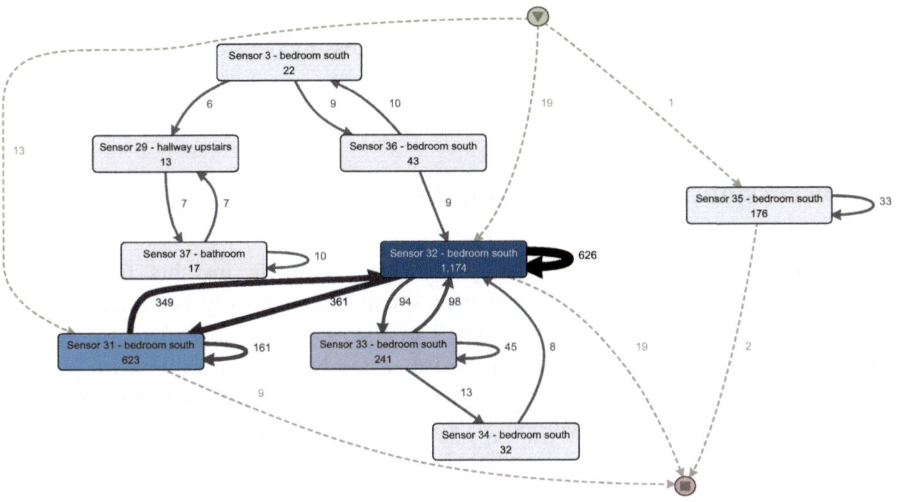

Fig. 7 Activity 2—sleeping (Kyoto-05)

preceded by activity 0, which involves sensors in the hallway and living room, as described in the previous paragraph.

The analysis shows that no perfect parameter combination exists. The parameter choice depends on the data set. This supports our approach of a feedback loop and automatic adjustment of parameters. Both concepts yield better results than a predefined parameter combination. It is important to note that the resulting process models are different from classic process models from business processes. Due to human behaviour, the resulting process models will always be more unstructured than business process models.

5 Related Work

Compared to closely related existing works in this field [3–5, 13–16] our paper is the first that covers the entire stack from sensor event data to process discovery and back. Partial steps of our approach have already been explored in [3, 4]. However, any qualitative feedback related to process model quality (e.g. fitness or recall) and the corresponding feedback loop is missing—several works [5, 14] target process mining from location sensors. Senderovich et al. [5] transform location sensor data into event logs but require domain knowledge from existing process models. In [14], user paths are discovered from indoor location sensors. Activities are discovered through clustering, but the benefits of process discovery, e.g. to identify concurrently executed process activities, still need to be explored. Similarly, the approach of [15] discovers indoor customer paths from sensor event data by applying fuzzy clustering. Process mining is used to increase the understandability of the clustering results. However, no overall process model from sensor event data is discovered.

6 Discussion, Implications and Future Work

Our evaluation results are promising to extract valuable insights from sensor location event data. We showed that user involvement is only necessary for the labelling of the activities at a later stage. The rest of the process of the framework can be automated. Our evaluation showed that combining clustering with process discovery algorithms is promising, and their isolated use would lead to fewer insights. Currently, we are developing a synthetic IoT event log generator allowing us to model multiuser IoT environments [17] with known ground truth allowing us to lift the single-occupancy constraint in the future. Our approach permits seeing activities as part of a process model rather than in isolation and, therefore, in context to other activities. The context allows the domain expert to make an informed decision and increases the accuracy of the labelling process. Although our evaluations rely on sensor location event data, it is possible to include other types of sensor event data in future iterations: our framework can easily integrate data sets with additional sensor information (e.g. temperature, humidity) and further improve the accuracy of discovered process models.

References

1. Janiesch, C., et al.: The internet of things meets business process management: a manifesto. IEEE Syst. Man Cybern. Mag. **6**(4), 34–44 (2020)
2. Janssen, D., Mannhardt, F., Koschmider, A., van Zelst, S.J.: Process model discovery from sensor event data. In: ICPM 2020. LNBIP, vol. 40. Springer, Berlin (2020)
3. Koschmider, A., Moreira, D.S.V.: Change detection in event logs by clustering. In: OTM 2018 Conferences, Proceedings, Part I, pp. 643–660 (2018)
4. Seiger, R., Zerbato,F., Burattin, A., García-Bañuelos, L., Weber, B.: Towards IoT-driven process event log generation for conformance checking in smart factories. In: EDOC Workshops 2020, pp. 20–26. IEEE, Piscataway(2020)
5. Senderovich, A., Rogge-Solti, A., Gal, A., Mendling, J., Mandelbaum, A.: The road from sensor data to process instances via interaction mining. In: Int. Conference on Advanced Information Systems Engineering, pp. 257–273. Springer, Berlin (2016)
6. Weijters, A., Ribeiro, J.: Flexible heuristics miner (FHM). In: CIDM 2011. IEEE, Piscataway (2011). https://doi.org/10.1109/cidm.2011.5949453
7. Berti, A., Zelst, S.J., Aalst, W.: Pm4py: bridging the gap between process-and data science (2019). arXiv preprint arXiv:1905.06169
8. Berti, A., Aalst, W.M.: Reviving token-based replay: Increasing speed while improving diagnostics. In: ATAED@ Petri Nets/ACSD, pp. 87–103 (2019)
9. Cook, D.J., Schmitter-Edgecombe, M.: Assessing the quality of activities in a smart environment. Methods Inform. Med. **48**(5), 480 (2009)

10. Harris, C.R., et al.: Array programming with NumPy. Nature 585, 357–362 (2020). https://doi.org/10.1038/s41586-020-2649-2
11. McKinney: Data Structures for Statistical Computing in Python. In: 9th Python in Science Conference. https://doi.org/10.25080/Majora-92bf1922-00a
12. Mannhardt, F., Leoni, M., Reijers, H.A.: Heuristic mining revamped: an interactive, data-aware, and conformance-aware miner. In: BPM (Demos). CEUR Workshop Proceedings, vol. 1920. CEUR-WS.org (2017)
13. Cameranesi, M., Diamantini, C., Storti, E.: Extraction of user daily behavior from home sensors through process discovery. IEEE Internet Things J. **7**(9), 8440–8450 (2020). https://doi.org/10.1109/JIOT.2020.2990537
14. Dogan, O., Martinez-Millana, A., Rojas, E., Sepúlveda, M., Munoz-Gama, J., ..., Fernandez-Llatas, C.: Individual behavior modeling with sensors using process mining. Electronics **8**(7) (2019). https://doi.org/10.3390/electronics8070766
15. Dogan, O., Öztaysi, B., Fernández-Llatas, C.: Segmentation of indoor customer paths using intuitionistic fuzzy clustering: process mining visualization. J. Intell. Fuzzy Syst. **38**(1), 675–684 (2020)
16. van Eck, M.L., Sidorova, N., van der Aalst, W.M.P.: Enabling process mining on sensor data from smart products. In: 2016 IEEE RCIS, pp. 1–12 (2016). https://doi.org/10.1109/RCIS.2016.7549355
17. Zisgen, Y., Janssen, D., Koschmider, A.: Generating synthetic sensor event logs for process mining. In: IIS - CAiSE Forum 2022, Leuven, Belgium, vol. 452, pp. 130–137. Springer, Berlin (2022). https://doi.org/10.1007/978-3-031-07481-3_15

Process Mining Meets Robotic Systems: Challenges and Opportunities

Flavio Corradini , Sara Pettinari , Barbara Re , Lorenzo Rossi , and Francesco Tiezzi

1 Introduction

A robotic system comprises one or more robots that use sensors, actuators, and management systems to interact with the environment, including humans or other robots, with the final goal of providing intelligent services [27]. Robots serve different sectors like agriculture, manufacturing, and health [13], offering benefits such as reducing manual operations, minimizing errors, and improving productivity and service quality [19]. Therefore, they are increasingly adopted to perform process activities of different domains, becoming integral elements in @systems operations [36]. Data collected by robots can enhance process models with real-world insights. Furthermore, their mission can be seen as a sequence of tasks enabling both robots' actions and inter-robot interactions. Thus, conceptualizing robots' missions as process models provides a high-level view of their behavior enabling the application of business process management techniques, such as process mining [8, 43, 45].

Integrating process mining and robotic systems is a natural consequence of the widely discussed application of process mining in the IoT field [3, 25]. However, unlike IoT, a robotic system exploits *context knowledge* to understand its surroundings and *behavioral abilities* to determine its decision-making processes and actions, such as maximizing efficiency and interacting effectively with humans or other robots to achieve shared goals. Therefore, analyzing robotic systems is

F. Corradini · S. Pettinari · B. Re (✉) · L. Rossi
School of Science and Technology, University of Camerino, Camerino, Italy
e-mail: flavio.corradini@unicam.it; sara.pettinari@unicam.it; barbara.re@unicam.it; lorenzo.rossi@unicam.it

F. Tiezzi
Dipartimento di Statistica, Informatica, Applicazioni, University of Florence, Firenze, Italy
e-mail: francesco.tiezzi@unifi.it

G. Fortino, M. Mecella (eds.), *Internet of Things Meets Business Process Management*, Internet of Things, https://doi.org/10.1007/978-3-031-90746-3_9

crucial for ensuring mission integrity and preventing system failures in dynamic environments [1]. Process mining applied within the robotic domain can enable the analysis of event logs generated during the execution of a robotic system. This application would address the complexity of understanding robots' behavior and interactions while ensuring system execution correctness [39]. Nevertheless, the literature is still in its early stages, lacking robust techniques tailored for this analysis. In this regard, data preparation techniques designed for these systems are required to abstract robotic data and capture the diverse and complex activities involved in their operations. Analysis techniques should be able to manage various behavioral abilities and multi-perspective data describing robots' context knowledge. Therefore, while using process mining for analyzing robotic systems can improve the understanding and optimization of processes, achieving this requires leveraging techniques capable of accurately analyzing the overall behavior.

Building upon current research in process mining and robotic systems, this chapter discusses the challenges and opportunities of using process mining techniques to analyze robotic systems. By outlining robots' characterization and possible directions for process mining techniques, we provide valuable insights for the research community regarding future process mining applications in the increasingly popular robotic domain. It is worth noting that, despite the name similarity, the application of process mining to robotic systems does not fall under the umbrella of robotic process automation, i.e., a collection of tools that operate on digital systems by emulating human activities [55].

The next sections are organized as follows. Section 2 specifies the characteristics of a robotic system both in terms of context knowledge and behavioral abilities. Section 3 presents and analyzes the applicability of process mining techniques to robotic systems. Section 4 focuses on the challenges associated with the event log preparation, process discovery, conformance checking, and process enhancement techniques. Finally, Sect. 5 concludes the work.

2 Robotic Systems Characterization

The execution of a robotic system is significantly shaped by its context knowledge and behavioral abilities. These two components highly influence a robotic mission's successful outcome and performance. This section first introduces a robotic system that is used as an example scenario. Then, it explores the context knowledge impacting the execution of a robotic system and presents the robot's behavioral abilities.

2.1 Reference Scenario in Logistic Domain

In the logistics domain, integrating robotic technologies has become crucial to enhance efficiency and reduce operational costs [47]. This domain comprises all the procedures, methods, and processes involved in moving people, raw materials, and goods along the supply chain and the transport system. By integrating robots that can navigate and carry out tasks independently, autonomy in transportation, like shipping orders, can be achieved. Additionally, robots can work alongside humans in storage and handling systems, such as warehousing operations.

The robotic warehouse scenario described below, depicted in Fig 1, comprises autonomous forklifts collaborating with humans and autonomous arms to automate order management and preparation. In the proposed scenario, human operators collaborate with multiple autonomous forklifts to gather goods for an order, while several autonomous robotic arms are responsible for order preparation. At system startup, a human operator triggers a forklift to begin collecting goods. The forklift is designed to locate the operator and adjust its actions accordingly. If one forklift encounters an issue, such as running out of battery, it can prompt another forklift to take over its tasks. Once a forklift is fully loaded with the required goods for an order, it navigates to one of the robotic arms. Subsequently, the robotic arm unloads

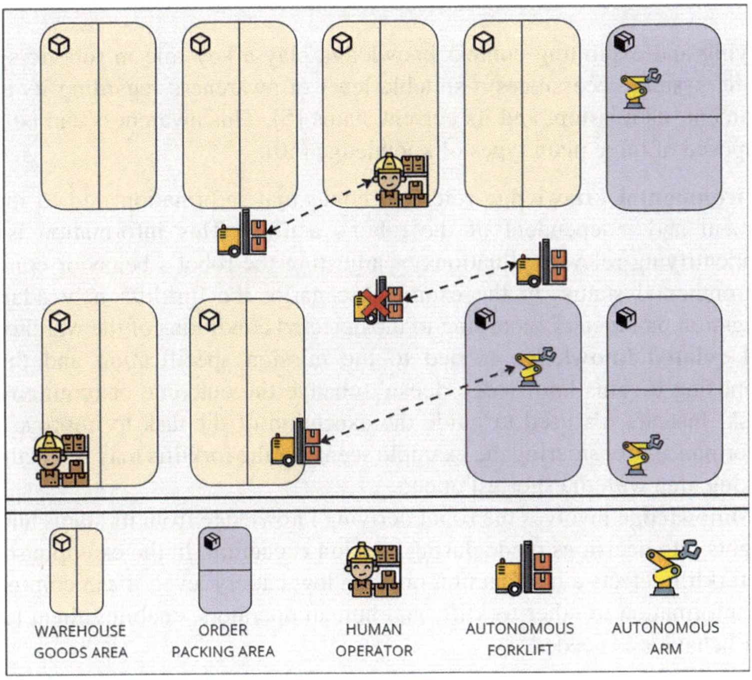

Fig. 1 Warehouse scenario

all the goods from the forklift and arranges them into a single order. The warehouse is divided into distinct areas, each serving a specific purpose. The **warehouse goods area** stores the goods and is dedicated to the collaborative operations of humans and forklifts for retrieving the goods, whereas the **order packaging area** is designated for autonomous arms operations, which are responsible for preparing the orders.

The mission comprises the collaboration of three main components:

- **Human operators** are assigned to a specific order. To fulfill its objective, each operator triggers a forklift and loads it with all the goods required for a specific order.
- **Autonomous forklifts** are triggered by a human operator. A forklift's behavior involves adjusting its behavior based on human actions and communicating with each other and the robotic arms. An autonomous forklift has several sensors for navigation and obstacle avoidance within the warehouse.
- **Autonomous arms** can collaborate with forklifts. Each arm is responsible for waiting for a forklift, unloading goods, and preparing orders. An autonomous arm has several sensors and a gripper mechanism to grasp and release goods properly.

2.2 *Robotic Systems Context Knowledge*

Identifying and exploiting context knowledge play a key role in robotic systems. A robotic system necessitates a suitable level of awareness regarding its external environment, its mission, and its current status [5]. This awareness can be viewed as composed of three main types of knowledge [50]:

- **Environmental knowledge** refers to contextual information tied to the environment and independent of the robot's actions. This information is useful for identifying relevant situations or adjusting the robot's behavior concerning environmental status. In the example scenario, the forklifts may adapt their navigation parameters according to the detected conditions of the warehouse.
- **Task-related knowledge** is tied to the mission specification and the tasks composing it. This knowledge doesn't change the outcome or requirements of a task. Instead, it's used to guide the execution of the task to improve system performance. Considering the example scenario, the forklifts may select the order packing area with the shortest queue.
- **Self-knowledge** involves the robot deriving knowledge from its status and offers insights into decisions made during mission execution. In the example scenario, if a forklift detects a malfunction or has a low battery level, it can communicate this information to other forklifts and human operators, enabling them to adjust their behavior as needed.

Such context knowledge impacting a robotic system execution is an identifiable configuration of features that are meaningful to characterize the world state and useful to influence the decision process of a robotic system [50].

2.3 Robotic Systems Behavioral Abilities

Considering the behavior of a robot, its execution depends on the capability to control several sensors and actuators in real time and on its interaction with the environment [8]. This capability shapes the context data accessible to the robot during its execution. In turn, context data can significantly influence its behavior, allowing it to react to context conditions not considered at the design time [30]. As presented in the *Robotics Multi-Annual Roadmap* [47], a robot can be characterized by several abilities that capture the system performance characteristics and are independent of any particular robot configuration or application domain. Among the identified abilities, four of these should be considered to provide a basis for setting performance metrics on the robot's behavior.

- **Decisional autonomy** refers to the ability of a robot to act autonomously. This autonomy can range from basic responses to sensor inputs to complete self-sufficiency in complex environments. Referring to the example scenario, the forklifts can choose their behavior in real time in response to humans' actions.
- **Interaction** is the ability of a system to interact physically, cognitively, and socially with humans, or other systems around it, including other robots. The ability to interact may be as simple as using a communication protocol, or as advanced as holding an interactive conversation in a social context. Referring to the example scenario, forklifts and autonomous arms can interact to complete the order packing.
- **Dependability** denotes the robot's ability to perform its given tasks without systematic errors. This ability influences the reliability and success of the deployment of robots in real life [7]. In the example scenario, forklifts can navigate autonomously without crashing against walls.
- **Adaptability** is the ability of the system to adapt itself to different work scenarios, environments, and conditions. Adaptation may take place over long or short time scales. It may relate to local control systems or actions, to the whole system, or to interaction. Referring to the example scenario, forklifts can optimize their navigation to optimize energy consumption.

All these abilities may coexist within the same system, or only a subset may be present. Furthermore, the presence and level of each ability can vary depending on the system's specific requirements.

3 Robotic Systems and Process Mining

Considering robots' characteristics, this section presents the complexity of data produced by robotic systems and how to manage them for process mining purposes effectively. Additionally, it discusses the importance of integrating multiple perspectives when preparing and analyzing robotic data. Moreover, the section explores how the behavioral abilities of robots can be correlated with process mining concepts. This analysis aids in comprehending how robotic abilities align with existing objectives and methodologies.

3.1 Robotic Event Data

Unlike traditional processes with standardized event logs, robotic systems lack uniform data structures and comprehensive logging mechanisms, making it challenging to capture the diverse and complex activities involved in robotic operations. The main challenge lies in obtaining robotic datasets in a format suitable for applying process mining techniques. Existing datasets primarily provide information in the form of collected images and sensor readings, with the aim of training machine learning algorithms to optimize the system [42, 58]. While these data are crucial for enhancing low-level robotic capabilities, their formats are not suitable for process mining-based analysis, which relies on high-level event logs capturing sequences of activities and decision points. However, robots typically generate a continuous stream of low-level context data without directly corresponding to meaningful activities. This poses a significant challenge in generating activity-centric robotic event data.

During its operation, a robot generates continuous event data in a *peripheral-centric* way, recording the data produced by sensors and actuators. A robotic peripheral-centric event typically includes the *time* when the event occurs, the name of the *peripheral* that triggered it, and additional attributes storing relevant data. For example, an event triggered by a laser sensor may include distance data as an attribute. These event data represent that the robot has accessed a peripheral functionality and enables the collection of information that shapes contextual knowledge. However, process mining techniques require event logs with an *activity-centric* structure, recording high-level activities performed to achieve objectives in the process model, such as movement actions to reach a destination.

To better clarify the difference between an event log produced by a robot and an event log suitable for process mining techniques, consider as an example part of the mission of a forklift from the reference scenario. The robot has to move around the warehouse (activity *Move*) following a human operator (activity *Detect Human*). The performed activities exploit several sensors and actuators; therefore, the event logs from both peripheral- and activity-centric perspectives are presented and discussed below. Table 1 shows an excerpt of a peripheral-centric event log of

Table 1 Peripheral-centric robotic event log

Time	Peripheral	Other data
09:33:20.147	Engine	{ linear_velocity: 1.5, angular _velocity: 0.0 }
09:33:20.147	Battery	{ level: 100 }
09:33:20.147	Camera	{ width: 960, height: 540, data: [...] }
09:33:20.149	Engine	{ linear_velocity: 0.8, angular _velocity: 0.5 }
09:33:21.239	gps	{ position: {1.0, 2.4, 0.0 }, orientation: {0.4, 1.1, 0.5 } }
09:33:25.550	Battery	{ level: 98 }
09:33:30.448	Battery	{ level: 97 }
09:33:30.941	Camera	{ width: 960, height: 540, data: [...] }
09:33:32.410	gps	{ position: {1.2, 2.8, 0.0 }, orientation: {0.5, 1.0, 1.2 } }
09:33:32.495	Laser	{ distance: 20.5 }
09:33:32.499	Laser	{ distance: 15.8 }
09:33:32:655	Camera	{ width: 960, height: 540, data: [...] }
09:33:33.003	Engine	{ linear_velocity: 1.5, angular _velocity: 0.0 }
09:33:33.110	Laser	{ distance: 31.7 }
09:33:33.118	Laser	{ distance: 30.6.5 }
09:33:33.120	Laser	{ distance: 29.4 }
09:33:33.220	gps	{ position: {3.2, 1.4, 0.0 }, orientation: {0.2, 1.4, 0.2 } }
09:33:33:240	Camera	{ width: 960, height: 540, data: [...] }
09:33:33.955	Laser	{ distance: 27.3 }
09:33:34.920	gps	{ position: {4.1, 1.8, 0.0 }, orientation: {0.3, 1.4, 0.2 } }
09:33:35.512	Battery	{ level: 92 }
...	...	...

one forklift execution, whereas Table 2 reports the same event log abstracted to an activity-centric point of view and related to multiple perspectives of interest. The peripheral-centric log describes the working flow of the robot peripherals, e.g., the first two rows of Table 1 indicate that the robot has activated the engines and has checked for its current battery status. On the other hand, the activity-centric log offers a high-level view of the robot's execution. For instance, the initial rows of Table 2 indicate that the robot has simultaneously initiated both the *Move* and *Detect Human* activities, specifying the corresponding spatial position and battery level.

Therefore, a robotic activity-centric event log should be produced to advance the research and application of process mining techniques and investigate their potentialities in this domain. Activity abstraction techniques, such as [34, 48, 57], could identify activities from low-level data. However, it is worth noting that in robotic systems, a low-level event may have been used to perform different activities; for instance, a forklift can exploit a camera sensor to move through the warehouse and detect the human operator. A different approach could be labelling the robotic code before its execution, like in [8, 26]. This approach enables a robotic expert to define the structure and the level of granularity of the event log before executing the system. Moreover, context knowledge should be incorporated into

Table 2 Activity-centric robotic event log

Time	Activity	Life cycle	Robot	Position	Battery
09:33:20.147	Move	Start	forklift_1	{position:{1.0, 2.4, 0.0}}	{level: 100}
09:33:20.147	Detect Human	Start	forklift_1	{position:{1.0, 2.4, 0.0}}	{level: 100}
09:33:33.240	Detect Human	Complete	forklift_1	{position:{3.2, 1.4, 0.0}}	{level: 97}
09:40:56.021	Move	Complete	forklift_1	{position:{4.1, 1.8, 0.0}}	{level: 92}
...	...	...	...	...	...

the activity-centric robotic event log, thus enabling a comprehensive analysis from multiple system perspectives [17, 32]. Several works have been proposed to address the combination of continuous data and event logs [8, 21, 31, 54]. Indeed, context data can be directly incorporated into the event log or stored in additional data models, which correlate this data with the events.

3.2 *Robotic Behavior*

The context variability of robotic systems, coupled with the diverse abilities that characterize them, generates event logs that can be utilized for various purposes. Indeed, by taking into account robots' behavioral abilities, the application of process mining techniques enables the discovery of whether a system has specific abilities, whether the robot has adhered to or deviated from these abilities, and how these abilities influence the system's execution.

Decisional Autonomy in Process Mining Based on contextual knowledge, a robot can autonomously choose how to perform its activities. This may create an event log where the robot records some infrequent behaviors.

Usually, in the process mining domain, the infrequent behaviors are usually handled as noise, i.e., discarded during the analysis [33]. However, especially in a robotic system, an infrequent behavior may reveal important insights into the process that can be derived from context variability. Indeed, deviations from the mission may be caused by decisions taken to deal with the context. Data-aware process mining techniques [11, 12, 33, 44] integrate the data perspective during the control-flow discovering, distinguishing infrequent activities from noise. These techniques allow for identifying process models where decision points correspond to data conditions. A robotic system would allow the identification of a choice made by the robot, such as performing a series of activities when it detects its battery level below a certain threshold.

Interaction in Process Mining The analysis of interacting processes poses additional challenges to be considered in the application of process mining. This is especially true in robotic systems where the interactions can be both one-to-one and many-to-many. Usually, no centralization point manages the delivery of a message [9]. Therefore produced event logs come from multiple sources and need to be correlated to analyze the behavior of the overall robotic system. This involves distinguishing interactions from the control flow to gain insights into each robot's behavior and the system functions as a whole.

Over the years, process mining researchers have partially considered these scenarios proposing approaches and techniques dealing with multiple event logs coming from different sources and containing message information [59]. These approaches fall within the umbrella of cross-organizational process mining [51], which focuses on processes executed across organizations and their respective units. The existing techniques mostly focus on privacy concerns [16, 52] rather than on issues concerning the overall behavior of interacting systems. Specifically, only a few works analyze the involved participants' behavior and interactions [9, 49, 60]. In a robotic system, employing these types of process mining techniques would facilitate the analysis of the impact of communication within the system. It would also enable the control of safe interactions between humans and robots.

Dependability in Process Mining The dependability ability ensures that a robotic system can carry out its tasks without encountering systematic errors. Therefore, the sequences of activities recorded in the log should be used to recognize goal achievement or identify potential anomalies.

Different process mining techniques can be used to ensure or recognize the dependability characteristic. Considering the analysis of system goals, goal-oriented process mining [10, 20, 23] integrates event logs with goal models to assess whether the executed behaviors align with predefined goals. This approach allows for the design of task objectives, providing insights into new requirements and potential process misalignment, whereas goal recognition techniques [41] aim at understanding the goals or intentions the system will accomplish. Alternative approaches for analyzing the dependability ability of a robotic system involve investigating system errors. Cancellation behavior and process deviation techniques [4, 28, 29] facilitate the identification and analysis of behaviors that deviate in an anomalous manner. Therefore, analyzing the dependability ability, by ensuring goal achievement or promptly detecting anomalies, would guarantee the reliability of a robotic system.

Adaptability in Process Mining The adaptability of a robot allows it to modify its activities over time. Consequently, a robot that can adapt produces event logs with variable sequences of activities that are context-dependent and rarely are repeated in the same way.

Various techniques are available in the process mining domain to extract process models and insights from event logs generated by adaptive systems. This makes it possible to analyze when and why changes occur [22]. Adaptive systems perform ad hoc changes, such as adding, removing, or relocating a process step, to adapt to changing contexts. These changes can be recorded in a change log, thus enabling the event log and the change log to be analyzed separately or together [22]. Whenever this additional log is not available, the identification of adaptive behaviors can be detected by finding the change points and the nature of changes given from the event log. The basic idea is to identify the drift in an event log [46] by examining successive sets of data values (of a specific size) to determine if there's a significant difference between the two sets [6, 14, 24, 35]. Analyzing robots' adaptability would facilitate the identification of behavioral changes over time and allow for comparing the impact of these changes relative to the initial behavior.

4 Challenges

When approaching a new domain, it is crucial to consider the three main classes of techniques within the process mining domain: process discovery, conformance checking, and process enhancement [53]. However, the potential of these techniques cannot be exploited without a suitable event log [15]. This section presents challenges that need to be addressed to foster the application of process mining techniques in analyzing robotic systems and to provide a basis for proposing or improving techniques tailored to robotic system analysis. The presented challenges arise from use cases and from drawbacks presented in the literature with current approaches for analyzing robotic systems, such as formal verification [30] and quantitative analysis [1].

C1. Specifying and Preparing a Structured Robotic Event Log
To produce a suitable event log, a data preparation phase is necessary. Notably, depending on the analysis objectives, robotic data can be processed for the application of traditional process mining or object-centric process mining techniques. Moreover, data stored in the event log and the resulting structure significantly influence the applicable techniques and the analysis outcomes.

Event logging is the first step to capturing robotic event logs. Robots' low-level data necessitates abstraction to extract meaningful activities and, subsequently, preprocessing to generate an activity-centric event log enriched with multiple perspectives related to robots' context knowledge. The analysis goals should guide the data preparation phase. Ideally, data collection and preparation should be planned before executing the system to obtain structured event logs with the necessary information and at the desired level of granularity. Otherwise, when analyzing already existing robotic data, incorporating domain knowledge into data preparation is essential to produce an event log useful for the analysis phase.

In addition to discrete events associated with the execution of robotic activities, context data should also be integrated into the robotic event log since it can be changed by a robotic activity or influence its behavior. Therefore, it is essential to keep track of all data collected between the start and completion of an activity to provide a comprehensive representation of the robot's environment and state throughout its operations.

C2. Improving Discovery Algorithms for Robotic Event Logs
The process discovery step aims at automatically deriving a process model describing the behavior observed in the event logs. Given the varying combinations of

robots' behavioral abilities, a comprehensive examination is necessary to determine whether existing techniques can be effectively applied to robotic systems.

The reconstruction of a robotic system execution via process models requires determining which discovery algorithms are best suited for analyzing the peculiarities of robotic event logs. Firstly, it is essential to determine which modeling notations can capture the behavioral abilities characterizing the robotic system. This includes selecting a notation capable of representing the output of control-flow discovery algorithms while considering the abilities characterizing the system. Notably, a robotic system is usually characterized by several behavioral abilities that should be discovered and analyzed altogether in the process model. Secondly, existing discovery algorithms must be tested to assess their suitability for handling the information recorded in the robotic event log. This may involve developing specialized algorithms or adapting existing techniques to accommodate the context knowledge collected by robotic systems.

C3. Improving Conformance Techniques to Deal with Robotic Event Logs
Conformance checking collects approaches for verifying how much the behavior of a process recorded in the event log differs from the expected behavior provided in the form of a process model. This enables a comparison between the intended behavior as depicted by the model and the real behavior captured in the event log. Given the autonomous nature of robotic systems, conformance checking techniques can facilitate evaluating whether the planned mission aligns with the executed one.

Directly related to the process discovery challenge, also conformance techniques may need to be adapted to the robotic domain. Existing techniques focused on the control-flow perspective can be exploited to assess the conformance between prescribed and actual robots' missions. Nevertheless, considering the impact of behavioral abilities on the produced process model, these techniques should be adapted to deal with the process model depicting the behavioral abilities of the robotic system. Even though behavioral abilities influence the control flow of a robotic system, context data directly affects system execution and should therefore be considered to evaluate the quality of a process discovery algorithm.

Focusing on the fitness and precision quality dimensions, context data must be incorporated into their calculations. In this regard, the robotic domain provides a foundational basis. Specifically, metrics related to robotic behavior aim to evaluate the system's effectiveness or efficiency in accomplishing its goals. These metrics may involve evaluating both the actions performed by the robot and how they are executed, taking into account the context influencing its execution. Alternatively, it may involve determining whether a task has been completed or not [38]. Hence, to compute context-based fitness and precision, conformance checking techniques should integrate or adapt formulas originating from the robotic domain. Moreover, since context data can influence robots' execution in various ways, analysts should have the option to select the degree of context data impact on the calculation of fitness and precision.

C4. Integrating New Data Visualizations Representing Context Data
The process enhancement step involves modifying or extending an existing process model based on the system mission depicted in the model and the event logs. Considering the context data generated during the execution of a robotic system, it is crucial to understand how to effectively integrate this data into the process model or represent it appropriately to provide additional insights.

Context data plays a key role in enhancing the process model. Multi-perspective techniques can be employed to enhance the process model by integrating multiple perspectives related to the control flow. The insights provided by this data can be used to directly show in the process model the impact of the context on the control flow. However, to prevent the process model from being overloaded with context data and to provide a clearer understanding of system analysis, it is possible to integrate additional visualizations alongside process models. The integration of process mining and visual analytics is gaining momentum [37]. Visual analytics offers interactive visualizations to understand, reason, and make decisions based on large and complex datasets. Combining it with process mining enables analysis based on process representation, supplemented by diagrams visualizing context data [2, 40, 56]. In the robotic domain, this integration would facilitate effective representation and analysis of robotic systems by offering a behavioral depiction and analysis of the system's control flow, enriched with new visualizations. This would require the identification of visualizations suitable to represent the dynamic context impacting robots' missions.

5 Conclusion

This chapter presented challenges and opportunities for integrating process mining techniques in robotic systems. The presented challenges range from data preparation to process enhancement and arise from use cases and the current state of the art concerning approaches for analyzing robotic systems. Following a discussion of each challenge, we have presented an overview of potential research opportunities. By addressing the challenges and exploring new research directions, robotic systems can be empowered by techniques belonging to the process mining discipline.

Notably, this chapter focused on the main classes of process mining techniques. However, the application of other techniques holds potential relevance in robotic systems. For instance, predictive process monitoring techniques [18] could enable the anticipation of future events or outcomes, thus presenting opportunities for enhancing robotic behavioral abilities.

References

1. Afzal, A., Le Goues, C., Hilton, M., Timperley, C.S.: A study on challenges of testing robotic systems. In: Software Testing, Validation and Verification, pp. 96–107. IEEE, Piscataway (2020)
2. Alman, A., Arleo, A., Beerepoot, I., Burattin, A., Di Ciccio, C., Resinas, M.: Tiramisù: making sense of multi-faceted process information through time and space (2024)
3. Bertrand, Y., Weerdt, J.D., Serral, E.: A bridging model for process mining and IoT. In: Process Mining Workshops. LNBIP, vol. 433, pp. 98–110. Springer, Berlin (2021)
4. Bezerra, F., Wainer, J., van der Aalst, W.M.: Anomaly detection using process mining. In: International Workshop on Business Process Modeling, Development and Support, pp. 149–161. Springer, Berlin (2009)
5. Bloisi, D.D., Nardi, D., Riccio, F., Trapani, F.: Context in robotics and information fusion. In: Context-Enhanced Information Fusion: Boosting Real-World Performance with Domain Knowledge, pp. 675–699 (2016)
6. Bose, R.J.C., van der Aalst, W.M., Žliobaitė, I., Pechenizkiy, M.: Handling concept drift in process mining. In: Advanced Information Systems Engineering, pp. 391–405. Springer, Berlin (2011)
7. Cavallo, F., Limosani, R., Fiorini, L., Esposito, R., Furferi, R., Governi, L., Carfagni, M.: Design impact of acceptability and dependability in assisted living robotic applications. Int. J. Interact. Design Manuf. **12**, 1167–1178 (2018)
8. Corradini, F., Pettinari, S., Re, B., Rossi, L., Tiezzi, F.: A methodology for the analysis of robotic systems via process mining. In: Enterprise Design, Operations, and Computing. LNCS, vol. 14367, pp. 117–133. Springer, Berlin (2023)
9. Corradini, F., Pettinari, S., Re, B., Rossi, L., Tiezzi, F.: A technique for discovering BPMN collaboration diagrams. Softw. Syst. Model. **23**, 1323–1343 (2024)
10. Dabrowski, J., Kifetew, F.M., Muñante, D., Letier, E., Siena, A., Susi, A.: Discovering requirements through goal-driven process mining. In: Requirements Engineering Conference Workshops, pp. 199–203. IEEE, Piscataway (2017)
11. De Leoni, M., van der Aalst, W.M.: Data-aware process mining: discovering decisions in processes using alignments. In: ACM Symposium on Applied Computing, pp. 1454–1461 (2013)
12. De Smedt, J., Hasić, F., vanden Broucke, S.K., Vanthienen, J.: Towards a holistic discovery of decisions in process-aware information systems. In: Business Process Management, pp. 183–199. LNCS, Springer, Berlin (2017)
13. de Silva, C.W.: Some issues and applications of multi-robot cooperation. In: Computer Supported Cooperative Work in Design, pp. 2–2. IEEE, Piscataway (2016)
14. de Sousa, R.G., Peres, S.M., Fantinato, M., Reijers, H.A.: Concept drift detection and localization in process mining: An integrated and efficient approach enabled by trace clustering. In: ACM symposium on applied computing, pp. 364–373 (2021)
15. De Weerdt, J., Wynn, M.T.: Foundations of process event data. In: Process Mining Handbook, pp. 193–211. LNBIP, Springer, Berlin (2022)
16. Elkoumy, G., Fahrenkrog-Petersen, S., Dumas, M., Laud, P., Pankova, A., Weildich, M.: Secure multi-party computation for inter-organizational process mining. In: Enterprise, Business-Process and Information Systems Modeling, vol. 387, pp. 166–181. Springer, Berlin (2020)
17. Fahland, D.: Process mining over multiple behavioral dimensions with event knowledge graphs. In: Process Mining Handbook, LNBIP, vol. 448, pp. 274–319. Springer, Berlin (2022)
18. Francescomarino, C.D., Ghidini, C.: Predictive process monitoring. In: Process Mining Handbook, LNBIP, vol. 448, pp. 320–346. Springer, Berlin (2022)
19. Gautam, A., Mohan, S.: A review of research in multi-robot systems. In: Industrial and Information Systems, pp. 1–5. IEEE, Piscataway (2012)
20. Ghasemi, M., Amyot, D.: From event logs to goals: a systematic literature review of goal-oriented process mining. Require. Eng. **25**(1), 67–93 (2020)

21. Goossens, A., De Smedt, J., Vanthienen, J., van der Aalst, W.M.: Enhancing data-awareness of object-centric event logs. In: Process Mining Workshops, pp. 18–30. Springer, Berlin (2022)
22. Gunther, C.W., Rinderle-Ma, S., Reichert, M., van der Aalst, W.M., Recker, J.: Using process mining to learn from process changes in evolutionary systems. Int. J. Bus. Process Integration Manag. **3**(1), 61–78 (2008)
23. Gurgen Erdogan, T., Tarhan, A.: A goal-driven evaluation method based on process mining for healthcare processes. Appl. Sci. **8**(6), 894 (2018)
24. Hassani, M.: Concept drift detection of event streams using an adaptive window. In: Conference on Modelling and Simulation, pp. 230–239 (2019)
25. Janiesch, C., Koschmider, A., Mecella, M., Weber, B., Burattin, A., Di Ciccio, C., Fortino, G., Gal, A., Kannengiesser, U., Leotta, F., et al.: The internet of things meets business process management: a manifesto. IEEE Syst. Man Cybern. Mag. **6**(4), 34–44 (2020)
26. Karnan, H., Nair, A., Xiao, X., Warnell, G., Pirk, S., Toshev, A., Hart, J., Biswas, J., Stone, P.: Socially compliant navigation dataset (scand): a large-scale dataset of demonstrations for social navigation. Robot. Autom. Lett. **7**(4), 11807–11814 (2022)
27. Kortenkamp, D., Simmons, R., Brugali, D.: Robotic systems architectures and programming. In: Handbook of Robotics, pp. 283–306 (2016)
28. Leemans, M., van der Aalst, W.M.: Modeling and discovering cancelation behavior. In: On the Move to Meaningful Internet Systems, pp. 93–113. Springer, Berlin (2017)
29. Leemans, S.J., Fahland, D., van der Aalst, W.M.: Exploring processes and deviations. In: Business Process Management Workshops, pp. 304–316. Springer, Berlin (2015)
30. Luckcuck, M., Farrell, M., Dennis, L.A., Dixon, C., Fisher, M.: Formal specification and verification of autonomous robotic systems: a survey. Comput. Surv. **52**(5), 1–41 (2019)
31. Mangler, J., Grüger, J., Malburg, L., Ehrendorfer, M., Bertrand, Y., Benzin, J.V., Rinderle-Ma, S., Serral Asensio, E., Bergmann, R.: Datastream XES extension: embedding IoT sensor data into extensible event stream logs. Future Internet **15**(3), 109 (2023)
32. Mannhardt, F.: Multi-perspective process mining (2018)
33. Mannhardt, F., De Leoni, M., Reijers, H.A., van der Aalst, W.M.: Data-driven process discovery-revealing conditional infrequent behavior from event logs. In: Advanced Information Systems Engineering, pp. 545–560. Springer, Berlin (2017)
34. Mannhardt, F., De Leoni, M., Reijers, H.A., Van Der Aalst, W.M., Toussaint, P.J.: From low-level events to activities-a pattern-based approach. In: Business Process Management, pp. 125–141. LNCS, Springer, Berlin (2016)
35. Martjushev, J., Bose, R.J.C., Van Der Aalst, W.M.: Change point detection and dealing with gradual and multi-order dynamics in process mining. In: Perspectives in Business Informatics Research, pp. 161–178. Springer, Berlin (2015)
36. Meyer, S., Ruppen, A., Magerkurth, C.: Internet of things-aware process modeling: integrating iot devices as business process resources. In: Advanced Information Systems Engineering, pp. 84–98. Springer, Berlin (2013)
37. Miksch, S.: Visual analytics meets process mining: challenges and opportunities. In: Process Mining, p. xiv. IEEE, Piscataway (2021)
38. Nelson, A.L., Barlow, G.J., Doitsidis, L.: Fitness functions in evolutionary robotics: a survey and analysis. Robot. Autonom. Syst. **57**(4), 345–370 (2009)
39. Nicoleta, T.C.: Process mining on a robotic mechanism. In: Software Testing, Verification and Validation Workshops, pp. 205–212. IEEE, Piscataway (2021)
40. Pini, A., Brown, R., Wynn, M.T.: Process visualization techniques for multi-perspective process comparisons. In: Asia Pacific Business Process Management. LNBIP, vol. 219, pp. 183–197. Springer, Berlin (2015)
41. Polyvyanyy, A., Su, Z., Lipovetzky, N., Sardina, S.: Goal recognition using off-the-shelf process mining techniques. In: Autonomous Agents and Multiagent Systems, pp. 1072–1080 (2020)
42. Pomerleau, F., Liu, M., Colas, F., Siegwart, R.: Challenging data sets for point cloud registration algorithms. Int. J. Robot. Res. **31**(14), 1705–1711 (2012)

43. Roldán, J.J., Olivares-Méndez, M.A., del Cerro, J., Barrientos, A.: Analyzing and improving multi-robot missions by using process mining. Autonom. Robots **42**(6), 1187–1205 (2018)
44. Rozinat, A., van der Aalst, W.M.: Decision mining in prom. In: Business Process Management, pp. 420–425. LNCS, Springer, Berlin (2006)
45. Rozinat, A., Zickler, S., Veloso, M., van der Aalst, W.M., McMillen, C.: Analyzing multi-agent activity logs using process mining techniques. In: Distributed Autonomous Robotic Systems 8, pp. 251–260. Springer, Berlin (2008)
46. Sato, D.M.V., De Freitas, S.C., Barddal, J.P., Scalabrin, E.E.: A survey on concept drift in process mining. Comput. Surv. **54**(9), 1–38 (2021)
47. SPARC: Robotics 2020—multi-annual roadmap (2017)
48. Tax, N., Sidorova, N., Haakma, R., van der Aalst, W.M.: Event abstraction for process mining using supervised learning techniques. In: SAI Intelligent Systems, pp. 251–269. Springer, Berlin (2018)
49. Tour, A., Polyvyanyy, A., Kalenkova, A., Senderovich, A.: Agent miner: an algorithm for discovering agent systems from event data. In: Business Process Management, pp. 284–302. LNCS, Springer, Berlin (2023)
50. Turner, R.M.: Context-mediated behavior for intelligent agents. Int. J. Hum.-Comput. Stud. **48**(3), 307–330 (1998)
51. van der Aalst, W.M.: Intra- and inter-organizational process mining: discovering processes within and between organizations. In: The Practice of Enterprise Modeling. LNBIP, vol. 92, pp. 1–11. Springer, Berlin (2011)
52. van der Aalst, W.M.: Federated process mining: exploiting event data across organizational boundaries. In: Smart Data Services, pp. 1–7. IEEE, Piscataway (2021)
53. van der Aalst, W.M.: Process mining: a 360 degree overview. In: Process Mining Handbook, pp. 3–34. LNBIP, Springer, Berlin (2022)
54. van der Aalst, W.M.: Experiences from the internet-of-production: using "data-models-in-the-middle" to fight complexity and facilitate reuse. In: Business Process Management Workshops, pp. 87–91. Springer, Berlin (2023)
55. van der Aalst, W.M., Bichler, M., Heinzl, A.: Robotic process automation. Bus. Inform. Syst. Eng. **60**(4), 269–272 (2018)
56. van der Aalst, W.M., de Leoni, M., ter Hofstede, A.H.: Process mining and visual analytics: Breathing life into business process models. BPM Center Rep. **17**, 699–730 (2011)
57. Van Eck, M.L., Sidorova, N., Van der Aalst, W.M.: Enabling process mining on sensor data from smart products. In: Research Challenges in Information Science, pp. 1–12. IEEE, Piscataway (2016)
58. Vayugundla, M., Steidle, F., Smisek, M., Schuster, M.J., Bussmann, K., Wedler, A.: Datasets of long range navigation experiments in a moon analogue environment on mount etna. In: International Symposium on Robotics, pp. 1–7. VDE (2018)
59. Vom Brocke, J., Rosemann, M.: Handbook on Business Process Management. Springer, Berlin (2014)
60. Zeng, Q., Sun, S.X., Duan, H., Liu, C., Wang, H.: Cross-organizational collaborative workflow mining from a multi-source log. Decis. Support Syst. **54**, 1280–1301 (2013)

NICE: The Native IoT-Centric Event Log Format for Process Mining

Yannis Bertrand, Silvestro Veneruso, Francesco Leotta, Massimo Mecella, and Estefania Serral

1 Introduction

As the utilisation of Internet of Things (IoT) devices in support of business processes (BPs) becomes more frequent, there is a growing recognition of the potential to leverage the data collected by these devices for process mining (PM). Most current PM methods that can incorporate IoT data follow a similar approach: the IoT data are preprocessed with event abstraction and event-case correlation techniques to be translated into a process event log in eXtensible Event Stream (XES) format (see [1–4]).

This method integrates IoT data into PM and allows for the application of existing control-flow and data-aware techniques. However, this method does not fully exploit the potential of IoT data. Often, the resulting high-level event log lacks contextual information (i.e. properties that can influence process execution, as explained in [1, 5]) that could be derived from the IoT data, or it has limited capability to incorporate such context information. Furthermore, by separating the abstraction phase from the analysis phase, the true potential of advanced algorithms to optimise both abstraction and model discovery together cannot be harnessed. For example, the development of an IoT-enhanced decision mining algorithm requires direct access to lower-level IoT data to learn the most relevant features directly from the source data, instead of relying on a delicate event abstraction step, which might leave important information behind, at a lower granularity level [6].

Y. Bertrand (✉) · E. Serral
Research Centre for Information Systems Engineering (LIRIS), KU Leuven, Brussels, Belgium
e-mail: yannis.bertrand@kuleuven.be; estefania.serralasensio@kuleuven.be

S. Veneruso · F. Leotta · M. Mecella
Dipartimento di Ingegneria Informatica, Automatica e Gestionale, Sapienza Università di Roma, Rome, Italy
e-mail: veneruso@diag.uniroma1.it; leotta@diag.uniroma1.it; mecella@diag.uniroma1.it

G. Fortino, M. Mecella (eds.), *Internet of Things Meets Business Process Management*, Internet of Things, https://doi.org/10.1007/978-3-031-90746-3_10

This shortcoming of existing approaches is to a large extent due to limitations of the most common event log standards, i.e. the XES [7] and the Object-Centric Event Log [8] (OCEL), and has been acknowledged in the IoT PM literature [9] and beyond [10]. In this paper, we discuss four challenges in IoT and process event data integration and list ten requirements for the storage of IoT-enhanced event logs and show that both XES and OCEL fail to meet more than half of these requirements. Building upon this observation, we present the Native IoT-Centric Event (NICE) log format, based on the conceptual model described in [1], which meets our requirements [6]. We then describe two event logs created following this format and show how it greatly facilitates more in-depth analyses of IoT-enhanced BPs.

The paper is organised as follows. Section 2 elaborates on the need for dedicated IoT-enhanced format. Section 3 introduces the previous research results that drove the development of the proposed event log format. Based on this, Sect. 4 describes four key challenges for IoT-enhanced logs, which are translated in ten concrete requirements in Sect. 6 [6]. In Sect. 5, we present the methodology we followed to create a new format, based on the requirements. Then, Sect. 7 describes the proposed format and shows that it meets the requirements. Section 8 provides a second validation of the model by applying it to generate logs for case studies in manufacturing and smart spaces. Section 9 discusses practical issues in NICE log creation, while Sect. 10 introduces relevant related works and compares them to the proposed format. Finally, Sect. 11 concludes the paper and outlines future works.

2 Motivation

To better understand the potential of IoT-enhanced PM, Table 1 provides a high-level overview of methods and techniques in the PM field. The table is structured vertically along the three main process mining task types. In the horizontal dimension, the evolution from a classical control-flow perspective to data-aware and IoT-enhanced approaches is depicted. The second and third column cover the lion's share of process mining research so far. However, an important gap is still present in terms of developing IoT-enhanced process mining techniques. The most prominent existing stream of research relates to activity mining, i.e. approaches to derive higher-level process events from low-level IoT data [3, 11–15]. Nonetheless, as illustrated by an only evocative set of so far unaddressed research opportunities, denoted in italics in Table 1, the widespread and grounded development of IoT-enhanced process mining techniques, covering all process mining tasks and possible applications, is far from realised yet. This is mainly due to the fact that the very source of the opportunities of IoT data is their fine granularity, which enables a deeper understanding of the process and the detection of finer patterns. As illustrated by a non-exhaustive set of examples in the table, there exist many potential new techniques that could be developed to take advantage of IoT data. For example:

Table 1 Positioning of IoT-enhanced techniques within the PM field, extending control-flow and data-aware approaches for all three main PM task types

		Control-flow	Data-aware	IoT-enhanced
90PM task types	Discovery	• Control-flow discovery [16–18]	• Data-aware process discovery [19, 20]	• Activity mining [3, 11–15, 21, 22]
		• Trace clustering [23, 24]	• Multi-perspective trace clustering [25]	• IoT context-aware trace clustering [26]
			• Multi-perspective process discovery [27]	• IoT-enhanced data-aware process discovery [28]
	Conformance checking	• Control-flow conformance checking [29]	• Data-aware conformance checking [30]	• IoT-enhanced deviation detection [31] • *IoT-driven root cause analysis*
	Extension		• Decision mining [32]	• IoT-enhanced decision mining [33]
			• Resource mining [34]	• *IoT-enhanced resource mining*
			• Predictive process monitoring [35]	• *IoT-enhanced predictive process monitoring*
			• Performance analysis [36]	• *Real-time predictive process monitoring*

- *IoT-driven root cause analysis*: delve into IoT measurements to discover the root cause of process deviations.
- *IoT-enhanced resource mining*: use IoT data (e.g. location, tags) to mine the fine-grained behaviour of resources in a process.
- *IoT-enhanced predictive process monitoring*: use IoT sensor data measurements to train predictive process monitoring models and facilitate real-time predictive process monitoring.

A common denominator of these envisioned IoT-enhanced PM techniques is the fact that simply abstracting fine-grained IoT data into a classical XES or OCEL event log considerably limits analysis possibilities. Accordingly, this paper proposes a new standard adapted to IoT-enhanced event logs, which is intended to give enough flexibility to develop new techniques taking advantage of the possibilities of IoT data.

3 Background

The definition of an event log suitable for IoT builds upon research and standardisation efforts carried on independently in the separated fields of BPM (especially PM) and IoT. In this section, we provide the background that influenced the proposed model.

3.1 Existing Standards for Event Logs

XES [7], the current standard event log model, is an XML-based model that mainly consists of the notions of event, case and log. It proposes standard attribute types to contextualise the events, e.g. the resource executing an activity and the cost of an activity. A standard activity life cycle is defined together with XES, based on which the status of an activity can be mapped with events relating to this activity. XES also allows the definition of new data attribute types through extensions.

Recently, the gain in maturity of the PM field has increased the urge to create alternative models. Multiple propositions that relax some assumptions of XES and allow for more flexibility in event data storage have been presented, e.g. in [8, 37, 38]. Among them, the OCEL [8] was designed to be more suitable for storing event logs extracted from relational databases and is widely considered as the main challenger of XES today. It introduces the concept of object, which generalises the notion of case by allowing one event to be linked with multiple objects instead of a single case. This removes the necessity to 'flatten' the event log by picking one case notion from the several potential case notions that often coexist in real-life processes.

3.2 Process Mining Using IoT Data

Challenges related to the application of PM to IoT data, and in particular to how event logs are represented, are reported in [6, 39]. The vast majority of the PM literature involving IoT data has focused on mining high-level events of the process from low-level IoT data to create XES event logs. Various frameworks to extract high-level logs from IoT data have been presented, e.g. [2, 3, 12, 14, 15, 40–43]. Traditional PM techniques can then be applied to these event logs to, e.g. discover control-flow models of the processes. A recent survey of process discovery to smart spaces, which represents a large chunk of the literature of PM applied to IoT, is provided in [44].

Although most of the existing literature is in IoT event abstraction, some other possible techniques have also been investigated. Banham et al. [45] propose to perform data-aware process discovery with IoT-based attributes. The framework proposed requires abstracting the IoT data to integrate them in an XES event log. A second work is proposed by Rodriguez-Fernandez et al. [31], who present an approach for IoT-enhanced deviation detection in the time series data directly (in a so-called time-series log). In a manufacturing context, [33] outlines an approach to derive decision rule patterns from TS sensor data by automatically featurising the sensor data and training a decision tree to learn the rules. Finally, a technique for multi-perspective trace clustering using IoT time series data was proposed in [26]. Remark that all these papers bumped into the limitations of traditional event logs and had to abstract the data first or to use the raw sensor data.

Another important research direction is represented by the automated segmentation of logs. Here, statistics-based techniques have been proposed [46], as well as techniques based on the structure of possibly mined processes [47].

3.3 *Existing Models Bridging IoT and Process Mining*

A first model for bridging IoT and PM was presented in [1] (reproduced in Fig. 1[1]). This model integrates IoT ontologies (like SOSA [48]) with typical PM concepts around the central notion of event, which is decomposed in three types of events:

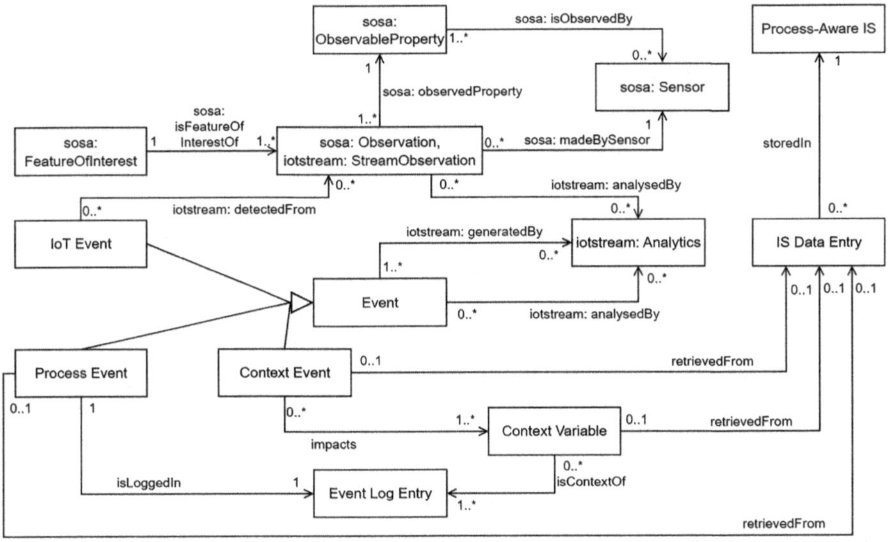

Fig. 1 Core of the bridging model for IoT and PM (taken from [1])

[1] In Figs. 1, 3, 4, arrowheads are used loosely to denote the direction of the associations as defined in the original ontologies, and not to represent the navigability between classes like in traditional UML.

(1) IoT events, which correspond to events as understood in IoT systems; (2) process events, which are events as understood in PM; and (3) context events, which correspond to changes in a parameter of the context of the process. For more information, see Sect. 7.1 or [1].

Next to this, a second conceptual model was described in [49]. This model, focusing on scenarios where the link between the process and the sensor data is known, introduces a new definition of events for IoT-enhanced BPs: CAIRO events, which have the specificity to be contextualised and observable.

4 Challenges

We pinpoint four key challenges of IoT data that make it difficult to integrate them in a traditional event log without having prior knowledge about the process or making non-trivial assumptions. These challenges are established based on an in-depth analysis of the literature (see Sect. 3) and based on the authors' experience in a number of case studies [50]. Observe that the literature also mentions other challenges than the ones listed here, e.g. data quality or data volume. However, we focused on the challenges that (1) have a direct influence on the mining and (2) pertain to the format of the event log.

4.1 C1: Granularity

The granularity of the sensor can be considered a topmost challenge [1, 3, 4, 11, 12, 51]. Process decisions are usually not made based on the raw sensor values, but rather depend on aggregations of the value of a sensor over a certain time or on combinations of the values of several sensors. Granularity is usually dealt with in the event abstraction step, i.e. it is considered as preprocessing. However, without prior knowledge, there is no way to know which aggregation or combination of sensor data should be applied to retrieve the relevant high-level context parameter that determines the decision, and the event abstraction is a mining step, which should be done based on the event log.

4.2 C2: Perspective Convolution

IoT data can relate to events of the control-flow or to the context of the process, or to both [4, 28, 52]. For example, accelerometer data could tell that crates of a product are being shaken during transport (context) or that the crates are being loaded in the truck (control-flow), a spike in fridge temperature can indicate that the fridge is opened to put in or take out crates (control-flow), etc. However, in existing event log standards, it is required to choose between using the IoT data as process events and attributes of a case or an event.

4.3 C3: Scope of Relevance

The relevance of the IoT data, especially when it relates to the context of a process, may not always be limited to an event or a case, but can relate to several events or several cases [11, 12, 52]. For example, the temperature in the truck does not necessarily impact only one crate of products, but usually many crates of different types of products transported in the same truck. And logging the value of temperature (and other variables) separately for each event of each product or crate potentially leads to duplicating huge chunks of data.

4.4 C4: Dynamicity

IoT data are inherently dynamic and not always synchronised with the process [31]. The dynamicity of context parameters is usually coped with by allowing data attributes to be updated in process events. However, IoT data are often loosely coupled with the process. Although there is no guarantee that the sensor data will have relevant values when process events are logged, this is usually the only point when context parameters can be updated. But without previous knowledge, it is not possible to know which sensor measurement(s) is or are relevant. For example, the observation of temperature that impacts the decision to keep a crate of fresh products or not can be made at any moment before the decision point, and there is no way to log it without an event. But this event (1) would not be linked with a process activity, and (2) this requires prior knowledge on the impact of temperature on the decision, which we assume we do not have, as it is the sort of effects that we would like to discover with PM.

5 Methodology

In this section, we outline the methodology we followed to create the NICE log format. It builds further upon the conceptual model presented by the authors in [1], which aimed at integrating IoT ontologies and event log models. To do so, multiple event types were defined to bridge the gap between the (physical) IoT world and the (digital) PM world. In this research, in line with design science methodology [53], we adopted an iterative approach, moving between design and evaluation through prototyping in order to continuously improve the quality of the event log format.

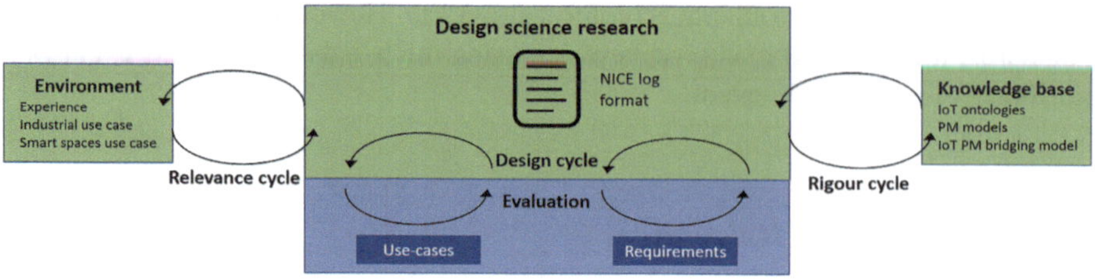

Fig. 2 Overview of our design science research approach

Concretely, as depicted in Fig. 2, the design science research methodology was implemented as follows:

- First, in the relevance cycle, we used our experience and analysed use cases to list the four major challenges presented in Sect. 4. Based on these four challenges, we derived the ten requirements detailed in Sect. 6. Next to this, the following evaluation criteria were defined: (1) fulfilling the ten requirements and (2) the ability to represent event logs from various types of IoT-enhanced BPs.
- Then, the rigor cycle brought two main types of related works: event log standards and models bridging the gap between IoT and PM. In particular, the bridging model presented by the authors in [1] was used as the main basis for the NICE log metamodel.
- Finally, the design cycle started from the conceptual model presented by the authors in [1]. To create the metamodel of the NICE format, we strived to harmonise the digital and physical parts of the model, adding new classes in the part of the model representing the digital world to align it with the classes of the physical world and creating new generic classes representing more abstract concepts shared by the digital and the physical worlds. Then, we translated the metamodel in an XML schema to create a standard format for NICE logs.

The evaluation step verified (1) that the metamodel fulfilled the requirements, (2) that all constructs and relationships of the metamodel were transcribed correctly in the XML schema and (3) that the format was capable of representing event logs from various types of IoT-enhanced BPs.

6 Requirements for an IoT-Enhanced Event Log Model

Based on the challenges identified previously, we put forward ten requirements that a suitable data model for IoT-enhanced event logs should meet.

- R1: Store high-level events.
- R2: Store low-level events.
- R3: Store intermediary/mixed-level events.
- R4: Enable traceability between high-level and low-level events.

- R5: Represent context at event level.
- R6: Represent context at activity level.
- R7: Represent context at case/object level.
- R8: Represent context at process level.
- R9: Update context parameters independently from process events.
- R10: Update context parameters at a higher frequency.

R1-4 connect to the ability to store data at different granularity levels (to meet C1). This means that data at either the typical low-level of raw IoT sensors, or the traditional high level of the process, or in-between levels, should be able to coexist in an IoT-enhanced event log. From this, it follows that the traceability of the IoT data should be guaranteed, to keep track of which raw low-level IoT data have been abstracted into which high-level process events or context parameters, and of possible in-between steps, e.g. by defining links between events at different levels. For instance, a change in the value of a sensor might indicate the crossing of a threshold, which can be a mixed-level event, located at a granularity level in-between raw IoT data and process events, and can be used to derive multiple higher-level events. Meeting these requirements would enable process mining to leverage IoT data to dig deeper into the working of the process.

Another effect of these requirements is that event abstraction is recognised as a mining step, and not just as a preprocessing step, since the low-level events are present in the event log itself. For example, a process analyst can more easily test different possibilities to derive high-level events from the IoT data and compare the resulting process models.

R5-10 relate to having a suitable representation of the (physical) context of the process. These requirements can be split in two subgroups. First, R5-8 state that context parameters should have different scopes (event, activity, case/object and process; see C3). Note that an event is understood as a change in the life cycle of an activity (e.g. start, complete); hence, the scope of an activity is broader than that of an event. For example, consider a process where many instances and multiple process activities take place in the same environment (i.e. in the same physical space). Some environmental phenomena are tracked by sensors s_1, s_2, ..., s_n. As such, these sensors may be collecting data that is relevant for different processes, different instances of a process, different activities within a process, etc. In a typical XES event log, the values of all the sensors (or of a parameter derived by abstracting them) would be logged for each event happening in this environment for each case (see Table 2). Better defining the scope of the IoT data could simplify the event log and help focus on the more relevant values of the sensors. Moreover, it could help to find links between different instances of the process that occur simultaneously.

Second, R9-10 stipulate that context parameters should be able to evolve with time at their own rhythm (C4), independently of control-flow events (C2). This aims at caring for situations where sensor data that are relevant for a decision can be observed at different moments during the execution of the process, i.e. typically at any point during the execution of an activity, and not only when events are logged. It also means that the sensor values can be irrelevant at the moment of logging.

Table 2 Example XES event log with sensor data being duplicated

Case ID	Event ID	Activity	Timestamp	Sensor 1	Sensor 2	...	Sensor n
1	1	A	12-01-22.09:28	42	0.2	...	10
2	2	A	12-01-22.09:28	42	0.2	...	10
1	3	B	12-01-22.09:30	42	0.3	...	10
3	4	A	12-01-22.09:31	42	0.3	...	10.1
3	5	B	12-01-22.09:31	42	0.3	...	10.1
1	6	C	12-01-22.09:32	40	0.2	...	10.1
2	7	B	12-01-22.09:34	40	0.2	...	10.1
4	8	A	12-01-22.09:34	40	0.2	...	10.1
...	...	...	...	...	...	...	...

7 Format Specification

In this section, we outline the format we propose for IoT-enhanced event logs and explain how we operationalised it to generate XML-based NICE logs.

7.1 Metamodel

The format consists of seven main concepts (in bold in Fig. 3), which are specialised to cope with the specificities of digital and physical data (to the left and the right of the figure, respectively). At its core, our data format contains lists of Data sources, Objects and Events. An Event is related to one or several Objects, which can be digital (Digital object, e.g. an order) or physical (Feature of interest, e.g. a fridge) or hybrid (e.g. a parcel). All Objects can have a collection of Properties, which can also be digital or physical, named Digital and Observable properties, respectively, and represent context parameters of the process, e.g. the total amount of the order, the temperature in the fridge, the weight and the value of the parcel. Note that Digital objects can only have Digital properties, while Features of interest can only have Observable properties. Events are derived from one or several Data entries or lower-level Events that are logged by a given Data source, which can either be a Process-Aware Information System (PAIS) or a Sensor. PAIS can record IS data entries, which typically correspond to rows in relational database, while Sensors make Observations, which represent individual measurements of an Observable property. We distinguish between three types of Events, which follow the hierarchy shown in Fig. 4:

- IoT event: an instantaneous change in a real-world phenomenon that is monitored by a Sensor, or derived from lower-level IoT Events, e.g. the temperature in a fridge decreasing.
- Process event: an instantaneous change of state in the transactional life cycle of an activity (corresponding to the usual notion of event in PM). This type of

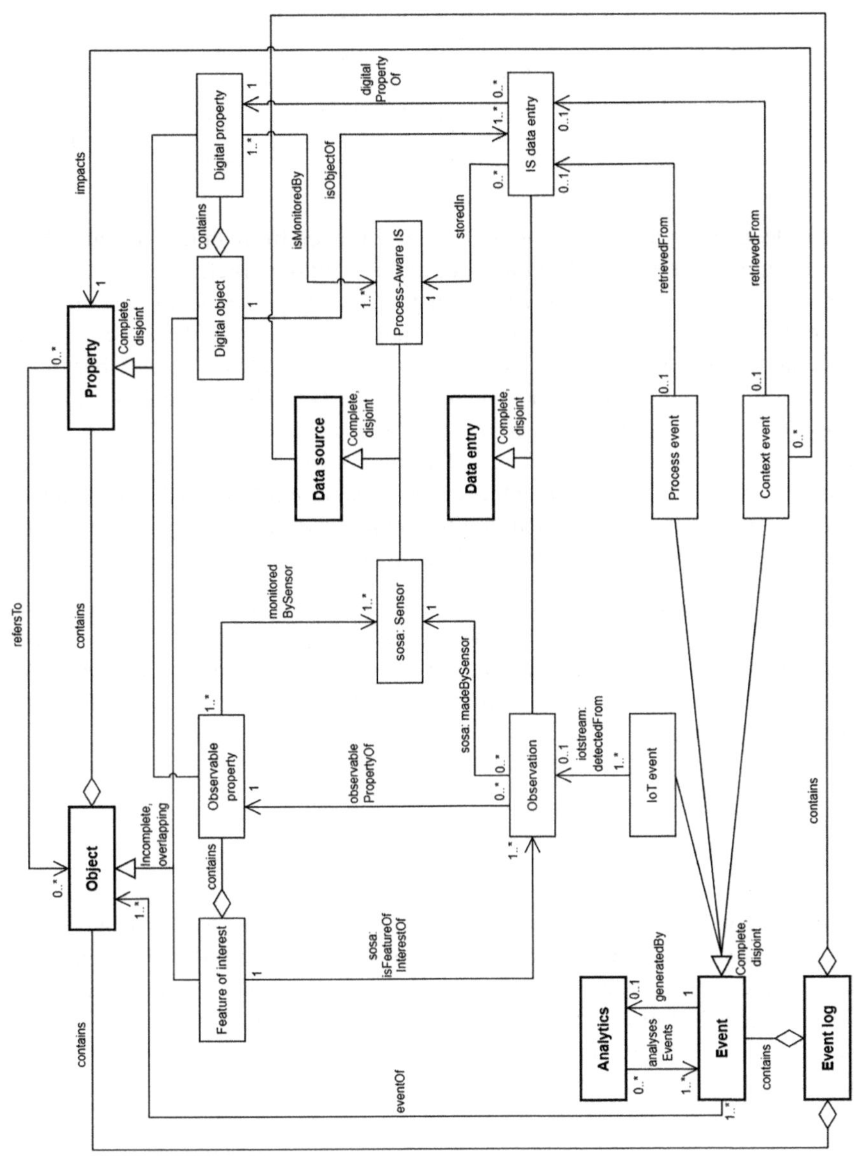

Fig. 3 Full metamodel of the NICE log format. Classes in bold are the main concepts of NICE logs

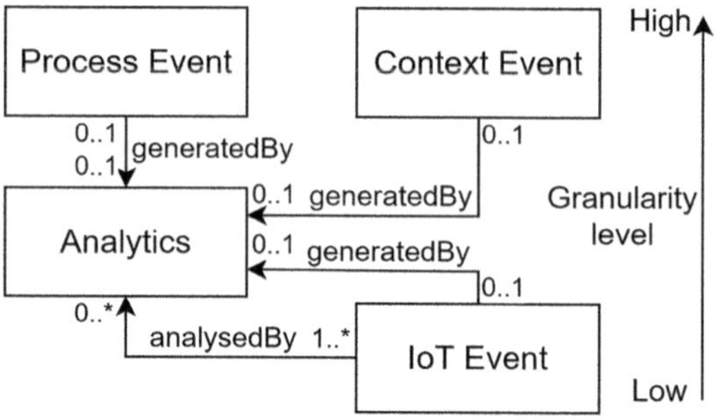

Fig. 4 Hierarchy of events in NICE logs (taken from [1])

event is deduced from one or more IoT events or taken from an IS data entry. For example, a product is taken from a fridge and loaded in a truck.

- Context event: an instantaneous change in a real-world phenomenon (i.e. observable property) or a digital property that has an impact on the execution of a specific process instance (i.e. it impacts a Property of an Object) but does not change its control-flow state. Such an Event typically influences the path followed in a process instance, how an activity is executed, etc. This type of event is deduced from one or more IoT events or taken from an IS data entry. For example, the pressure in a tank exceeds a maximum threshold, prompting a human intervention in a chemical production process; the amount of a claim is updated by a customer in a claim handling process.

Higher-level Events can be derived from lower-level Events via the notion of Analytics, which represents any technique used for event abstraction (e.g. rule-based reasoning techniques like CEP and data-driven models). IoT events can cascade until a deduced event has a direct relationship with the process, i.e. it is a Process event or a Context event.

7.2 XML Implementation

In this section, we describe how we translated the model in Fig. 3 into an XML format for representing NICE logs. XML was chosen for its flexibility and its popularity, though the model could also be transcribed in other languages such as JavaScript Object Notation (JSON) or YAML Ain't Markup Language (YAML). The full schema is available in the following online repository: https://github.com/ybertran/NICELogFormat/blob/main/NICELogSchema.xsd.

7.2.1 Overall Structure

The XML schema of the processed log is structured as follows: the `EventLog` tag is the root and contains three main elements, identified respectively by the tags:

1. `ObjectList`. This element contains a list of Objects. In IoT-enhanced BPs, different Objects interact and an Event can involve a combination of Objects (e.g. users, locations, machines). Objects also contain `Property` elements, which can be updated by Events. Three types of Objects can be stored, depending on the types of properties they have: *(i)* `DigitalObjects`, which only have `DigitalProperties`; *(ii)* `FeaturesOfInterest`, which only have `ObservableProperties`; and *(iii)* hybrid Objects, which can have both DigitalProperties and ObservableProperties.
2. `DataSourceList`. This element contains a list of available `DataSources`. We distinguish between two types of DataSources: *(i)* physical DataSources, typically `Sensors` that monitor ObservableProperties, e.g. the opening of a door or the temperature, and *(ii)* digital DataSources, named `ProcessAwareInformationSystems` (PAISs), that record ProcessEvents and the value of DigitalProperties.
3. `EventList`. This element contains a list of `Events`. There are three types of events: IoTEvent, ContextEvent and ProcessEvent. `IoTEvent` is used to represent low-level types of Events, such as raw sensor measurements, and, if relevant, mixed-level types of Events, such as the aggregation of lower-level IoTEvents into *actions*, i.e. atomic interactions with the environment or a part of it (e.g. sitting on a chair, opening the fridge). A `ContextEvent` represents a change in the value of a Property of an Object, which impacts the execution of the process (e.g. influence a decision). It is related to an Object and the Property it updates and contains this Property's new value. Finally, `ProcessEvent` is used to represent high-level types of Events, e.g. groups of atomic interactions with the environment that are performed with a common goal. It represents an execution of a business activity (e.g. prepare dinner, register an order).

   Each Event is associated with an identifier, a timestamp, a value and one or more Objects via their ID. Note that the meaning of the value attribute depends on the type of Event: for IoTEvents and ContextEvents, it represents the new value of the Property they relate to; for ProcessEvents, it is used to store the life cycle of the activity logged (e.g. start, complete). An IoTEvent has an additional optional child element called `Observation` that contains the raw sensor value, if available, while ProcessEvent and ContextEvent can have an optional child element `ISDataEntry` representing the source database entry from which it is detected. Next to this, ProcessEvent has one additional attribute that refers to the label of the related activity.

   These three levels of Events are connected through the `Analytics` element that references the lower-level Events that generate a higher level of Event and the `Method` used. Analytics elements are the physical counterpart of ISDataEntry, as they also indicate the source of a higher-level Event, but are used when this Event is derived from a lower-level IoTEvent. For instance, the mixed-level IotEvent called 'open the fridge' in its Analytics has references to the lower-level Events related to the opening of the fridge door and the triggered motion sensors near the fridge. The higher-level ProcessEvent related to the activity 'cook', in its Analytics, has the references of the mixed-level Events that all together compose the activity ‘cook’, including the mixed-level IotEvent ‘open the fridge’. Figure 9 shows an example of an Analytics element used to derive ContextEvents in the first case study, and Fig. 10 gives examples of mixed-level IoTEvents and ProcessEvents derived by Analytics in the second case study.

7.2.2 Specialisations

As outlined in Sect. 7.1, the main elements presented in the previous section have specialisations corresponding to digital and physical versions of the concepts. The specialisations of element types are implemented in XSD by means of extensions. Generic element types are defined with the attributes and elements common to their specialised types, and extensions define more particular attributes and elements of the specialised types. For instance, the generic Property type defines the attributes ID, propertyName, dataType and initValue, which are common to the specialisations DigitalProperty and ObservableProperty, which extend the generic type with references to a PAIS and a sensor, respectively (see Fig. 5 for the XSD definition of the Property and ObservablePorperty types).

There is one exception to this approach: the generic Object type was defined with all attributes and elements of Objects, and specialisations of Object were specified via restrictions to the types of Properties that they contain with respect to the generic Object type. This is due to the fact that Objects can be instantiated as such, unlike other generic types that are only instantiated via specialised types. In this way, DigitalObjects and FeaturesOfInterest can only have DigitalProperties or ObservableProperties, respectively, while still allowing ‘hybrid’ Objects to have Properties of both types.

```
  ...
  <xs:complexType name= "Property">
      <xs:attribute name= "ID" type= "xs:string"/>
      <xs:attribute name= "propertyName" type= "xs:string"/>
      <xs:attribute name= "dataType" type= "xs:string"/> <!-- refers to the data
          type: numerical (discrete/continuous), binary, etc. -->
      <xs:attribute name= "initValue" type= "xs:string"/>
  </xs:complexType>

<xs:complexType name= "ObservableProperty">
  <xs:complexContent>
    <xs:extension base= "Property">
      <xs:attribute name= "monitoredBySensor" type= "xs:string"/> <!-- refer to
          Sensor -->
    </xs:extension>
  </xs:complexContent>
</xs:complexType>
  ...
```

Fig. 5 Example declaration of generic and specialised types in the schema

7.2.3 Consistency and Identity Constraints

To ensure the coherence of the data stored in NICE logs, identity constraints are defined over elements that need to be unique and referred to by other elements. More specifically, keys are declared over Event IDs, Object IDs, DataSource IDs and Property IDs and their specialisations. See Table 3 for a summary of the keys defined and the elements referring to them.

Figure 6 shows an example of such an identity constraint. Lines 2–5 define the constraint that the ID attribute of Sensor elements needs to be unique, i.e. two Sensors cannot have the same ID. Then, lines 7–10 specify that the monitoredBy-Sensor attribute of ObservableProperty needs to refer to the ID of a Sensor element, ensuring that all ObservableProperties are monitored by an existing Sensor in the log.

Table 3 Overview of the keys and references defined for NICE logs

Key	Referenced by
Event ID	Analytics
Object ID	Event, Observation, ISDataEntry
FeatureOfInterest ID	Observation
DigitalObject ID	ISDataEntry
Sensor ID	Observation, ObservableProperty
PAIS ID	ISDataEntry, DigitalProperty
Property ID	ContextEvent
ObservableProperty ID	Observation
DigitalProperty ID	ISDataEntry

```
1   ...
2   <xs:key name= ”sensorIDKey”>
3       <xs:selector xpath= ”.//DataSourcesList/Sensor”/>
4       <xs:field xpath=”@ID”/>
5   </xs:key>
6   ...
7   <xs:keyref name= ”observablePropertySensorInObjectIDRef” refer= ”sensorIDKey”>
8       <xs:selector xpath= ”.//ObjectsList/Object/ObservableProperty” />
9       <xs:field xpath= ”@monitoredBySensor”/>
10  </xs:keyref>
11
12  <xs:keyref name= ”observablePropertySensorInFeatureOfInterestIDRef” refer= ”
        sensorIDKey”>
13      <xs:selector xpath= ”.//ObjectsList/FeatureOfInterest/ObservableProperty” />
14      <xs:field xpath= ”@monitoredBySensor”/>
15  </xs:keyref>
16  ...
```

Fig. 6 Example declaration of identity constraint and references in the schema. This XSD excerpt specifies that the IDs of Sensors have to be unique and that the monitoredBySensor attribute of ObservableProperties has to refer to the ID of a Sensor. Note that two reference constraints have to be declared, as ObservableProperties can be found in Objects and in FeatureOfInterests

7.3 Theoretical Requirement Fulfilment

In this section, we evaluate our format by confronting it with the requirements outlined in Sect. 6:

- R1 (Store high-level events, C1): Obtained with process and context events.
- R2 (Store low-level events, C1): Obtained with IoT events.
- R3 (Store intermediary events, C1): Obtained with IoT events derived from others.
- R4 (Enable traceability between high-level and low-level events, C1): Achieved with the concept of Analytics, which links derived events with the events they are derived from
- R5 (Represent context at event level, C3): Done with the properties of the objects linked to an event.
- R6 (Represent context at activity level, C3): Done with properties of the objects linked to the events of an activity; an additional activity object could link events together for convenience (link the events to the activity object, and the context to the activity object).
- R7 (Represent context at case/object level, C3): Context is represented at object level with the object properties; case is a special object type.
- R8 (Represent context at process level, C3): This requirement can be completed with a process-wide object (e.g. the house in a smart home log).
- R9 (Update context parameters independently from process events, C2): Context parameters (Properties) are updated by context events and not by process events.
- R10 (Update context parameters at a higher frequency (than process events), C4): This requirement is achieved together with R9, as context events can happen at any rate, distinguished from process events.

8 Case Studies

To showcase the usability of the data format, we present two case studies. In each case study, we outline the process, explain how we constructed a NICE log from it and present a short analysis facilitated by the use of the NICE format.

8.1 Real-Life Manufacturing Use Case

8.1.1 Use Case Description

The first use case is the real-life process of a partner company active in the production of chemical products. Their production process can be summarised in four main steps:

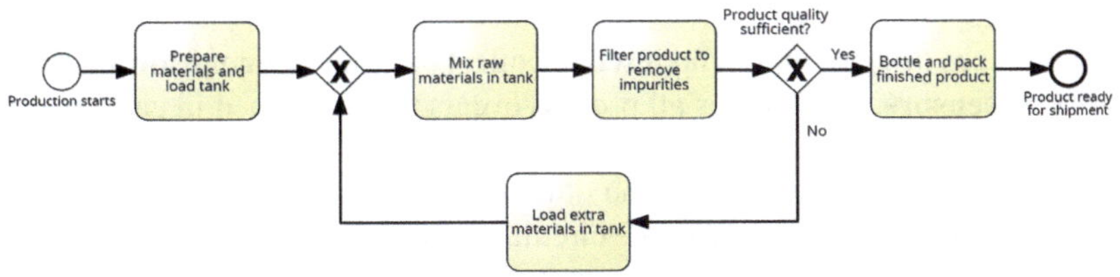

Fig. 7 High-level model of the process analysed in the experiment

1. Preparing raw material and loading the tank
2. Mixing the raw material in the tank
3. Circulating the product through filters to remove impurities
4. Bottling and packing the finished product

Sometimes, the quality of the product is not high enough after filtering, i.e. some characteristics of the product do not meet the specifications. In this case, an adjustment is applied by loading additional raw materials into the tank and repeating steps 2 and3, resulting in the high-level production process depicted in Fig. 7.

This seemingly simple process has to be executed with extreme precision and care as the slightest presence of impurities in the finished product greatly diminishes its quality. This is why the company is interested in analysing production logs and TS sensor data together to find out variation in process execution. In particular, the flow of the product has to remain as stable as possible during the circulation step, as surges in the flow could push impurities through the filters

Two main data sources were used to create the NICE log: (1) logs from the production system, which contain the sequences of activities executed for each process instance and trace attributes, and (2) TS data from sensors tracking the flow of the product in the four tanks and in the pipes leading through the filters every second. The data represent 161 complete process instances, which correspond to 199.4 million rows of sensor data.

8.1.2 NICE Log Creation

Objects

Two main object types were extracted to create the NICE log: **batches** and **tanks**. Each batch represents one instance of the process, and as such are digital objects that correspond to the case notion of the process. Batches gather various properties like the *recipe* followed, which is used to retrieve the planned activity sequence and the amount and type of raw materials to use. Next to this, batches also refer to the *tank* physical object in which the batch was produced. Tanks have the property *capacity*, which denotes the maximum quantity of product that can be created in one batch, and several properties that are tracked by sensor data sources, such as the *pump circulation flow* and *tank pressure*. Finally, tanks also have the property *flow status*, which is derived from the observations of pump circulation flow via context events.

Data Sources

Two main types of data sources are used to collect data: an **information system (IS)** and several **sensors**. The IS logs all process events that happened in each execution of the process and stores some properties of each batch, such as the *recipe* and *types of raw materials used*. Sensors, on the other hand, are attached to tank objects and tracks various properties related to the circulation of the product in the tank, like the *pump circulation flow* and the *pressure* at different points in the production line.

Events

Events of each type are recorded during the execution of the process. First, **IoT events** are extracted from the raw sensor observations as changes of more than 1% in the value of successive observations. This yields 1.1M IoT events, which corresponds to an average of around 7000 IoT events per batch, compared to more than 1M observations per batch.

Then, **context events** are derived from the flow IoT events to keep track of the flow status property of the product. More specifically, this property follows a cycle between each start and stop of the pump, as follows:

- Flow initialisation (flow is raising and unstable)
- Flow stable
- Flow drifting (indicates that the flow is unstable)
- Flow termination (flow rapidly drops to zero)

Figure 8 shows the time series plot of a pump run, with Table 4 containing the corresponding derived context events pertaining to the pump flow status property; remark that multiple drifts can follow each other in sequence.

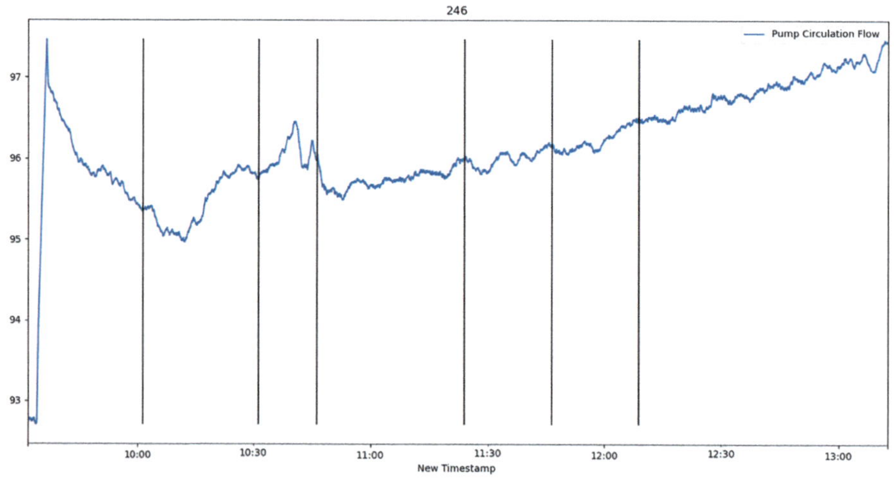

Fig. 8 Example time series plot of the pump circulation flow property. Black vertical lines indicate the start timestamp of drift context events

Table 4 Context events derived from the time series depicted in Fig. 8

Timestamp	Property	Value
t_1	Short-term flow status	Flow initialisation
t_2	Short-term flow status	Stable flow
t_3	Short-term flow status	Short mild upward flow drift
t_4	Short-term flow status	Short mild downward flow drift
t_5	Short-term flow status	Long weak upwards flow drift
t_6	Short-term flow status	Short weak upward flow drift
t_7	Short-term flow status	Short weak upward flow drift
t_8	Short-term flow status	Long weak upward flow drift
t_9	Short-term flow status	Flow termination

Drift context events are derived from the time series data of the pump circulation flow using a sliding window to compare the average values of the flow in consecutive time windows to detect rising and decreasing trends. Moreover, various types of drift context events are distinguished based on their duration (i.e. long or short) and intensity (i.e. strong, mild or weak) to enable a fine-grained analysis of the behaviour of the flow. This step is recorded in Analytics elements, such as the one in Fig. 9, which gives an example of the analytics behind the derivation of a drift context event. In total, 31,741 context events were generated.

Finally, 3631 process events marking the start and the end of activities are recorded by the manufacturing execution system. These process events are at a finer granularity level than the activities depicted in Fig. 7 and also encompass some pre- and post-processing of the batches, e.g. placing the filters before loading the tank and cleaning the tank at the end of the process.

8.1.3 Analysis

This NICE log helps us in the analysis of the flow and its interplay with the rest of the process. For instance, it can be used to identify batches with similar pump flow status life cycles, in which to look for similar batch characteristics that might be the root cause of the pump flow status sequence. This can be particularly valuable for batches of lower or higher quality and could help improve the designed process at a low level, e.g. by refining the control limits of the sensor parameters. Comparing pump flow statuses could also help identify unfrequent sequences of context events, which could point to anomalous batches.

A complementary analysis is to start from batches that are known to be of lower quality and search for a root cause in the pump flow status, as it is believed that a more unstable flow leads to a worse final product quality.

```
1  ...
2  <IoTEvent id=”62” timestamp=”2020−01−01T00:26:26” value=”OFF”>
3      <EventObjectRelationship objectID=”objID_1”/>
4      <EventObjectRelationship objectID=”objID_2”/>
5      <EventObjectRelationship objectID=”objID_26”/>
6      <Observation timestamp= ”2020−01−01T00:26:26” value= ”OFF” sensor= ”s14”
           observablePropertyInFeatureOfInterest= ”propID2”>
7          <FeatureOfInterest>objID_26</FeatureOfInterest>
8      </Observation>
9  </IoTEvent>
10
11 <IoTEvent id=”63” timestamp=”2020−01−01T00:31:58” value=”ON”>
12     <EventObjectRelationship objectID=”objID_1”/>
13     <EventObjectRelationship objectID=”objID_2”/>
14     <EventObjectRelationship objectID=”objID_26”/>
15     <Observation timestamp= ”2020−01−01T00:31:58” value= ”ON” sensor= ”s14”
           observablePropertyInFeatureOfInterest= ”propID2”>
16         <FeatureOfInterest>objID_26</FeatureOfInterest>
17     </Observation>
18 </IoTEvent>
19 ...
20 <IoTEvent id=”65” timestamp=”2020−01−01T00:31:58” value=”ON”>
21     <EventObjectRelationship objectID=”objID_1”/>
22     <EventObjectRelationship objectID=”objID_2”/>
23     <EventObjectRelationship objectID=”objID_24”/>
24     <Observation timestamp= ”2020−01−01T00:31:58” value= ”ON” sensor= ”s8”
           observablePropertyInFeatureOfInterest= ”propID4”>
25         <FeatureOfInterest>objID_24</FeatureOfInterest>
26     </Observation>
27 </IoTEvent>
28 ...
29 <ContextEvent ID=”66” timestamp=”2020−01−01T00:31:58”
       impactsObservablePropertyInObject=”propID5” value= ”short mild upwards flow
       drift”>
30     <EventObjectRelationship objectID= ”objID_0” />
31     <EventObjectRelationship objectID= ”objID_1” />
32     <Analytics>
33         <AnalysesEvent>62</AnalysesEvent>
34         <AnalysesEvent>63</AnalysesEvent>
35         <AnalysesEvent>65</AnalysesEvent>
36         <Method name=”sliding window”>
37             <Parameter name= ”window size” value= ”900”/>
38             <Parameter name= ”variation threshold” value= ”0.01”/>
39             <Parameter name= ”direction” value= ”upwards”/>
40         </Method>
41     </Analytics>
42 </ContextEvent>
43 ...
```

Fig. 9 Example analytics element in the log (lines 32–41)

8.2 Smart Spaces Use Case

8.2.1 Use Case Description

For our second case study, we generated a log simulating the behaviour of two inhabitants living in a smart home for 4 weeks. To do so, we used the smart spaces simulator presented in [54], which we adapted to generate NICE logs. More specifically, the two users were programmed to execute daily life activities such as *Cook_and_eat, Do_the_dishes, Drink, Eat_cold, Eat_warm, Exercise, Go_work, Sleep, Rest, Shop, Use_Computer, Watch_tv* and *Wc*. Moreover, the simulator was programmed so that during 1 week, the users show a different behaviour, i.e. they stay at home the whole day instead of going to work in the morning and coming back home in the evening during weekdays. This change in behaviour is due to extremely high outside temperatures during that week, which are tracked by temperature sensors and logged as IoT events.

8.2.2 NICE Log Creation

Objects
Three types of objects are present in this use case. First, **users** represent the two inhabitants living and executing activities of daily life in the smart home. Then, **locations** are the various rooms in the simulated house, necessary to pinpoint the movements of the users, detected by motions sensors. Finally, every **day** of the simulation is also represented by an object. This object type is used in combination with users to define cases, each case being one day for one user.

Data Sources
This use case also comprises the two main types of data sources typical of NICE logs: **sensors** and an **IS**. In particular, 30 motion sensors capture the presence of users in various locations in the home, and one temperature sensor keeps track of the inside temperature. Next to this, an IS supporting the routines of the inhabitants keeps track of the status of the air conditioning system.

Events
Two types of events are recorded directly by the data sources. First, motion sensors log the presence of users in their vicinity as **IoT events**. Similarly, changes in temperature detected by the temperature sensor are also logged as IoT events. Second, the status of the air conditioning controlled by the IS is recorded as **context events** indicating when it is turned on and off. Next to this, a series of higher-level events are derived from the low-level IoT events mentioned above. These consist in mixed-level IoT events that represent simple actions performed by the users (e.g. going to bed) and process events logging activities of daily life composed of multiple actions (e.g. sleeping). Figure 10 shows an example of derivation of a process event from lower-level IoT events.

```
...
<IoTEvent ID="66" timestamp="2020-01-01T00:31:58" value= "Go_bathroom_sink" >
    <EventObjectRelationship objectID="objID_1"/>
    <EventObjectRelationship objectID="objID_2"/>
    <Analytics>
        <AnalysesEvent>62</AnalysesEvent>
        <AnalysesEvent>65</AnalysesEvent>
        <Method name="CEP"/>
    </Analytics>
</IoTEvent>
...
<IoTEvent ID="86" timestamp="2020-01-01T05:36:18" value="Go_bed">
    <EventObjectRelationship objectID="objID_1"/>
    <EventObjectRelationship objectID="objID_2"/>
    <EventFeatureOfInterestRelationship objectID= "objID_26"/>
    <Analytics>
        <AnalysesEvent>68</AnalysesEvent>
        <AnalysesEvent>70</AnalysesEvent>
        <AnalysesEvent>75</AnalysesEvent>
        <AnalysesEvent>78</AnalysesEvent>
        <AnalysesEvent>80</AnalysesEvent>
        <AnalysesEvent>85</AnalysesEvent>
        <Method name="CEP"/>
    </Analytics>
</IoTEvent>
...
<IoTEvent ID= "88" timestamp= "2020-01-01T05:36:22" value= "OFF">
    <EventObjectRelationship objectID= "objID_1"/>
    <EventObjectRelationship objectID= "objID_2"/>
    <EventFeatureOfInterestRelationship objectID= "objID_23"/>
    <Observation timestamp= "2020-01-01T05:36:22" value= "OFF" sensor= "s3" observablePropertyInObject= "propID1">
        <FeatureOfInterest>objID_23</FeatureOfInterest>
    </Observation>
</IoTEvent>

<IoTEvent ID= "89" timestamp= "2020-01-01T05:40:52" value= "ON">
    <EventObjectRelationship objectID= "objID_1"/>
    <EventObjectRelationship objectID= "objID_2"/>
    <EventFeatureOfInterestRelationship objectID= "objID_23"/>
    <Observation timestamp= "2020-01-01T05:40:52" value= "ON" sensor= "s3" observablePropertyInObject= "propID1">
        <FeatureOfInterest>objID_23</FeatureOfInterest>
    </Observation>
</IoTEvent>

<IoTEvent ID="90" timestamp="2020-01-01T05:40:52" value="Go_in_bed">
    <EventObjectRelationship objectID="objID_1"/>
    <EventObjectRelationship objectID="objID_2"/>
    <EventFeatureOfInterestRelationship objectID= "objID_23"/>
    <Analytics>
        <AnalysesEvent>88</AnalysesEvent>
        <AnalysesEvent>89</AnalysesEvent>
        <Method name="CEP"/>
    </Analytics>
</IoTEvent>
```

Fig. 10 (continued)

```

<ProcessEvent ID="91" timestamp="2020−01−01T05:40:52" label= "Sleeping" value
    = "complete">
    <EventObjectRelationship objectID= "objID_1" />
    <EventObjectRelationship objectID= "objID_2" />
    <Analytics>
        <AnalysesEvent>66</AnalysesEvent>
        <AnalysesEvent>86</AnalysesEvent>
        <AnalysesEvent>90</AnalysesEvent>
        <Method name="CEP"/>
    </Analytics>
</ProcessEvent >
...
```

Fig. 10 Example of events connected by the analytics element. In particular, in this portion of the log, there are five IoT events and one process event. The mixed-level IoT event with ID 90, which represents the atomic action *go_in_bed*, is generated by the analysis of the low-level IoT events with IDs 88 and 89, with the method called CEP. The high-level process event with ID 90, which completes the event *sleeping*, is generated by the analysis of the mixed-level IoT events with IDs 66, 86, and 90, with the method called CEP. The derivation of events with IDs 66 and 86 follows the same principle, but the lower-level IoT events from which they are derived are not shown for brevity

8.2.3 Analysis

Based on these IoT events, a context event is derived when the temperature exceeds the maximum acceptable temperature for work (set to 32.5 degrees for office workers in Belgium) and sets a property 'Temperature suitable for work' to False. When the temperature decreases below the threshold again, another context event is generated, which sets 'Temperature suitable for work' back to True. Figure 11 contrasts directly-follows graphs (DFGs), while the 'Temperature suitable for work' property was False (Fig. 11a) and over the whole log (Fig. 11b). From the figure, we can see that the behaviour of the users differs in both circumstances, as the activity Go_work is not performed during the heatwave period and more activities are recorded at home.

These context events make the change in the behaviour of the users easy to explain, as the days of anomalous behaviour are flagged with context events and characterised by a value of 'Temperature suitable for work' equal to True. This way, all relevant information is stored in the log and is easily traceable, as lower-level IoT events are still present. Moreover, this approach makes it possible to set or mine different thresholds for 'Temperature suitable for work', e.g. for users working in different sectors, and to create one context parameter for each user, with different rules, which would be much more difficult in traditional event log formats. Once clearly identified with the context events, deviating behaviour can be removed from the log or treated separately simply by removing the events recorded between the two context events. Finally, storing the low-level data in the log enables the use of, e.g. decision mining or trace clustering techniques to mine the finer-grained rules behind the break from work from the low-level data directly, which is not possible in a XES or OCEL log.

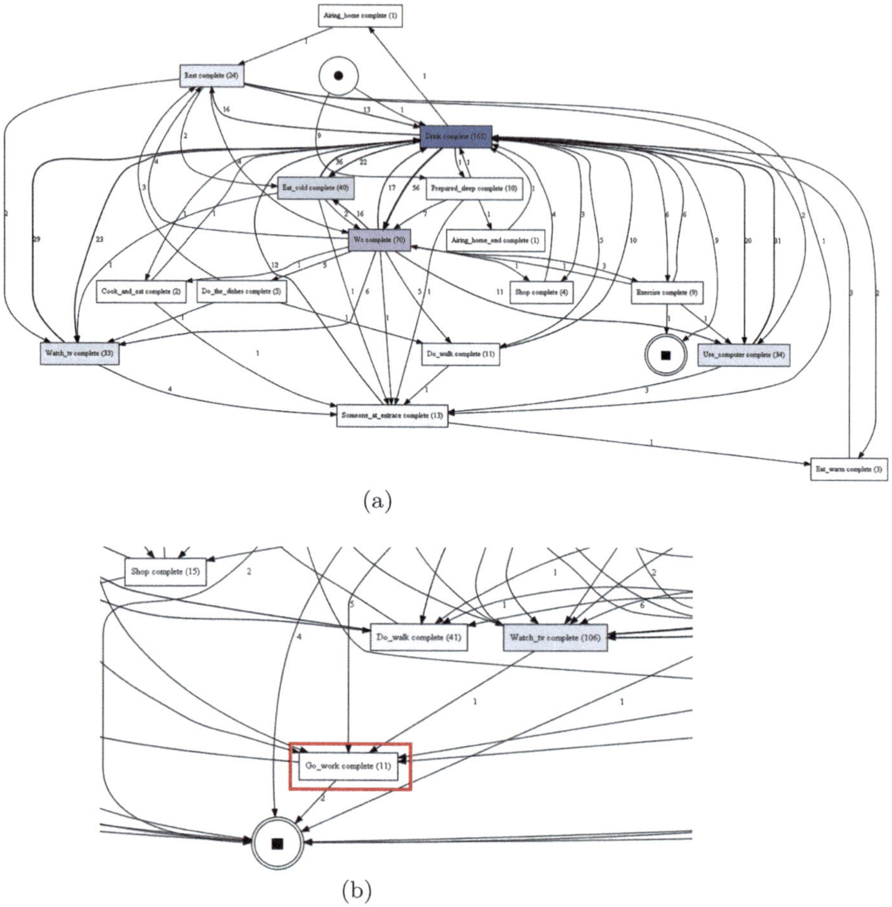

Fig. 11 DFGs of the behaviour of the users in the simulated log. (**a**) DFG of the behaviour of the users during the heatwave. (**b**) Zoom in on the ‘Go_work’ activity in the DFG of the behaviour of the users over the whole log

9 Discussion

In this paper, we presented a new format that enables a wider look at the possible uses of IoT data in process mining. While the existing literature largely tackles the issue of extracting process events from IoT data (so-called activity mining), other potential uses of IoT data listed in Table 1 remain almost entirely unexplored. One of the main hurdles complicating the development of new techniques is the restrictions that are imposed by the current main standards for event logs, namely, XES and OCEL. All the data have to be stored as high-level events or attributes of high-level events or cases (objects in the case of OCEL), leaving context as a second-class citizen.

In contrast with existing standards, the NICE log format is able to store low-, mixed- and high-level data and integrates control-flow, context data and IoT observations meaningfully, as indicated by its meeting the requirements listed in Sect. 6. The usability and usefulness of the format are demonstrated in two case studies illustrating different features of NICE logs. First, the manufacturing case study shows its use in a complex real-life scenario, where the IoT time series data need to be filtered and abstracted into elaborate context events to unleash valuable insights. In addition to this, the traceability inherent to NICE logs allows us to combine the abstracted and raw data, making it easy to dive further into the IoT events from which context events were derived and generate new context events as the analysis goal evolves. Second, the smart spaces case study showcased the possibility to decompose the abstraction step by extracting multiple levels of events, forming a potentially complex hierarchy of activities and actions. The analysis showed how these events can then be used, e.g. to detect concept drifts. Moreover, with the exception of the AC context events, the whole log consists of raw and abstracted sensor data, which exemplifies the ability of NICE logs to be entirely (or almost entirely) based on IoT data.

Some prior knowledge is typically required to derive context events from IoT data. Indeed, context events are by definition directly relevant to the high-level process, contrarily to individual IoT observations, which means they have to be carefully defined to capture the most important happenings from the parameters tracked by sensors. This being said, in cases where such knowledge is scarce, generic rule and pattern discovery approaches could be followed to find patterns indicating relevant context events from the IoT data.

The XML schema clearly defines the structure of the NICE format and can guide the generation of more NICE logs in the future. Next to this, although the NICE log model was designed for IoT-enhanced logs combining IoT and traditional IS-based event data, the model also allows storing IoT data only (with IoT events), and even IS data only.

Finally, while OCEL 2.0 as-is was evaluated as unable to meet the requirements in Sect. 7.3, a model combining the newly presented standard for machine data [55] to OCEL 2.0 could potentially meet more of the requirements.

10 Related Works

In the last years, several data formats have been adapted to IoT-enhanced event logs [31, 54, 56–58]. In this section, we present these new alternatives and compare them to our data format with respect to the requirements identified in [6].

Of these models, the least tailored ones to IoT-enhanced logs are D-OCEL [59], iDOCEM [58] and OCEL 2.0 [60]. D-OCEL is an extension of the OCEL [8], which introduces dynamic attributes, thereby solving one of the main issues faced by OCEL for IoT-enhanced event logs. Then, iDOCEM [58] integrates D-OCEL with the MERODE conceptual modelling framework [61] to improve data collection in

BPs, resolving some of the issues related to the granularity of events. OCEL 2.0 [60] is the most recent version of the OCEL standard, which also enables attribute value updates.

In [57], authors present the XES DataStream extension, which aims at facilitating the storage of IoT data in XES logs. DataStream introduces, among others, the notions of *point* and *datastream*, which respectively represent individual sensor observations and group together points that belong to a same event or trace. This enables the storage of all IoT data in the log, together with the impacted events or traces.

Another format based on XES is described in [54] and represents IoT-enhanced logs generated by the smart home simulator used in this paper. It generates multiple files: a segmented high-level event log, an unsegmented high-level event log and a sensor log.

Finally, the TS log, a model which is completely independent from existing standards like XES and OCEL, was presented in [31]. This model is designed exclusively for TS data, which are collected around a process and can be used to derived the execution of activities (i.e. process events). A TS log therefore contains a set of variable names, a domain function for each variable, an index and a function recording the value of each variable at each timestamp.

Comparing our model to related works, the NICE log format is the most flexible and balanced and the only one fulfilling all requirements expressed in [6] (see Table 5). The original OCEL did allow neither data attribute updates nor traceability and only has one event type. XES, D-OCEL and OCEL 2.0 have the same limitations, except that they allow for data attribute updates, but only with (process) events. XES with DataStream does much better, but still requires all sensor data points to be linked with one process event, and the support for mixed-level events is uncertain. The simulation tool does not support all context representation and enables only limited traceability (and information is scattered in multiple files). Finally, TS logs focus fully on low-level data and cannot store traditional process data.

Compared to the conceptual model presented by the authors in [1], the format of the NICE logs differs in several aspects. Primarily, NICE logs rely on the concept of object instead of the concept of case. This is because objects offer more flexibility and can be used to precisely define the scope of context parameters as object properties. Then, while the original conceptual model gave a more detailed description of IoT concepts than of process concepts, NICE logs harmonise both perspectives with overarching concepts. Finally, attention has been given to reducing the redundancy between some concepts and attributes.

Table 5 Comparison of different solutions with respect to the requirements

		XES [7]	OCEL [8]	D-OCEL [59]	iDO-CEM [58]	DS [57]	Simulator [54]	TS log [31]	OCEL 2.0 [60]	NICE log
Requirements	R1: Store high-level events	✓	✓	✓	✓	✓	✓		✓	✓
	R2: Store low-level events				✓	✓	✓	✓		✓
	R3: Store intermediary/mixed-level events						✓			✓
	R4: Enable traceability between high-level and low-level events				✓	✓	✓			✓
	R5: Represent context at event level	✓	✓	✓	✓	✓	✓		✓	✓
	R6: Represent context at activity level			✓	✓	✓			✓	✓
	R7: Represent context at case/object level	✓	✓	✓	✓	✓			✓	✓
	R8: Represent context at process level	✓		✓	✓		✓		✓	✓
	R9: Update context parameters independently from process events						✓	✓		✓
	R10: Update context parameters at a higher frequency				✓	✓	✓	✓		✓

11 Conclusion

In this paper, we presented a new format for IoT-enhanced event logs. This format combines the different event types presented in [1] with the notion of object from OCEL to form a very flexible format, capable of integrating IoT data with process data with minimal information loss. The format was evaluated in Sects. 7.3 and 8, where we gave evidence of its theoretical robustness and its practical usefulness, respectively. In Sect. 10, we also compared our new format with existing alternatives and showed that it is the only one fulfilling the requirements stated in Sect. 6. Moreover, a first practical application demonstrated how generating a log following our format enables process analyses integrating IoT data.

Some limitations of our work include the limited availability of NICE logs to experiment with so far and the potential amplifying effect of multiple event abstraction steps on sensor data quality issues. To cope for these limitations, in future works, we first plan to publish NICE logs from real-life processes and further develop the format by creating more tool support to generate and manipulate logs. Next to this, we would like to design new analysis techniques taking advantage of the capabilities of the new format, e.g. for IoT-enhanced trace clustering, IoT-enhanced predictive process monitoring with context events. Also, looking into the compatibility of NICE logs with the OCEL 2.0 format could help leveraging existing techniques and tools. Finally, we consider investigating the similarities of NICE logs with other existing IoT-enhanced log formats (e.g. DataStream) to further advance interoperability and facilitate data creation and exchange in the community.

Acknowledgments The work of Yannis Bertrand and Estefanía Serral was supported by the Flemish Fund for Scientific Research (FWO) with grant number G0B6922N. The work of Francesco Leotta and Silvestro Veneruso was supported by the Sapienza University of Rome with grant number RM120172B3B5EC34.

References

1. Bertrand, Y.,De Weerdt, J.,Serral, E.: An expert-validated bridging model for IoT process mining. Bus. Inform. Syst. Eng. **66**, 737–756 (2024)
2. Janssen, D., Mannhardt, F., Koschmider, A., Zelst, S.J.: Process model discovery from sensor event data. In: ICPM 2020, pp. 69–81 (2020)
3. Seiger, R., Zerbato, F., Burattin, A., Garcia-Banuelos, L., Weber, B.: Towards IoT-driven process event log generation for conformance checking in smart factories. In: EDOCW, pp. 20–26 (2020)
4. Soffer, P., et al.: From event streams to process models and back: challenges and opportunities. Inform. Syst. **81**, 181–200 (2019)
5. Serral, E., De Smedt, J., Vanthienen, J.: Making business environments smarter: a context-adaptive petri net approach. In: UIC 2014, pp.343–348 (2014)
6. Bertrand, Y., De Weerdt, J., Serral, E.: Assessing the suitability of traditional event log standards for IoT-enhanced event logs. In: International Conference on Business Process Management, pp. 63–75 (2022)

7. Günther, C.W., Verbeek, E.: XES standard definition (2014)
8. Ghahfarokhi, A.F., Park, G., Berti, A., Aalst, W.M.P.: OCEL: A Standard for Object-Centric Event Logs. CCIS, vol. 1450, pp. 169–175 (2021)
9. Janiesch, C., Koschmider, A., Mecella, M., Weber, B., Burattin, A., Di Ciccio, C., Fortino, G., Gal, A., Kannengiesser, U., Leotta, F., The internet of things meets business process management: a manifesto. IEEE Syst. Man Cybern. Mag. **6**(4), 34–44 (2020)
10. Beerepoot, I., Di Ciccio, C., Reijers, H.A., Rinderle-Ma, S., Bandara, W., Burattin, A., Calvanese, D., Chen, T., Cohen, I., Depaire, B.: The biggest business process management problems to solve before we die. Comput. Ind. **146**, 103837 (2023)
11. Eck, M.L., Sidorova, N., Aalst, W.M.P.: Enabling process mining on sensor data from smart products. In: RCIS, pp. 1–12(2016)
12. Koschmider, A., Janssen, D., Mannhardt, F.: Framework for process discovery from sensor data. In: EMISA, p. 8 (2020)
13. Mannhardt, F., Bovo, R., Oliveira, M.F., Julier, S.: A taxonomy for combining activity recognition and process discovery in industrial environments (2018)
14. Trzcionkowska, A., Brzychczy, E.: Practical aspects of event logs creation for industrial process modelling. MAPE **1**(1), 77–83 (2018)
15. Valencia-Parra, A., Ramos-Gutierrez, B., Varela-Vaca, A.J., Gomez-Lopez, M.T., Bernal, A.G.: Enabling process mining in aircraft manufactures: extracting event logs and discovering processes from complex data, pp. 166–177 (2019)
16. Aalst, W., Weijters, T., Maruster, L.: Workflow mining: discovering process models from event logs. TKDE **16**(9), 1128–1142 (2004)
17. Leemans, S.J.J., Fahland, D., Aalst, W.M.P.: Discovering Block-Structured Process Models from Event Logs Containing Infrequent Behaviour. LNBIP, vol.171, pp. 66–78 (2014)
18. Weijters, A., Aalst, W.M.P., Medeiros, A.K.: Process mining with the heuristics miner algorithms (2006)
19. Leoni, M., Aalst, W.M.P.: Data-aware process mining: discovering decisions in processes using alignments. In: SAC '13, p. 1454 (2013)
20. Mannhardt, F., Leoni, M., Reijers, H.A., Aalst, W.M.P.: Data-Driven Process Discovery - Revealing Conditional Infrequent Behavior from Event Logs. LNCS, vol. 10253, pp. 545–560 (2017)
21. Seiger, R., Malburg, L., Weber, B., Bergmann, R.: Integrating process management and event processing in smart factories: a systems architecture and use cases. J. Manuf. Syst. **63**, 575–592 (2022)
22. Ahmadi, Z., De Weerdt, J., Serral, E.: A novel contextualization method for process discovery using activity specialization hierarchies. In: BPMDS'2024 (2024)
23. Song, M., Günther, C.W., Aalst, W.M.P.: Trace Clustering in Process Mining. LNBIP, vol. 17, pp. 109–120 (2009)
24. De Weerdt, J., Broucke, S., Vanthienen, J., Baesens, B.: Active trace clustering for improved process discovery. TKDE **25**(12), 2708–2720 (2013). https://doi.org/10.1109/TKDE.2013.64
25. Jablonski, S., Rögliner, M., Schönig, S., Wyrtki, K.M.: Multi-perspective clustering of process execution traces. EMISAJ, 2–122 (2019)
26. Bertrand,Y., De Weerdt,J., Serral, E.: A novel multi-perspective trace clustering technique for IoT-enhanced processes: a case study in smart manufacturing. In: International Conference on Business Process Management, pp. 395–412. Springer, Berlin (2023)
27. Mannhardt, F.: Multi-perspective process mining. Ph.D. Thesis, Eindhoven University of Technology (2018)
28. Banham, A., Wynn, M.T.: XPM: a framework for process mining with exogenous data, 12 (2021)
29. Dunzer, S., Stierle, M., Matzner, M., Baier, S.: Conformance checking: a state-of-the-art literature review. In: S-BPM ONE '19 Proceedings, pp. 1–10 (2019)

30. Werf, J.M.E.M., Verbeek, H.M.W., Aalst, W.M.P.: Context-Aware Compliance Checking. LNCS, vol. 7481, pp. 98–113 (2012)
31. Rodriguez-Fernandez, V., Trzcionkowska, A., Gonzalez-Pardo, A., Brzychczy, E., Nalepa, G.J., Camacho, D.: Conformance checking for time-series-aware processes. IEEE TII **17**(2), 871–881 (2021). https://doi.org/10.1109/TII.2020.2977126
32. Bazhenova, E., Buelow, S., Weske, M.: Discovering Decision Models from Event Logs. LNBIP, vol. 255, pp. 237–251 (2016)
33. Scheibel, B., Rinderle-Ma, S.: Decision mining with time series data based on automatic feature generation. In: International Conference on Advanced Information Systems Engineering, pp. 3–18. Springer, Berlin (2022)
34. Nakatumba, J.: Analyzing resource behavior using process mining. In: BPI'09, pp. 69–80 (2010)
35. Di Francescomarino, C., Ghidini, C., Maggi, F.M., Milani, F.: Predictive process monitoring methods: which one suits me best? arXiv:1804.02422 [cs] (2018)
36. Bozkaya, M., Gabriels, J.:Process diagnostics: a method based on process mining, 7 (2009)
37. Popova, V., Fahland, D., Dumas, M.: Artifact lifecycle discovery. arXiv:1303.2554 [cs] (2013)
38. Fahland, D., Montali, M., Lebherz, J., van der Aalst, W.M.P., van Asseldonk, M., Blank, P., Bosmans, L., Brenscheidt, M., Di Ciccio, C., Delgado, A., et al.: Towards a Simple and Extensible Standard for Object-Centric Event Data (OCED)—Core Model, Design Space, and Lessons Learned (2024). arXiv preprint arXiv:2410.14495
39. Leotta, F., Mecella, M., Mendling, J.: Applying process mining to smart spaces: perspectives and research challenges. In: CAiSE 2015 Workshops, pp. 298–304. Springer, Berlin (2015)
40. Zelst, S.J., Mannhardt, F., Leoni, M., Koschmider, A.: Event abstraction in process mining: literature review and taxonomy. Granular Comput. **6**, 719–736 (2021)
41. Koschmider, A., Mannhardt, F., Heuser, T.: On the Contextualization of Event-Activity Mappings. LNBIP, vol. 342, pp. 445–457. Springer, Berlin (2019)
42. Dimaggio, M., Leotta, F., Mecella, M., Sora, D.: Process-based habit mining: Experiments and techniques. In: UIC 2016, pp. 145–152. IEEE, Piscataway (2016)
43. Leotta, F., Mecella, M., Sora, D.: Visual process maps: a visualization tool for discovering habits in smart homes. JAIHC **11**(5), 1997–2025 (2020)
44. Bertrand, Y., Abbeele, B., Veneruso, S., LeottalF., Mecella, M., Serral, E.: A survey on the application of process discovery techniques to smart spaces data. EAAI **126**, 106748 (2023)
45. Banham, A., Leemans, S.J., Wynn, M.T., Andrews, R., Laupland, K.B., Shinners, L.: XPM: enhancing exogenous data visibility. AI Med. **133**, 102409 (2022)
46. Leoni, M., Pellattiero, L.: The benefits of sensor-measurement aggregation in discovering IoT process models: a smart-house case study. In: BPM 2021, pp. 403–415 (2021)
47. Esposito, L., Leotta, F., Mecella, M., Veneruso, S.: Unsupervised segmentation of smart home logs for human habit discovery. In: IE'22, pp. 1–8. IEEE, Piscataway (2022)
48. Janowicz, K., Haller, A., Cox, S.J.D., Le Phuoc, D., Lefrançois, M.: Sosa: a lightweight ontology for sensors, observations, samples, and actuators. J. Web Semantics **56**, 1–10(2019)
49. Franceschetti,M., Seiger, R., Weber, B.: An event-centric metamodel for IoT-driven process monitoring and conformance checking (2023)
50. Torres, V., Serral, E., Valderas,P., Pelechano, V., Grefen, P.: Modeling of IoT devices in business processes: a systematic mapping study. In: CBI 2020, vol. 1, pp. 221–230 (2020)
51. Zerbato, F., Seiger, R., Federico, G.D., Burattin, A., Weber, B.: Granularity in process mining: can we fix it? 40–44 (2021)
52. Janiesch, C., Koschmider, A., Mecella, M., Weber, B., Burattin, A., et al.: The internet-of-things meets business process management: a manifesto. IEEE Syst. Man Cybern. Mag. **6**(4), 34–44 (2020). arXiv: 1709.03628
53. Hevner, A.R., March, S.T., Park, J., Ram, S.: Design science in information systems research. MIS Q. 75–105 (2004)
54. Veneruso, S., Bertrand, Y., Leotta, F., Serral, E., Mecella, M.: A model-based simulator for smart homes: enabling reproducibility andstandardization. JAISE **15**(2), 143–166 (2023)

55. Unterberg, L.T.G., Koren, I., Aalst, W.M.: Maximizing reuse and interoperability in industry 4.0 with a minimal data exchange format for machine data. Modellierung 2024, 103 (2024)
56. Goossens, A., De Smedt, J., Vanthienen, J., Aalst, W.M.: Enhancing data-awareness of object-centric event logs. In: ICPM 2022, pp. 18–30. Springer, Berlin (2022)
57. Mangler, J., Grüger, J., Malburg, L., Ehrendorfer, M., Bertrand, Y., Benzin, J.-V., Rinderle-Ma, S., Serral Asensio, E., Bergmann, R.: Datastream xes extension: embedding IoT sensor data into extensible event stream logs. Future Internet **15**(3), 109 (2023)
58. Verbruggen, C., Goossens, A., De Smedt, J., Vanthienen, J., Snoeck, M.: idocem. In: Software and Systems Modeling, pp. 1–25 (2024)
59. Goossens, A., De Smedt, J., Vanthienen, J., Aalst, W.M.: Enhancing data-awareness of object-centric event logs. In: Process Mining Workshops: ICPM 2022 International Workshops, Bozen-Bolzano, Italy, October 23–28, 2022, Revised Selected Papers, pp. 18–30. Springer, Berlin (2023)
60. Berti, A., Koren, I., Adams, J.N., Park, G., Knopp, B., Graves, N., Rafiei, M., Liß, L., Unterberg, L.T.G., Zhang, Y., et al.: Ocel (object-centric event log) 2.0 specification (2024). arXiv preprint arXiv:2403.01975
61. Snoeck, M.: Enterprise information systems engineering. The MERODE Approach (2014)

The Internet of Production: Accelerating the Optimization of Production Processes via Models-in-the-Middle

Leah Tacke Genannt Unterberg, István Koren, and Wil M. P. van der Aalst

1 Introduction

The chances and challenges connected to Industry 4.0 are a current hot topic in research and industry alike. While it promises data-based and AI-driven technological advances, a major challenge arises from the subsequent need for effective data management and utilization strategies. The former calls for data interoperability, while effective data utilization in turn requires close interdisciplinary and inter-domain cooperation. In multidisciplinary engineering environments, this is further complicated, as even the ability to communicate clearly is not guaranteed due to domain-dependent usage of terms. For both aspects, siloization, i.e., lack of interaction and exchange, leads to wasted and duplicated efforts, as well as missed opportunities.

On the one hand, data producers/owners can benefit from access to analytics software solutions of peers who are using similar data. On the other hand, data science researchers can benefit from access to varied data sets that their peers working with conceptually similar data may be using. The gap lies in recognizing which data is conceptually similar and actually making it interoperable. To bridge it, we propose the creation of intermediate meta-models that can serve as common exchange formats. These *Models-in-the-Middle* (MitM) facilitate not just practical data exchange but also enable implementation of configurable algorithms and transformations. Choosing the appropriate level of abstraction is difficult, as models that are too low (too close to source data) do not provide much value because they cannot cover many data sources, while models too high up need complicated and

L. Tacke Genannt Unterberg · I. Koren · W. M. P. van der Aalst (✉)
Chair of Process and Data Science, RWTH Aachen University, Aachen, Germany
e-mail: l.tgu@pads.rwth-aachen.de; koren@pads.rwth-aachen.de; wvdaalst@pads.rwth-aachen.de

G. Fortino, M. Mecella (eds.), *Internet of Things Meets Business Process Management*, Internet of Things, https://doi.org/10.1007/978-3-031-90746-3_11

unique mappings from the said source data—defeating their point as intermediate models.

We focus on OCED (Object-Centric Event Data) and MAED (Measurement and Event Data), which tackle two very relevant types of data in regard to "BPM meets IoT." OCED is the evolution of the traditional event data formalism employed in process mining. MAED is intended to closely correspond to the actual outputs of machines and other networked IoT devices such as sensors. These assets are also commonly referred to as cyber-physical systems (CPS). As pointed out in the *The Internet of Things Meets Business Process Management Manifesto* [48], bringing such data into the BPM discipline is of great importance, considering the availability and potential impact.

The rest of the paper is structured as follows. First, we discuss related work in Sect. 2. Next, we elaborate on the idea and necessity of *Models-in-the-Middle* in Sect. 3 and present the first proposals for OCED and MAED in Sect. 4 and Sect. 5, respectively. We present the potential of connecting MitM with each other in Sect. 6, leading into a discussion of their benefits in Sect. 7. Lastly, we provide an outlook and conclude in Sect. 8.

2 Related Work

Some related work lies within the process mining domain, as it is a field that has from its beginnings focused on operationalizing existing event data for specialized algorithms. There are also various standardization initiatives for IoT data coming from industry.

Meta models for the integration of IoT data with higher-level business process event data have been proposed by Bertrand et al. [26, 27] and Mangler et al. [63]. The former introduce an XML-based format and attempt to model the connection between low-granularity IoT data and coarse event data at the business process level. The model combines an object-centric and data-centric view. However, it is also very event-centric, in that it treats sensor readings as *IoT events* without considering them as connected time series. This view can miss out on time series-specific analyses, e.g., interpretation in the frequency domain. Though their concept of *analytics* covers this conceptionally, it requires further, unspecified conversions to implement.

The latter propose DataStream as a XES extension, thereby intending to enrich traditional event data with sensor data. Their data model implementation is originally XML-based but they additionally define a YAML-based representation for performance reasons. Still, their model treats measurement data as individual, untyped JSON data points. This complicates automated algorithmic usage as the *schema* of the data streams has to be inferred and is not enforced to be consistent.

Potential applications of integrating time series data into process mining are numerous. The authors of [17] propose a structure to link exogenous time series data to process events via linking, slicing, and transformation functions. They motivate the necessity of this data integration in the medical field. Other application domains are in smart homes [42] and smart manufacturing [40].

The mentioned existing approaches focus on the process mining side of things and treat IoT data as a source of derived events and process context information. These efforts can be understood as MitM between MAED and OCED. For one, they often still require non-standardized raw data to be transformed and enriched with process-level context. Additionally, the introduction of BPM-specific concepts makes them unsuitable as low-level exchange formats for general IoT data.

Other relevant research lies in the field of data management/access and semantic web technologies (SWT). These, often ontology-based, technologies focus on making richly semantically annotated data from heterogeneous sources accessible via common ontologies. Research in cyber-physical production systems (CPPS) has pointed to SWT as a facilitator for unified data access [28]. Ekaputra et al. survey different ontology-based approaches to integrate separate data sources in multidisciplinary engineering environments (MDEE) in [41]. They categorize approaches into classes based on how unifying/higher-level ontologies are used in the creation/connection of separate data sources. Our proposal of meta-models fits into this by enabling domain-specific ontology-based instantiation and thus semantically interoperable concrete models.

IoT data standardization itself appears mostly driven by practice rather than research. Major industry initiatives are the OPC UA companion specifications,[1] IDTA[2] asset administration shell, and Automation ML e.V.[3] The first two propose highly detailed data models/specifications for relevant types of IoT data, while Automation ML is itself a modeling language for hierarchical industrial IoT systems. The three players vow to collaborate instead of duplicating their work [39].

As a result of having to appeal to a vast number of stakeholders, these standardization efforts are less concept-driven and instead heavily focus on being comprehensive. Trading algorithmic accessibility and simplicity for detailed representational power positions such very wide models closer to, or even *as*, raw data. Their complexity makes it very difficult to design configurable (research) software for them. As MitM, these data models could rather serve as input for MAED and OCED. Technologies connecting these specifications to formal data models [54, 72] may help in achieving this in a semiautomated manner.

3 Toward Models-in-the-Middle

A large project in service of Industry 4.0 is the Internet of Production (IoP) at RWTH Aachen University. It is an ambitious research project that envisions the creation

[1] https://opcfoundation.org/about/opc-technologies/opc-ua/
[2] https://industrialdigitaltwin.org/
[3] https://www.automationml.org/

of a network of interconnected production systems akin to a "World Wide Lab." The project brings together over 35 research groups from engineering disciplines and computer science. About 200 researchers are involved, representing expertise in various manufacturing processes like textile processing, injection molding, metal forming, additive manufacturing, milling, etc., as well as materials science, production technology, and data science. The participating institutes contribute with not only leading research but also extensive industry-grade physical assets: each represented manufacturing domain has shopfloors of machines and equipment. The focus ranges from singular machines to entire assembly lines. This is already a source for one vocabulary confusion: the notion of *process*. Researchers with the former focus may think of the discrete or continuous operation that their machine performs on a workpiece, while people who consider assembly lines and production chains understand *the process* as all operations performed to produce a product.

The project employs the concept of *Digital Shadows* [22, 49] as a means to optimize and connect production processes over various stages of the product life cycle. They can be regarded as the part of *Digital Twins* that mirrors the state of real-world assets. Digital Shadows are intended to be task-specific and purpose-built collections of data and models for analyses at different "ranges"—from machine tool optimization to improvement of entire value streams or processes.

To generate knowledge with Digital Shadows, raw data from various sources has to be transformed, refined, unified, and ingested by data science techniques like statistical methods, machine learning, AI, and specialized approaches such as process mining (PM). It is clear that interoperable data models and shared understanding of concepts are pivotal in achieving this efficiently. To chart the data landscape in the IoP, a survey about what data researchers use was conducted among the IoP members. Over 80 data models and descriptions were collected and analyzed. Submissions were tagged with high-level categories in an effort to identify their commonalities. Category frequencies are shown in Fig. 1. The most prevalent categories are *Machine* (54/82) and *Process* (54/82), followed by *Material* (39/82) and *Simulation and Optimization* (24/82). The *Machine* category encompasses machine configuration/master data but predominantly consists of live IoT data collected during machine operation.

While submissions within the categories share concepts to a great extent, almost all of them use unique, use-case-specific, data models. On top of that, the data sources are extremely heterogeneous in syntax, i.e., using application-specific data formats and storage. Standardization is typically not applied. As a consequence, there is a lack of data interoperability. Most use cases use ad hoc data transformations, custom tooling, and in-house implementations. Reuse of implemented approaches or systems is almost nonexistent.

We envision that intermediate meta-models, so-called *Models-in-the-Middle* (MitM), can help to efficiently connect data to method. Equipping such data models with concrete exchange formats tackles the data interoperability gap, while the selection of minimal but powerful relevant concepts connects languages, both natural and programming, of cooperating disciplines. Figure 2 exemplifies the contrasting situations. Each arrow represents a data transformation from source data

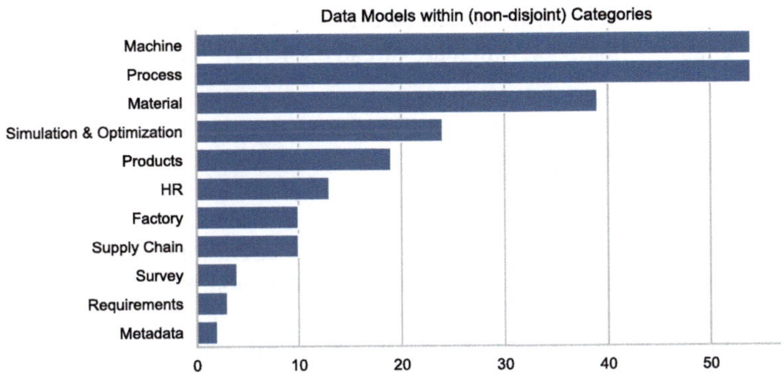

Fig. 1 The number of data models from the IoP project-wide survey within each top-level category. Submissions can belong to multiple categories

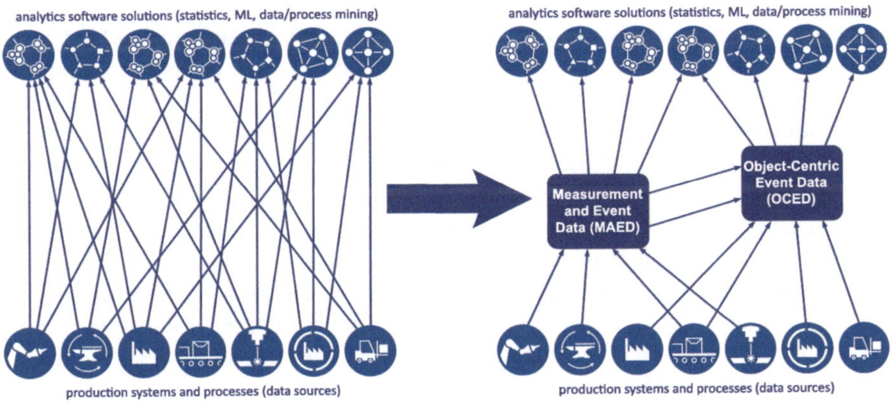

Fig. 2 Connecting source data and data science (AI, ML, statistical, etc.) approaches. The status quo with mostly one-to-one pipelines on the left versus the intended state on the right where approaches developed for meta-models can be effectively reused. Presented with the two exemplary meta-models for MAED (Measurement and Event Data) and OCED (Object-Centric Event Data).

to suitable input for an analytics technique. Even very similar data sets require unique connections which is why MitM intend to bundle the connections and provide configurable and reusable connections to specific data science approaches at the top. In the following, we present our proposals for the first two MitM, OCED and MAED.

4 Object-Centric Event Data (OCED)

In this section, we discuss Object-Centric Event Data: its roots in traditional event logs, why it is necessary, how it can be represented, and what can be done with it.

In particular, we present our proposed MitM for OCED and illustrate the types of analyses it enables—many of which are already software-supported.

4.1 The Success of XES

Process mining has since its infancy focused on operationalizing existing event data for algorithmic usage. The availability of shareable data sets and a unified understanding of *what event data is* was pivotal in the development of data-based algorithms like process discovery. An agreement on the minimal concepts that make up event data—traditionally *case id*, *time*, and *activity*—allowed researchers to develop *compatible* and *reusable* approaches. Algorithms could easily be applied to event logs from different original data sources and even be chained, e.g., to facilitate generic pre-processing.

This is how the standardization of XES (eXtensible Event Stream) [44] as an XML-based exchange format for event data helped the research software platform ProM (Process Mining Workbench) to flourish. The ProM ecosystem has supported the creation of, at this point, thousands of PM algorithms. It served as a development test bed for methods that were eventually picked up by commercial vendors, such as classically Disco by Fluxicon (a direct spin-off) and nowadays highly relevant Celonis, Software AG, and SAP Signavio. More modern community-driven software frameworks such as pm4py [25] have since been created, but they still rely on the standardization of the inputs and outputs. Pondering the current state of the art had there not been a shared understanding of what minimally constitutes event data, that was underpinned by concrete formalized data format, may be interesting for the reader to consider.

Adoption of PM in industry has been increasing significantly. Fortune Business Insights reported that the process mining software market was valued at USD 1.13 billion in 2022 and projected cumulative aggregated growth rates (CAGR) of 40–50% until 2030 [2]. Similarly, Gartner has predicted CAGR of above 30% between 2021 and 2025 [1]. Generally, the application and impact of the discipline are expected to grow. Recently and also increasingly, an object-centric view on processes has gained consideration. The perspective shift from *traditional* event data to *Object-Centric Event Data* also requires a more powerful data model.

4.2 Introducing the OCED Meta-Model

The demand for OCED stems from the observation and concession that in real life, processes are typically never isolated but rather interact and influence each other via the involved entities (objects). This idea is not new [5, 66]; however, the importance of this perspective has become more apparent as the goals shift toward more holistic process analyses. Beerepoot et al. allude to this with *expansive BPM* as one of the key challenges to tackle by the BPM discipline [20]. Objects can be thought of abstractly as potentially anything, but practically they are often already present in ERP systems—the main data source for PM, e.g., requests, orders, invoices, and deliveries. In a production setting, they could refer to machines, workpieces, robots, operators, materials, etc. It is clear that when looking at these settings broadly, there is no clear "most important" case notion. Essentially, instead of neat independent sequences of events, the event data is a graph of objects and events.

Note that it is important to distinguish between instance and instance type, e.g., objects and object types, just like event and activity (event type). Each object can be considered as a potential case identifier, i.e., used to identify events "that belong together." Object types can then be regarded as case notions. However, analyzing a general graph of events and objects by projecting on individual paths generated from this case notion, which is commonly referred to as *flattening*, does not have sufficient representational power to not introduce discrepancies. In the literature, two primary issues were coined: *convergence* and *divergence* [3]:

Convergence Events referring to multiple object instances of the selected type are duplicated in multiple cases and can as such be "double-counted," leading to misleading statistics.

Divergence Events of the same type referring to the same object of the selected type but differing with regard to other object types cannot be differentiated.

For a more detailed discussion of these issues, we refer the reader to related literature [3, 4].

The recently proposed OCEL 2.0 [24] serves as a meta-model for OCED. We reprint it here in Fig. 3 in UML-like notation.

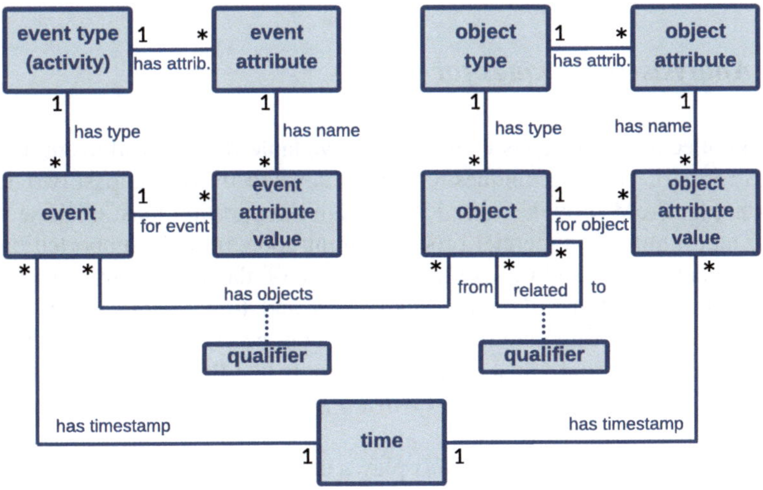

Fig. 3 The proposed meta-model for OCED

The meta-model describes how the main concepts can be represented relationally. The two squares of event (top left)/object (top right), type, attribute, and attribute value can be regarded as "gadgets" (template graph structures) that amalgamate the schema, i.e., type information, together with instances. These cycles connect events/objects to their type that in turn defines which named attributes they may possess. While these "typing gadgets" make the diagram seem more complex than it is, the essential concepts are *events*, *objects*, their *relationships*, *attributes*, and *time*.

Events always have an associated timestamp and can refer to objects via a qualified relation. That means an event can have different *types* of relations to objects. As a simple example, an event of type *transport* may be connected to two objects of type *location*, which are qualified as *origin* and *destination*, respectively. Without the qualification, a transport event would be completely ambiguous. Next, objects can be related to each other via a qualified relation. This allows modeling how the objects themselves are statically related. Objects also have a dynamic aspect, in that their attribute values can have a timestamp associated with them. With this, it is possible to model that the state of an object, as expressed by its attributes, can change over time, independently of recorded events.

The representational power of this data model is very high, as highly dynamic and complex data can be mapped to it. At the same time, it is minimal with respect to its core concepts. This makes it broadly applicable in regard to ease of *creation* of data sets. Furthermore, the typing makes it very accessible for automated algorithmic usage and thus facilitates efficient and expressive *utilization*. As it strikes a good balance between these two aspects, it is ideal as a MitM. To further elaborate on the algorithmic and semantic accessibility of data with these concepts, we present an overview over possible analysis techniques.

4.3 Analysis Techniques for OCED

A large collection of analysis techniques is available for classical event data (e.g., in XES format). These techniques have been developed over the past two decades. For example, ProM provides over 1500 plug-ins supporting most of these process mining algorithms. Commercial process mining tools initially supported only few of these analysis capabilities. Often the focus of these tools was on Directly-Follows Graphs (DFGs) showing frequencies and times. However, the process mining market is rapidly maturing with many commercial systems providing more advanced features initially just supported by tools like ProM. The next step in this evolution is the adoption of *Object-Centric Process Mining* (OCPM) using OCED as input.

Next to event data, *process models* play a central role in process mining. These models can be descriptive (de facto) as well as normative (de jure). The "real process" is considered as the actual target of analyses with event data and process models serving as proxies for it. Based on two decades of process mining research, there exists a common categorization of methods into [5]:

- *Discovery*
- *Conformance*
- *Process enhancement*
- *Operational support*

Discovery uses data to infer a model, conformance quantifies or highlights discrepancies between data and a model, process enhancement transfers insights from data and model to the real-life process in terms of adaptations, while operational support does this in an online fashion and focuses on prediction and next-action recommendation.

Moving from classical event logs to OCED and object-centric process models, all traditional techniques need to be "reinvented" to deal with multiple object types. There are two key ingredients in OCED that make things markedly different: the qualified *event-to-object* (E2O) relationships that allow for events referring to multiple objects rather than a single case and the qualified *object-to-object* (O2O) relationships that can be used to model object relations, for example, a "part of" relationship [4].

There exist many *process discovery* techniques that can be applied to *flattened* event logs, i.e., [7, 14, 15, 31, 55, 82]. These can be leveraged to deal with multiple object types. The baseline approach first discovers a process model per object type and then folds the different models. When creating an event log per object type, we need to "flatten" the events, i.e., events are replicated for each object of the selected type that is involved in the event. The resulting flattened event log can be used as input for any discovery algorithm that does not create duplicate activities, i.e., it is essential that there is just one modeled activity for each activity in the event log. Most process-discovery algorithms have this property, e.g., the alpha algorithm [7], DFG-based algorithms, and the family of inductive discovery algorithms [55]. Therefore, this is no real limitation. After discovering a process model per object type, we need to merge these models using the joint activities. To do this, there are two challenges. First of all, the frequencies are no longer correct due to convergence problems [4]. Therefore, we need to transform the activity frequencies back to the original frequencies recorded in the event log. This is possible by simply reporting the original frequencies and enriching connections to reflect the number of objects involved. In a DFG or Petri net, connections typically model a single case following through the model, and now we need to show that there may be a variable number of objects involved. This is done by showing cardinalities on the connections, e.g., one or more objects of a given type are involved. Second, we need to merge the models in such a way that the semantics are preserved. This is ensured by having unique activities in the model and assuming that the different object types are synchronized. This was problematic in earlier approaches considering multiple objects that communicated instead of being involved in shared events. See, for example, the Proclet-based and artifact-centric approaches that are more involved and require input that is often not available [43]. Currently, there are various object-centric discovery approaches producing object-centric DFGs [23], object-centric Petri nets [6], and object-centric BPMN models. Note that the inductive mining approach produces process trees that can be mapped onto Petri nets or BPMN models having unique activities. This is often used in the background. For example, the Celonis Process Adherence Manager (formerly known as Process Sphere) used the inductive mining approach to create object-centric BPMN models.

For *conformance checking*, we follow the same approach but in the other direction. Here, we start from an object-centric process model. This could be an object-centric DFG, an object-centric Petri net, or an object-centric BPMN model. To check conformance, we first flatten both the event logs and process model per object type and then use standard conformance checking approaches such as token-based replay and alignments [71]. Next to this, it is possible to check cardinality constraints directly on the event log. For example, if the model says that an activity a always involves one object of type x, one or more objects of type y, and at most two objects of type z, then this can be checked directly on the OCED. It is also possible to use more sophisticated techniques as described in [61]. This approach creates alignments for multiple objects also considering their relations.

In OCED, there is a not a single case notion and therefore the *scope of a process* is not so clear. In this new setting, a process can be seen as a selection of event types (activities) and object types, i.e., a view. Moreover, the notion of a *process instance* (i.e., a case notion) is missing. However, it is possible to define new process instance notions also called process execution that relate to connected events and objects [10].

For *process enhancement*, we want to enrich process models with additional information, e.g., timing information or decision logic. This involves using event attributes and object attributes (see meta-model). Here, it is possible to use the same strategy: simply flatten event data and process model to separate the object types and use the existing process enhancement techniques. For example, token-based replay and alignments can be used to relate event attributes and object attributes (i.e., case data) to model elements. The tricky thing is again in the merging of models and relating the different object types. However, resources can now be seen as objects rather than attributes, which allows for more realistic models also incorporating capacity constraints and resource availability. Most of the approaches to lift analysis techniques to multiple object types are still baseline approaches. See, for example, the approach to create object-centric simulation models [52].

Operational support refers to techniques used to influence running process executions, for example, the detection of performance and compliance problems such as an order being late or a check being skipped. When this is done at runtime, it is possible to generate an alert or create an intervention. This can be combined with predictive techniques and recommender systems. Having information about multiple objects helps to improve the quality of operational support. In this context, we also talk about action-oriented process mining or digital twins [67].

Object-Centric Process Mining (OCPM) broadens the horizons of traditional process mining. Unlike traditional methods that focus on a single case (notion), OCPM facilitates viewing all operational activities from any perspective, employing a single, consistent source of truth. Changing viewpoints does not necessitate additional data extractions and offers the flexibility of on-demand process mining views. It also helps in avoiding traditional distortions like divergence and convergence while identifying issues involving multiple object types, often found at the intersection points of processes and organizational units. For instance, delays in customer orders could be attributed to various factors such as production issues, logistic challenges, procurement problems, or understaffing.

The OCEL 2.0 standard provides a systematic way of storing OCED. Moreover, there are several open-source tools supporting the above techniques, some of which we apply in the following application. Also commercial tools such as Celonis have adopted OCPM by supporting the discovery of object-centric DFGs, object-centric BPMN models, object-centric conformance checking, and object-centric performance analysis. However, OCED and OCPM are limited to discrete events. Therefore, we would like to leverage our experiences with OCEL/OCED Models-in-the-Middle and also create similar models for processes with continuous data.

4.4 Example Application

Various tooling has already been created for the accompanying data format of OCEL 2.0,[4] which supports JSON, XML, and *sqlite* file formats. The popular process mining software pm4py [25] includes support for data import/export as well as object-centric discovery. There is also a fully web-based visualization tool[5] that provides a quick glimpse into OCEL 2.0 instances. Additionally, there are ocpa [9] and ocpi [8], which implement some of the previously mentioned methods.

As an example, we use a publicly accessible synthetic event log about a logistics process [51]. Figure 4 gives an overview of the participating object types in the form of a UML-like diagram. The edges show the static object-to-object (O2O) relations (directed edges) as well as their dynamic interactions, which are derived from co-occurrence in events (undirected edges). The diagram also lists the attributes associated with the object types, like the *Amount* and *Status* of a *Transport Document*. The latter is actually a dynamic attribute, as the status changes over the course of a transport document's life cycle, as exemplified in Table 1 (reproduced output of *ocelot* tooling).

To get some insight into how these objects interact, we employ *discovery*. A mined colored Petri net is shown in Fig. 5. It describes the interactions between *customer orders*, *transport documents*, and *vehicles*.

[4] https://www.ocel-standard.org/

[5] https://ocelot.pm/

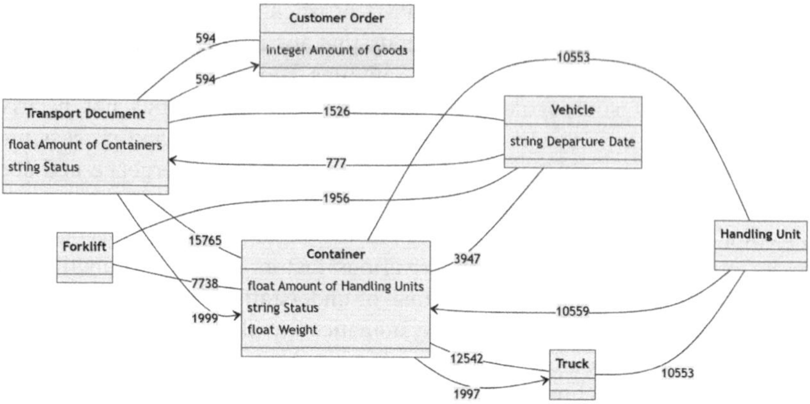

Fig. 4 A UML-like diagram of object-to-object (O2O) relations. Static ones are directed arcs, while derived dynamic co-occurrence relations are represented by undirected edges

Table 1 The status change history of a transport document

Time	Value
2023-05-23T08:22:17	
2023-05-30T08:58:48	In transit
2023-06-09T09:00:00	Shipped

We can then employ a new type of *conformance checking* on, e.g., the E2O multiplicities. Table 2 shows the results on the considered log (reproduced from *ocpm.info* tooling). Drilling down into the last listed deviation, we can find the departures, and by extension, vehicles, containers, and transport documents where there were abnormally few or many of the latter. This allows identifying under which circumstances vehicles are under- or overloaded.

Another common basic analysis technique in PM is *variant analysis*. Traditionally, they summarize similar cases, i.e., typically via the control-flow perspective. Lifted to the object-centric paradigm, they aggregate different patterns of interactions between object types [11]. Figure 6 shows an example based on the analyzed log. We can see that in over half of the "cases" involving customer orders, transport documents, and vehicles, there are no vehicles involved: we might infer that these orders have not been fulfilled yet. Similarly, the next 16% of instances do not even have a transport document attached to them, indicating that these orders remained completely unhandled at the point of export. The third most common variant then shows what one may expect as a more typical flow: a customer order is first created and registered, then a transport document is created for it, a vehicle is booked for that order, and it is loaded with multiple containers. Other variants show instances where multiple vehicles were booked to fulfill an order.

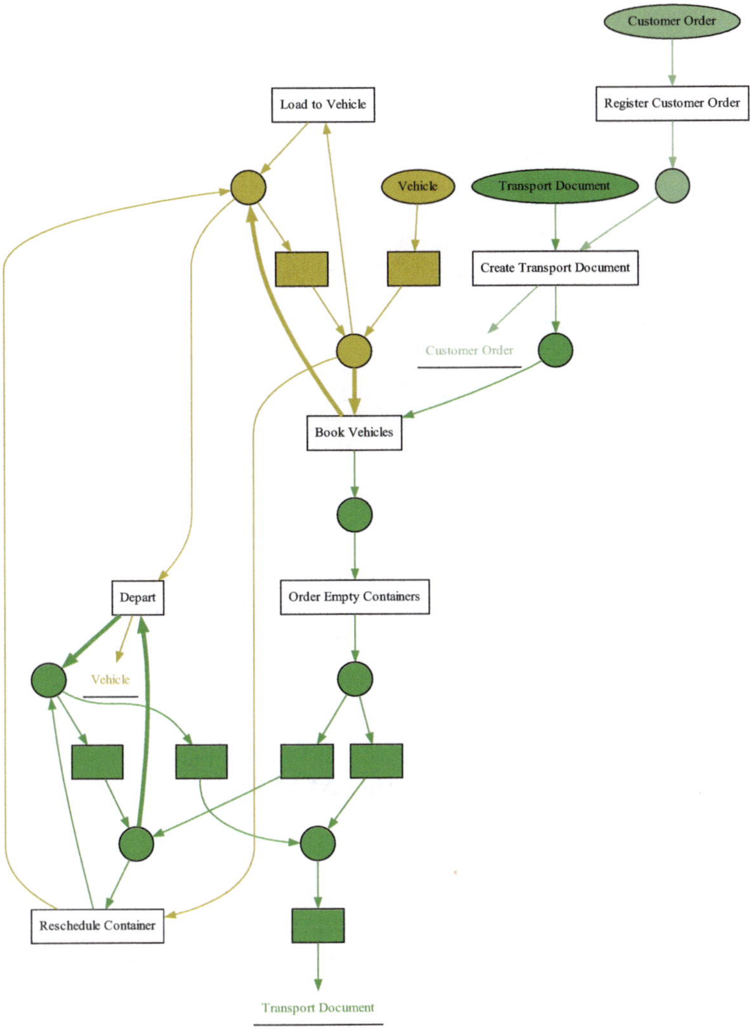

Fig. 5 An object-centric Petri net discovered for a simulated logistics handling process

4.5 Transferring the Idea

While the importance and impact of a standardized MitM have been recognized and its advantages have been enjoyed by the process mining community, other fields have not had that luxury. Our analysis of data usage in the IoP had shown that even though over half the submitted data models used time series data collected from machines and sensors, no commonly understood exchange format is used. As this

Table 2 Deviations of E2O multiplicities

Activity	Object type	Avg objects	Std objects	#Deviations
Book vehicles	Vehicle	1.953	0.212	6
Order empty containers	Container	3.813	1.101	1009
Depart	Container	18.787	7.106	618
Depart	Transport document	7.487	2.605	146

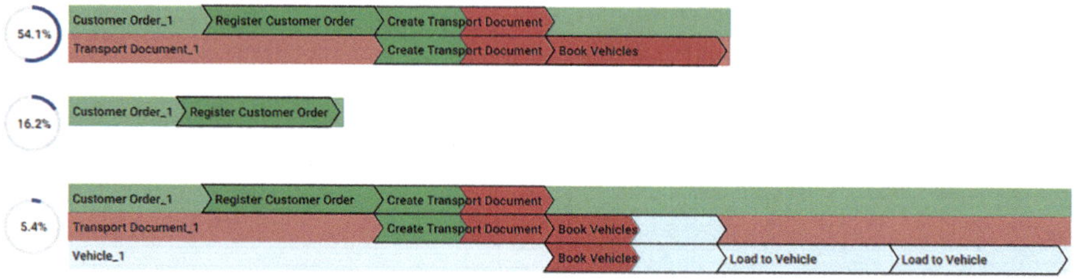

Fig. 6 Object-centric variant visualization of interactions between customer orders, transport documents, and vehicles

kind of IoT data is also extremely relevant for the BPM community [48], we looked at the commonalities in this type of data and conceptualized *Measurement and Event Data*.

5 Measurement and Event Data (MAED)

In trying to transfer our experiences with event data conceptualization, we also focused on the minimal ingredients necessary to facilitate both broad applicability and rich data science accessibility. In contrast to OCED, this model is not as mature and some details are subject to change. Still, we trust in the usefulness of the current state of MAED.

5.1 Introducing the MAED Meta-Model

MAED is intended to be very close to the raw data that machines and sensors produce, which makes it easier to map original data sources. Raw data producers are usually unaware of their context, e.g., surrounding process(es). Devices may provide metadata for data provenance and schema information, but especially the data from high-frequency sensors may just end up as millions of rows of timestamps and associated values. In contrast to PM-focused proposed models and formats [27, 63] that integrate IoT and classic event data, we target a lower level of abstraction.

The main concepts are *Time*, *Observation*, and *Object*. Observations represent digitally recorded data points with a timestamp and associated values. They are differentiated into *Events* and *Measurements*, which have syntactically the same structure but differ semantically. Events are discrete and possibly spontaneous and typically emerge from the system, indicating state changes or transitions. Measurements, on the other hand, are actively sampled in (usually) regular intervals. While the occurrence of an event already provides information regardless of its attributes, measurements are expected, so their mere occurrence is not valuable information. Measurements describe the state of a system in terms of their attribute values. Ideally, they can be regarded as multivariate time series. In practice, some pre-processing like interpolation [56] and resampling may be necessary to use them effectively. Lastly, objects in this model are rather simplistic compared to OCED as they are only used as an identifier for the physical or/virtual asset/entity that observations are related to.

Similar to OCED, MAED is intended to be a container format that captures more than a single isolated process. With the object identifier and typing of events and measurements, instances can already contain multiple individual (time) series that correspond to, e.g., multiple processes in a production chain. What is missing, however, and this was realized by considering *data utility*, is a division in time. This is because the assumption that measurement data is quasi-continuously measured holds only in a narrow view: within an individual process execution. Generally, in IoT settings, devices may only produce data while they are active, e.g., switched on. Furthermore, to facilitate comparative analyses on the same object, the extraction of sub-time series is necessary. This is one of the main use cases in the production setting: to assess how varying conditions affect production processes. Especially in a research setting such as the IoP, experiments with different materials, machine configurations, etc., are routinely conducted and later comparatively analyzed.

So, in addition to these low-level concepts, we introduce *Segments*. They are sequences of time intervals and correspond to a kind of case notion of event data. Even if this kind of annotation is often not part of the raw data itself, it is usually present in the surveyed data models. In the simplest case, there may just be one kind of segment that delineates the active from the dead times for each measurement and object combination. Such a trivial segmentation could even be inferred via the gaps in observations. As a technical note, we use segments as simply a sequence of time intervals without any restrictions. This contrasts possible intuition of requiring complete coverage of the time domain, nonoverlapping intervals, or monotonicity along increasing indices. We chose this unrestricted notion because this level of flexibility is necessary to fully model how such notions of context can be related to time.

As an example, consider the concept of *workpiece*, which is often central in discrete production processes, like milling, and chains thereof. A segmentation that tracks the durations of various objects, here, e.g., CNC machines, interact with a sequence of workpieces, has to support re-ordering and gaps. Taking into account that some machines may interact with multiple workpieces at the same time, like a heat-treatment oven, overlap is also necessary. Explicitly including the concept here

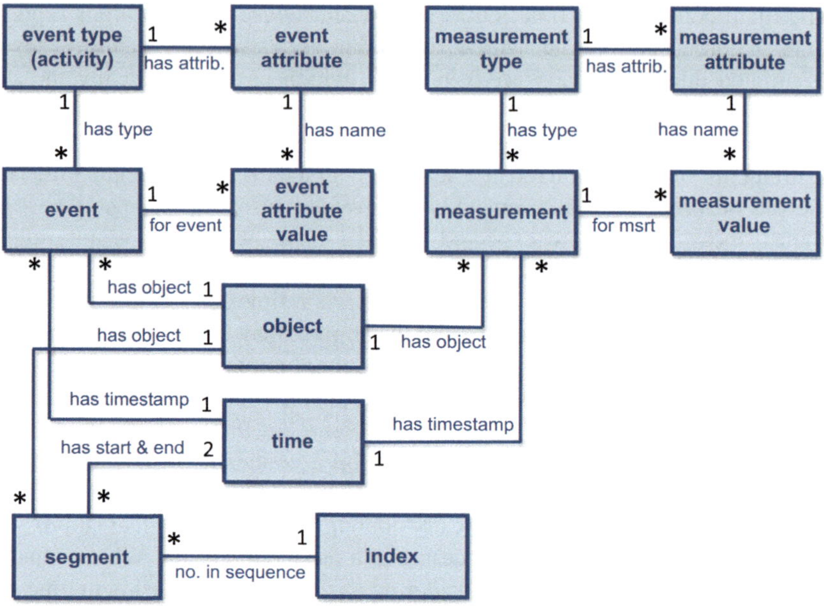

Fig. 7 The proposed meta-model for MAED

rounds out the analytic accessibility of MAED, which we expand upon in Sect. 5.2. It also partially bridges the gap to OCED in providing a model element for context, which has been amply discussed in prior work on IoT in PM [26, 63].

These concepts were chosen after considering all relevant submissions in the IoP data model survey and consulting with data owners in the project about how they collect and store their data. Due to the great heterogeneity, we favor this high degree of simplicity. Figure 7 depicts our meta-model of the concepts in UML-like fashion. We use similar "typing gadgets" as in the OCED model to ensure that observations belonging to a certain time series have the same features.

Along with this minimal conceptualization, we provide a concrete CSV file-based implementation and mathematical formalization. For details, refer to [75]. Note that we are still evolving the model, hence the meta-model discrepancy with the cited work. Equipped with the model, we present a collection of approaches/methods that are applicable to it.

5.2 *Analysis Techniques for MAED*

As MAED was conceived to represent IoT data, or more generally, cyber-physical systems data, perspectives involve the union, and particularly intersection, of discrete and continuous data. Our focus here lies on the latter, as this is the less explored domain within PM. In contrast to (business process) event data, which is commonly only narrowly considered as the workhorse of the (extended) BPM discipline, the usage of time series data is ubiquitous across all natural and life sciences, as well as data science and ML. Scoping out the range of possible analyses over such a landscape is nearly impossible, but there are certainly some common baseline tasks.

For one, the central role of models in PM is also present in many research areas dealing with this combination of event and time series data. Particularly, disciplines that study physical systems in some manner typically make use of continuous state models. *Hybrid Models*, for example, employed in CPS [37], even combine such continuous models with discrete ones. Other applications of models to observed quantified phenomena may be more focused on discovering the interactions of variables, understanding how one evolving process affects the other, or predicting/forecasting future values.

Starting out, in MAED, we have *Events*, *Measurements*, and *Segments*. As these are indexed over *Objects* and *Time*, we can view them as sets of series. Then, we interpret measurements as time series (TS), i.e., samples of an unknown function that is continuous over compact intervals of time. Without further processing, these may be highly irregular due to gaps as mentioned in Sect. 5.1. This will be addressed in the following. We present possible connections of the above data concepts among each other and with *models* in Fig. 8. Note that tripled connections, e.g., *discovery* from *Event Series* to *Discrete Models* on the left, represent (typically) set-level transformations in contrast to individual element transformations. The intent is to differentiate the required multiplicity, as some methods, like time series clustering, natively work on sets of TS, while others such as regression work on individual time series. In the named example, we can thus represent that discrete model discovery from events typically requires multiple traces of events—not just one long one.

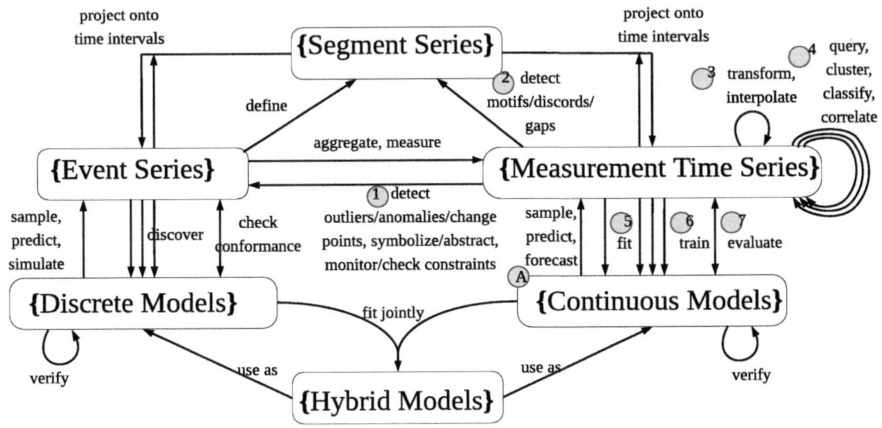

Fig. 8 A schematic overview of analysis techniques that are applicable to MAED. We focus particularly on the time series side

With that said, consider the top row of the figure. Segment series can be used to divide event/measurement series into multiple separate subseries by projecting onto their time intervals. This allows integrating analysis results in next iterations of analysis, as regions of interest "can be zoomed in on." The segments themselves can be directly defined by events, i.e., some event type signifies the start, while another signifies the end. Additionally, segments can be derived from measurement TS via detection of motifs (frequent patterns) [33, 60, 64], discords/anomalous sequences (unusual patterns) [60, 65, 70], or gaps.

Then, there are reciprocal connections between events and measurement TS. Events can be aggregated, e.g., their occurrence frequency over time, or measured in other more complex ways via their attributes or interplay. Going back from TS to events, techniques such as outlier/anomaly detection [29, 35, 73] and change point detection (drift in the assumed underlying generating process) [13] can be applied. Furthermore, symbolizing/abstracting or discretizing TS [58, 59] can be used to characterize a TS in terms of state transitions between finitely many states, i.e., discrete events. Another form of event extraction comes from monitoring or checking constraints on TS, e.g., in the form of temporal logic expressions [18, 32]. Whenever constraints are violated, an event can be raised to signify noncompliance. Such derived events can then be used to close the loop via defining segments along which TS can be further divided.

Briefly considering the left side around event series, we can reproduce traditional process mining tasks such as *discovery* and *conformance checking*. (Discrete) models, such as Petri nets and finite state automata, can then be used for verification and to predict or simulate, and sample if they are stochastic, new events.

There are two looping connections on the TS node. First, there are a multitude of transformations of TS to TS. They range from pre-processing steps such as resampling and interpolation [56] to methods from classical time series analysis like differencing, lagging, and detrending to yield stationarity [45]. Furthermore, signal processing techniques such as filtering and smoothing can be used to create new TS to work with. Applications of sophisticated methods, such as the *Kalman filter* [79] and its extensions, or *Fourier* and wavelet transforms [30, 77], into (mixed) frequency domain may be regarded as explicit continuous models instead.

Second, there are methods applicable to sets of TS that broadly also yield TS. There is querying and indexing [38] of TS, e.g., for database retrieval of "the most similar previously recorded TS." These fall under the umbrella of distance-based methods, like TS clustering [12, 50]. Related are methods for TS classification [16, 47], e.g., classifying which TS represent an intended behavior from those that do not. Particularly, machine learning methods in this area often rely on feature extraction [34], which typically involves breaking down a TS of indeterminate length into a fixed set of descriptive features. Another type of analysis that we can consider here is correlative analysis between TS. Methods range from faceted/paneled/overlaid visualizations with a human eye to detect correlations, over cross-correlation analysis [68], to *Granger causality* [74], which is a measure of how well one TS can predict another. These methods enable deriving some relations between TS and provide insight into how different measurement types, or measurements from different objects, interact.

Next, we go into the connections to *continuous models*. We differentiate *fitting* and *training* to emphasize the multiplicity difference. With fitting, we typically refer to traditional/structural TS forecasting models such as the AR/MA (auto-regressive, moving average) family [45] or other regression models that are parameterized on a singular TS. More sophisticated AR/MA variants such as multiseasonal (S)ARIMA(X) models [46], optionally with modeling of exogenous factors (X), and methods incorporating the frequency domain for seasonality such as TBATS [36] have also been proposed. The inclusion of exogenous variables draws attention to the fact that our measurements are generally *multivariate*, i.e., have a vector of features/variables. Many time series approaches are designed for univariate TS and are then later lifted to the multivariate setting. Examples are VARMAX and multivariate ARCH [62].

For *training*, we specifically refer to methods that parameterize a model based on a multitude of samples. Among these fall the ever-growing *machine learning* models [57]. They either use feature extraction as mentioned before or directly model sequential data with recurrent neural networks, or, recently, transformers [81]. To effectively train deep learning models on sometimes limited data, data augmentation [80] may be practically relevant.

A different suite of continuous models comes in the form of systems of differential equations that are very common in domains with strong modeling priors such as mechanical engineering. In these domains, models are commonly hand-designed, e.g., by making use of descriptions of natural laws, and only their parameters or initial conditions are *fitted* or *solved*. This is often with the goal of *verification* [37] in mind. Particularly in engineering, such physics-based models can be parts of Digital Shadows and Twins [22, 49]. Differential equations come in many types such as linear, nonlinear, stochastic, ordinary, or partial. The literature, particularly bordering statistical physics, is extensive [69]. Worth mentioning here as well are stochastic and Markov processes [78, 83], which on their own come with entire fields of more mathematical research.

With fitted/trained forecasting models, one can create new measurements via auto-regressive multistep forecasting. Many of these continuous models make underlying assumptions about stochasticity, e.g., about error terms. Some are also explicitly probabilistic in their predictions and representation and can thus be sampled. Particularly, models trained on sets of TS may be able to simulate or predict the entire TS at once.

Lastly, models typically need to be *evaluated*. Time series models can make use of regression metrics for goodness of fit like standard prediction error metrics, such as mean squared error (MSE) and mean absolute percentage error (MAPE). There also exist variants for stochastic models, such as pinball loss and the continuous rank probability score.

A standard methodology of evaluation of predictive models is k-fold cross-validation where the data is split into a train and test set and a model is fitted and scored within each fold. In time series, independence assumptions that give

Table 3 List of analysis techniques for MAED connected to measurement time series

Number in figure	Technique	References
1	Detect outliers/anomalies	[29, 35, 73]
	Detect change points	[13]
	Symbolize/abstract	[58, 59]
	Monitor/check constraints	[18, 32]
2	Detect motifs/discords/gaps	[33, 60, 64, 65, 70]
3	Transform	[45, 77]
	Interpolate	[56]
4	Query	[38]
	Cluster	[12, 50]
	Classify	[16, 47]
	Correlate	[68, 74]
5	Fit	[36, 45, 46]
6	Train	[47, 57, 80, 81]
7	Evaluate	[21]

cross-validation its theoretical validity are explicitly violated. Therefore, the basic standard is a held-out interval of test data at the end of the time series, or fully out-of-sample (OOS) approaches, i.e., scoring predictive power on a different TS. Still, variants of cross-validation, including rolling and expanding windows, or fully splitting the TS, can be used in practice [21].

As the last node in the diagram, we have *Hybrid Models* [37], which are particularly interesting for MAED. They usually generalize other models by regarding parameters as dynamic instead of static and can be regarded as an application of *superstatistics* [19]. Hidden Markov models [84] combine a latent discrete state space with continuous emissions that depend on the states. They represent one commonly used technique for time series change point detection.

We summarize the annotated outgoing connections of measurement time series in Table 3.

With all of these techniques at one's disposal, and particularly, by making use of chained applications, the possibilities of working with MAED are seemingly endless. Our goal is to develop tooling that facilitates simplified application of such approaches by providing transformations/extractions into the exact required input formats. Next, we return to our setting of specifically IoT data in the form of *machine data* and present an exemplary application.

5.3 Example Application

Currently, we provide some basic tooling for loading/exporting/converting and visualizing MAED instances. As a proof of concept, we apply our tooling and techniques from Sect. 5.2 to showcase some possible analyses on a data set from the IoP project, specifically data from a CNC milling machine that has been sampled to be handleable.

To start, we use an overview visualization that shows at a glance *when* events and measurements took place. Refer to Fig. 9a. As human analysts, we can see that there are two events with the complementary suffixes "_start" and "_end". They denote NC program starts and ends and are thus ideal candidates to segment the time frame into separate milling operations. For each time interval, here, *(machine) cycle*, we get an individual time series of measurements.

We can now apply comparative approaches on the set of TS. Consider Fig. 10, an aggregated visualization of the distribution of three selected measurement features over the machine cycles. The typical distribution of values for these features can be eyeballed on the vertical axis. In machining, stresses and cutting forces, aggregated over a milling operation, are valuable indicators of potential misconfigurations and tool wear.

Via visual inspection, we might want to specifically check the TS of cycle 16, as the *xc* values were lower than usual. Figure 9b shows the values of *xc* over cycle 16. Marked in red are outliers as detected via [76], which could lead to further investigations into these particularly low and high values. Then there is a segment, marked in light gray, that appears to be periodically oscillating. Vibrations are a common occurrence and their analysis is common practice.

Taking a technique from signal processing, we can perform a spectral analysis on this sub-TS, which is presented in Fig. 11. The spectrogram shows that there is a dominant oscillation at 1Hz for the first 20 seconds (within the gray shaded area), when a jump in frequency and widening in dominant frequencies occurs that is afterward split off into two overlapping drifting signals. A domain expert might want to investigate further, by correlating the found patterns with the machine settings and workpiece and drill bit in use.

This was quite a shallow manual exploration; however, the presented techniques, applied from Sect. 5.2, are actually being used in the creation of Digital Twins/Shadows, e.g., in tool wear prediction. Our data model supports these domain-specific use cases we encountered in the IoP project well. However, MAED and also OCED can be used outside of their respective "domains of inspiration." The reach of MitM can be further extended by *connecting* them.

6 Connecting Models-in-the-Middle

Models-in-the-Middle individually are already useful by serving as standardized exchange formats. However, when considering a collection of MitM, we can increase their combined utility even more. Creating connections between MitM via transformations, extractions, and enhancement algorithms, essentially and effectively, increases the "reach" of data. Data that was initially collected in one domain and treated with a domain-specific understanding of its typical analyses may find

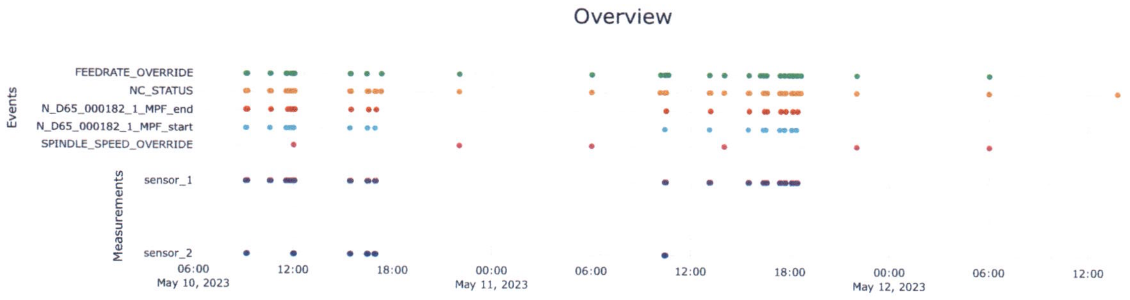

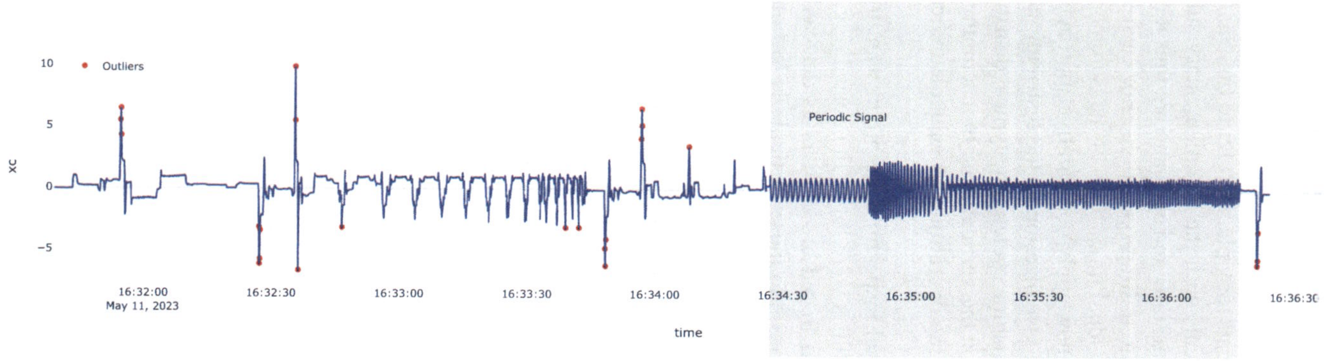

Fig. 9 (**a**) Overview over *when* events/measurements occur in a dataset. Useful for visually checking co-occurrence patterns, especially events. (**b**) Individual time series visualization with auto-detected outliers [76] and a further annotation of a segment with periodic values

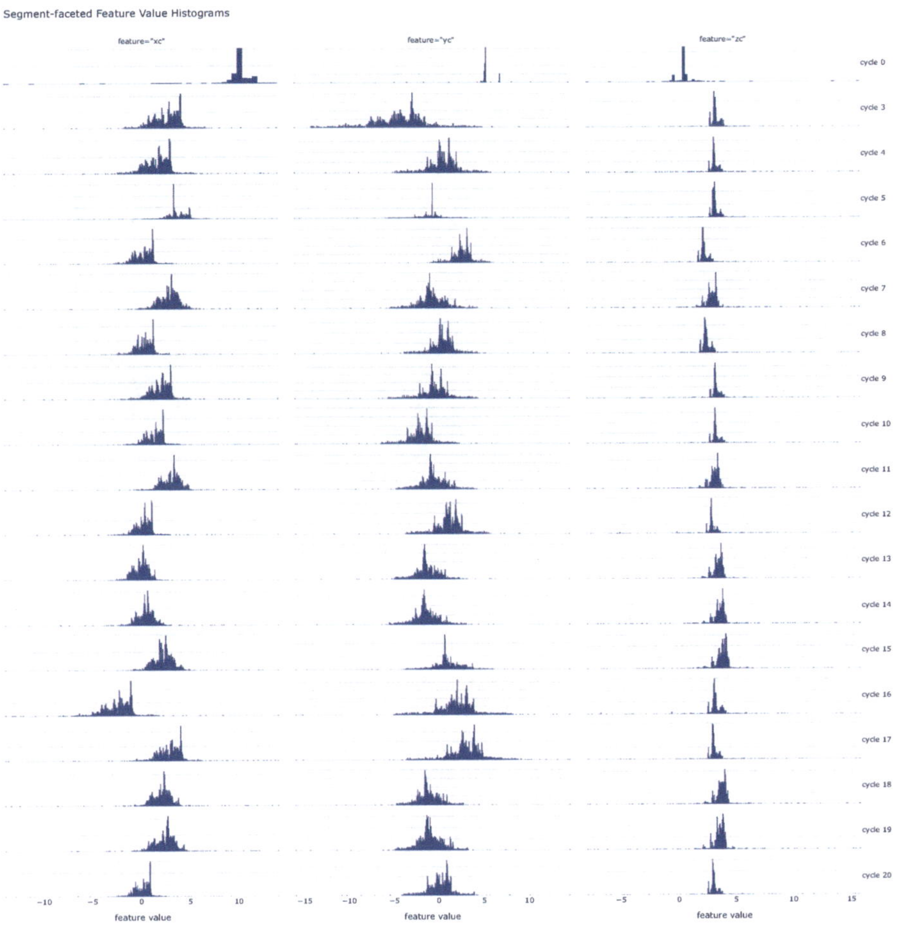

Fig. 10 Histogram of feature values for the machine cycle segmentation

a new life in another domain. The transformation algorithms between MitM may range from trivial, i.e., just converting the data on a syntactic level, to rich semantics-imbuing approaches. Such rich transformations may be created by deriving and inferring higher-level concept present in some type of data, represented in a MitM. For example, deriving event notions from continuous measurement data and transferring them into OCED, or even traditional event data, would open them up to the whole wealth of PM methods. Note that even just conversions on a syntactic level can be useful in connecting data to algorithms and implementations of another MitM. In addition to making the data more interoperable, such connections could also foster other interdisciplinary cooperation, like IoT data integration into BPM.

To make this more specific, we discuss possible connections between our proposed MitM, OCED and MAED. One may consider MAED to be specialized to

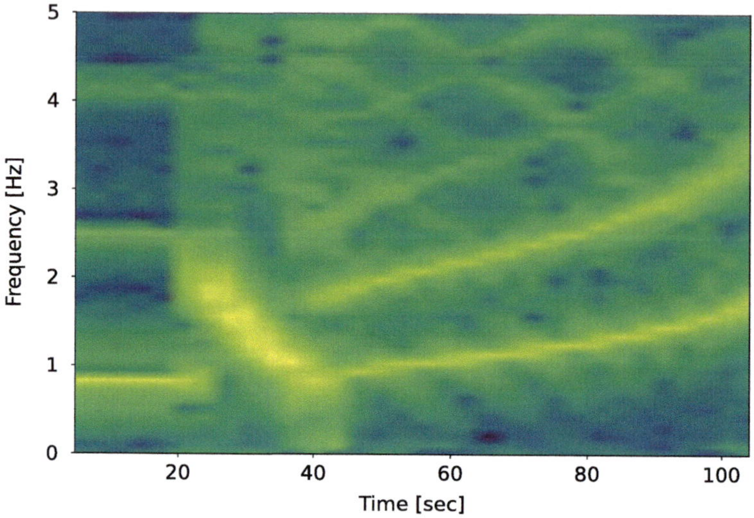

Fig. 11 Spectrogram of the interval with identified periodic oscillations

the low-level discrete/continuous time series domain, while OCED is more focused on structure and interactions among objects over discrete points in time. A good way to play into the strengths of their respective foci is by transforming conceptually between discrete events and continuous measurements. This corresponds to the arrows between event series and measurement time series in Fig. 8.

To get from OCED to MAED, we can *aggregate* and *measure*. As an example, we recall the event log about container logistics from the sample OCED application in Sect. 4.4. As basic aggregations, we could calculate the number of event and object occurrences over time, optionally per respective type. In addition to this, arbitrarily complex patterns could be counted such as the occurrence of customer orders with unusually many vehicles booked for their transport document. We may measure durations between such patterns, or base measurements on object and event attributes, for example, the revenue per container, or costs per vehicle, per quarter/month/week/day. We can thus generate time series of all kinds of interesting KPIs. By faceting such calculations per object/event type, or another grouping, we can easily make them multivariate. At this point, we can apply the whole machinery applicable to, and ideally implemented for, MAED, such as transforming, segmenting, correlating, fitting, and forecasting. Once we have gained some insights, or, e.g., fitted a forecasting model, this can again be played back into OCED. New object types for certain types of identified motifs, outliers, or anomalies can be created, and events generated for detected change points (potential *concept drift* in the process) and instances of violated constraints on KPIs. Once back in "process mining land" with OCED, we may apply discovery on such derived events and analyze object co-occurrences and relations for particular motif types. At this point, we can close the cycle to again derive MAED datasets.

Such a back-and-forth between MitM, working with increasingly higher-level semantics and fully utilizing the strengths and existing configurable software for each type of data, is evidently a very powerful analysis technique. The challenge lies in creating a framework that supports doing so efficiently and effectively.

7 Advantages of Models-in-the-Middle

Models-in-the-Middle do not just facilitate a common understanding of a kind of data but also provide the much-needed basis for *data interoperability* that enables developing generic and reusable algorithms and applications. The success of XES is that it facilitated the usage of a common process mining research software framework, specifically ProM. Transferring the concept, we envision the creation of a counterpart of this framework for MAED.

In our case study of the IoP project, we noted that many use cases use the same time series-based methods but are still implemented completely independently. To realize the full potential of MitM, reusable and generic implementations that can be configured via APIs or UIs are necessary. Similar to how domain-agnostic visualization like the "Dotted Chart" in ProM can be applied to extracted, simulated, generated, and refined event data, analogous tools could be developed for MAED. As a proof of concept, we implemented some initial tooling in the form of a web app.[6,7] Two basic visualizations it provides are shown in Fig. 12. Data owners in the IoP were enthusiastic about the concept of such generic tooling to alleviate duplicated efforts and to make implementations more shareable.

8 Conclusion

Effective data management as well as utilization is necessary to efficiently and effectively work on digitalizing and optimizing (production) processes. In this paper, we presented how intermediate data models can foster data interoperability and cross-domain and cross-discipline cooperation by providing a common middle ground for data exchange and implementation reuse. Specifically, we introduced two example MitM that are especially relevant for the Internet of Production research project and the BPM discipline: OCED and MAED. We showed their applicability using initial tool support and by collaborating with data owners in the IoP.

While our work was inspired by the need to bridge the gap between data source and data consumer in the context of digitalized production using Digital Shadows, the potential application of the methodology is much larger. Other such

[6] https://git-ce.rwth-aachen.de/machine-data/mdata_app

[7] https://mdata.cluster.iop.rwth-aachen.de/

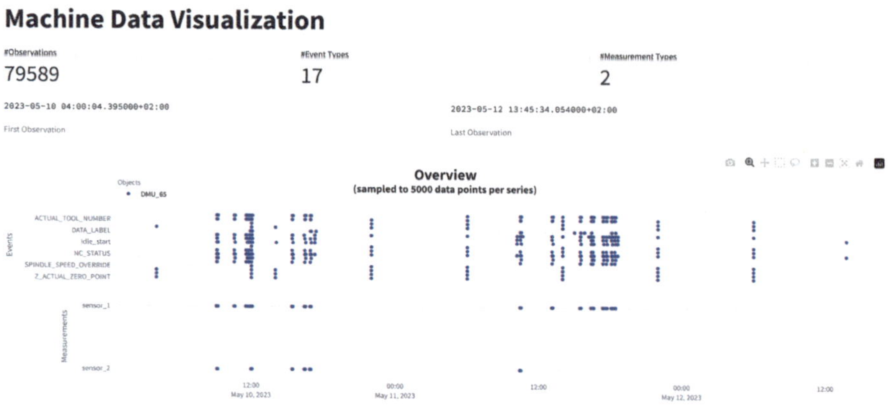

(a)

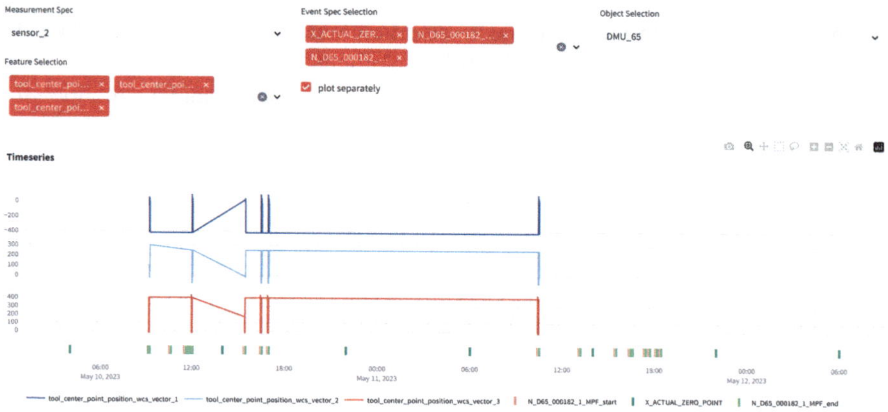

(b)

Fig. 12 Two example generic visualizations based on MAED input. (**a**) Overview visualization of a MAED instance. It shows *when* events and measurements were recorded. (**b**) Measurement time series of selectable features together with events

meta *Models-in-the-Middle* could benefit completely different areas of research. Creating relations between MitM could impose a hierarchical structure upon them. Lastly, concrete domain-specific instantiations of these meta-models could provide standardization on a lower level. Using ontology-based methods, activity and object types, or event/measurement types and their relation could be "preconfigured" to meet the needs of automated semantic-aware analyses. This would be akin to ontology-based data access [41, 53].

Acknowledgments Funded by the Deutsche Forschungsgemeinschaft (DFG, German Research Foundation) under Germany's Excellence Strategy—EXC-2023 Internet of Production—390621612. We also thank the Alexander von Humboldt (AvH) Stiftung for supporting our research.

References

1. Forecast Analysis: Process Mining, Worldwide. https://www.gartner.com/en/documents/4007520
2. Process Mining Software Market Size, Share & Growth [2030]. https://www.fortunebusinessinsights.com/process-mining-software-market-104792
3. van der Aalst, W.M.P.: Object-centric process mining: dealing with divergence and convergence in event data. In: Ölveczky, P.C., Salaün, G. (eds.) Software Engineering and Formal Methods, pp. 3–25. Lecture Notes in Computer Science. Springer, Berlin (2019). https://doi.org/10.1007/978-3-030-30446-1_1
4. van der Aalst, W.M.P.: Object-centric process mining: unraveling the fabric of real processes. Mathematics **11**(12), 2691 (2023). https://doi.org/10.3390/math11122691
5. van der Aalst, W.M.P.: Process mining—data science in action, 2nd edn. Springer, Berlin (2016). https://doi.org/10.1007/978-3-662-49851-4
6. van der Aalst, W.M.P., Berti, A.: Discovering object-centric petri nets. Fundamenta Informaticae **175**(1–4), 1–40 (2020). https://doi.org/10.3233/FI-2020-1946
7. van der Aalst, W.M.P., Weijters, T., Maruster, L.: Workflow mining: discovering process models from event logs. IEEE Trans. Knowl. Data Eng. **16**(9), 1128–1142 (2004). https://doi.org/10.1109/TKDE.2004.47
8. Adams, J.N., van der Aalst, W.M.P.: OCπ: object-centric process insights. In: Application and Theory of Petri Nets and Concurrency: 43rd International Conference, PETRI NETS 2022, Bergen, Norway, June 19–24, 2022, Proceedings, pp. 139–150. Springer, Berlin (2022). https://doi.org/10.1007/978-3-031-06653-5_8
9. Adams, J.N., Park, G., van der Aalst, W.M.P.: Ocpa: a Python library for object-centric process analysis. Softw. Impacts **14**, 100438 (2022). https://doi.org/10.1016/j.simpa.2022.100438
10. Adams, J.N., Park, G., van der Aalst, W.M.P.: Preserving complex object-centric graph structures to improve machine learning tasks in process mining. Eng. Appl. Artif. Intell. **125**, 106764 (2023). https://doi.org/10.1016/j.engappai.2023.106764
11. Adams, J.N., Schuster, D., Schmitz, S., Schuh, G., van der Aalst, W.M.: Defining cases and variants for object-centric event data. In: 2022 4th International Conference on Process Mining (ICPM), pp. 128–135 (2022). https://doi.org/10.1109/ICPM57379.2022.9980730
12. Aghabozorgi, S., Seyed Shirkhorshidi, A., Ying Wah, T.: Time-series clustering A decade review. Inform. Syst. **53**, 16–38 (2015). https://doi.org/10.1016/j.is.2015.04.007
13. Aminikhanghahi, S., Cook, D.J.: A survey of methods for time series change point detection. Knowl. Inform. Syst. **51**(2), 339–367 (2017). https://doi.org/10.1007/s10115-016-0987-z
14. Augusto, A., Conforti, R., Dumas, M., La Rosa, M., Polyvyanyy, A.: Split miner: automated discovery of accurate and simple business process models from event logs. Knowl. Inform. Syst. **59**(2), 251–284 (2019). https://doi.org/10.1007/s10115-018-1214-x
15. Augusto, A., Conforti, R., Dumas, M., Rosa, M.L., Maggi, F.M., Marrella, A., Mecella, M., Soo, A.: Automated discovery of process models from event logs: review and benchmark. IEEE Trans. Knowl. Data Eng. **31**(4), 686–705 (2019). https://doi.org/10.1109/TKDE.2018.2841877
16. Bagnall, A., Lines, J., Bostrom, A., Large, J., Keogh, E.: The great time series classification bake off: a review and experimental evaluation of recent algorithmic advances. Data Mining Knowl. Discov. **31**(3), 606–660 (2017). https://doi.org/10.1007/s10618-016-0483-9
17. Banham, A., Leemans, S.J.J., Wynn, M.T., Andrews, R.: xPM: A framework for process mining with exogenous data. In: Munoz-Gama, J., Lu, X. (eds.) Process Mining Workshops, pp. 85–97. Lecture Notes in Business Information Processing. Springer, Berlin (2022). https://doi.org/10.1007/978-3-030-98581-3_7
18. Bartocci, E., Deshmukh, J., Donzé, A., Fainekos, G., Maler, O., Ničković, D., Sankaranarayanan, S.: Specification-based monitoring of cyber-physical systems: a survey on theory, tools and applications. In: Bartocci, E., Falcone, Y. (eds.) Lectures on Runtime Verification: Introductory and Advanced Topics, pp. 135–175. Lecture Notes in Computer Science. Springer, Berlin (2018). https://doi.org/10.1007/978-3-319-75632-5_5

19. Beck, C., Cohen, E.G.D.: Superstatistics. Phys. A: Stat. Mech. Appl. **322**, 267–275 (2003). https://doi.org/10.1016/S0378-4371(03)00019-0
20. Beerepoot, I., Di Ciccio, C., Reijers, H.A., Rinderle-Ma, S., Bandara, W., Burattin, A., Calvanese, D., Chen, T., Cohen, I., Depaire, B., Di Federico, G., Dumas, M., van Dun, C., Fehrer, T., Fischer, D.A., Gal, A., Indulska, M., Isahagian, V., Klinkmüller, C., Kratsch, W., Leopold, H., Van Looy, A., Lopez, H., Lukumbuzya, S., Mendling, J., Meyers, L., Moder, L., Montali, M., Muthusamy, V., Reichert, M., Rizk, Y., Rosemann, M., Röglinger, M., Sadiq, S., Seiger, R., Slaats, T., Simkus, M., Someh, I.A., Weber, B., Weber, I., Weske, M., Zerbato, F.: The biggest business process management problems to solve before we die. Comput. Ind. **146**, 103837 (2023). https://doi.org/10.1016/j.compind.2022.103837
21. Bergmeir, C., Hyndman, R.J., Koo, B.: A note on the validity of cross-validation for evaluating autoregressive time series prediction. Comput. Stat. Data Anal. **120**, 70–83 (2018). https://doi.org/10.1016/j.csda.2017.11.003
22. Bergs, T., Gierlings, S., Auerbach, T., Klink, A., Schraknepper, D., Augspurger, T.: The concept of digital twin and digital shadow in manufacturing. Procedia CIRP **101**, 81–84 (2021). https://doi.org/10.1016/j.procir.2021.02.010
23. Berti, A., van der Aalst, W.M.P.: OC-PM: Analyzing object-centric event logs and process models. J. Softw. Tools Technol. Transf. **25**(1), 1–17 (2023). https://doi.org/10.1007/s10009-022-00668-w
24. Berti, A., Koren, I., Adams, J.N., Park, G., Knopp, B., Graves, N., Rafiei, M., Liß, L., Tacke Genannt Unterberg, L., Zhang, Y., Schwanen, C., Pegoraro, M., van der Aalst, W.M.P.: OCEL (object-centric event log) 2.0 specification (2024)
25. Berti, A., van Zelst, S.J., van der Aalst, W.M.P.: Process mining for Python (PM4Py): bridging the gap between process- and data science (2019). https://doi.org/10.48550/arXiv.1905.06169
26. Bertrand, Y., De Weerdt, J., Serral, E.: A Bridging Model for Process Mining and IoT. In: Munoz-Gama, J., Lu, X. (eds.) Process mining workshops, pp. 98–110. Lecture Notes in Business Information Processing. Springer, Berlin (2022). https://doi.org/10.1007/978-3-030-98581-3_8
27. Bertrand, Y., Veneruso, S., Leotta, F., Mecella, M., Serral Asensio, E.: NICE: the native IoT-centric event log model for process mining. In: Lecture Notes in Business Information Processing. Springer, Germany (2023). https://lirias.kuleuven.be/4111092
28. Biffl, S., Lüder, A., Gerhard, D. (eds.): Multi-Disciplinary Engineering for Cyber-Physical Production Systems. Springer, Berlin (2017). https://doi.org/10.1007/978-3-319-56345-9
29. Blázquez-García, A., Conde, A., Mori, U., Lozano, J.A.: A review on outlier/anomaly detection in time series data. ACM Comput. Surv. **54**(3), 56:1–56:33 (2021). https://doi.org/10.1145/3444690
30. Brigham, E.O.: The Fast Fourier Transform and Its Applications. Prentice-Hall, Englewood Cliffs (1988)
31. Carmona, J., Cortadella, J., Kishinevsky, M.: A region-based algorithm for discovering petri nets from event logs. In: Dumas, M., Reichert, M., Shan, M.C. (eds.) Business Process Management, pp. 358–373. Lecture Notes in Computer Science. Springer, Berlin (2008). https://doi.org/10.1007/978-3-540-85758-7_26
32. Chen, G., Liu, M., Kong, Z.: Temporal-logic-based semantic fault diagnosis with time-series data from industrial Internet of Things. IEEE Trans. Ind. Electron. **68**(5), 4393–4403 (2021). https://doi.org/10.1109/TIE.2020.2984976
33. Chen, Y., Nascimento, M.A., Ooi, B.C., Tung, A.K.H.: SpADe: On shape-based pattern detection in streaming time series. In: 2007 IEEE 23rd International Conference on Data Engineering, pp. 786–795 (2007). https://doi.org/10.1109/ICDE.2007.367924
34. Christ, M., Braun, N., Neuffer, J., Kempa-Liehr, A.W.: Time series FeatuRe extraction on basis of scalable hypothesis tests (tsfresh a Python package). Neurocomputing **307**, 72–77 (2018). https://doi.org/10.1016/j.neucom.2018.03.067
35. Cook, A.A., Mı sırlı, G., Fan, Z.: Anomaly detection for IoT time-series data: a survey. IEEE Internet Things J. **7**(7), 6481–6494 (2020). https://doi.org/10.1109/JIOT.2019.2958185

36. De Livera, A.M., Hyndman, R.J., Snyder, R.D.: Forecasting time series with complex seasonal patterns using exponential smoothing. J. Am. Stat. Assoc. **106**(496), 1513–1527 (2011). https://doi.org/10.1198/jasa.2011.tm09771
37. Deshmukh, J.V., Sankaranarayanan, S.: Formal techniques for verification and testing of cyber-physical systems. In: Al Faruque, M.A., Canedo, A. (eds.) Design Automation of Cyber-Physical Systems, pp. 69–105. Springer, Berlin (2019). https://doi.org/10.1007/978-3-030-13050-3_4
38. Ding, H., Trajcevski, G., Scheuermann, P., Wang, X., Keogh, E.: Querying and mining of time series data: experimental comparison of representations and distance measures. Proc. VLDB Endow. **1**(2), 1542–1552 (2008). https://doi.org/10.14778/1454159.1454226
39. Drath, R., Mosch, C., Hoppe, S., Faath, A., Barnstedt, E., Fiebiger, B., Schlögl, W.: Interoperabilität mit der Verwaltungsschale, OPC UA und AutomationML. https://opcfoundation.org/wp-content/uploads/2023/04/Diskussionspapier-Zielbild-und-Handlungsempfehlungen-fur-industrielle-Interoperabilitat-5.3-protected.pdf
40. Ehrendorfer, M., Mangler, J., Rinderle-Ma, S.: Clustering raw sensor data in process logs to detect data streams. In: Sellami, M., Vidal, M.E., van Dongen, B., Gaaloul, W., Panetto, H. (eds.) Cooperative Information Systems, pp. 438–447. Lecture Notes in Computer Science. Springer, Cham (2024). https://doi.org/10.1007/978-3-031-46846-9_25
41. Ekaputra, F., Sabou, M., Serral, E., Kiesling, E., Biffl, S.: Ontology-based data integration in multi-disciplinary engineering environments: a review. Open J. Inform. Syst. **4**, 1–26 (2017)
42. Esposito, L., Leotta, F., Mecella, M., Veneruso, S.: Unsupervised segmentation of smart home logs for human habit discovery. In: 2022 18th International Conference on Intelligent Environments (IE), pp. 1–8 (2022). https://doi.org/10.1109/IE54923.2022.9826776
43. Fahland, D.: Describing behavior of processes with many-to-many interactions. In: Donatelli, S., Haar, S. (eds.) Application and Theory of Petri Nets and Concurrency, pp. 3–24. Lecture Notes in Computer Science. Springer, Berlin (2019). https://doi.org/10.1007/978-3-030-21571-2_1
44. Günther, C.W., Verbeek, H.M.W.: XES—standard definition (2014). https://pure.tue.nl/ws/portalfiles/portal/3981980/692728941269079.pdf
45. Hamilton, J.D.: Time Series Analysis. Princeton University Press, Princeton (2020)
46. Hyndman, R.J., Athanasopoulos, G.: Forecasting: Principles and Practice. OTexts (2018)
47. Ismail Fawaz, H., Forestier, G., Weber, J., Idoumghar, L., Muller, P.A.: Deep learning for time series classification: a review. Data Mining Knowl. Discov. **33**(4), 917–963 (2019). https://doi.org/10.1007/s10618-019-00619-1
48. Janiesch, C., Koschmider, A., Mecella, M., Weber, B., Burattin, A., Di Ciccio, C., Fortino, G., Gal, A., Kannengiesser, U., Leotta, F., Mannhardt, F., Marrella, A., Mendling, J., Oberweis, A., Reichert, M., Rinderle-Ma, S., Serral, E., Song, W., Su, J., Torres, V., Weidlich, M., Weske, M., Zhang, L.: The Internet of Things meets business process management: a manifesto. IEEE Syst. Man Cybern. Mag. **6**(4), 34–44 (2020). https://doi.org/10.1109/MSMC.2020.3003135
49. Jarke, M., Schuh, G., Brecher, C., Brockmann, M., Prote, J.P.: Digital shadows in the Internet of Production. **2018**(115) (2018). https://ercim-news.ercim.eu/en115/special/2104-digital-shadows-in-the-internet-of-production
50. Javed, A., Lee, B.S., Rizzo, D.M.: A benchmark study on time series clustering. Mach. Learn. Appl. **1**, 100001 (2020). https://doi.org/10.1016/j.mlwa.2020.100001
51. Knopp, B., Graves, N.: Container Logistics Object-centric Event Log (2023). https://doi.org/10.5281/zenodo.8428084
52. Knopp, B., Pourbafrani, M., van der Aalst, W.M.P.: Discovering object-centric process simulation models. In: 2023 5th International Conference on Process Mining (ICPM), pp. 81–88 (2023). https://doi.org/10.1109/ICPM60904.2023.10271944
53. Kontchakov, R., Rodríguez-Muro, M., Zakharyaschev, M.: Ontology-based data access with databases: a short course. In: Rudolph, S., Gottlob, G., Horrocks, I., van Harmelen, F. (eds.) Reasoning Web. Semantic Technologies for Intelligent Data Access: 9th International Summer School 2013, Mannheim, Germany, July 30 August 2, 2013. Proceedings, pp. 194–229. Lecture Notes in Computer Science. Springer, Berlin (2013). https://doi.org/10.1007/978-3-642-39784-4_5

54. Lee, B., Kim, D.K., Yang, H., Oh, S.: Model transformation between OPC UA and UML. Comput. Stand. Interfaces **50**, 236–250 (2017). https://doi.org/10.1016/j.csi.2016.09.004
55. Leemans, S.J.J., Fahland, D., van der Aalst, W.M.P.: Scalable process discovery and conformance checking. Softw. Syst. Model. **17**(2), 599–631 (2018). https://doi.org/10.1007/S10270-016-0545-X
56. Lepot, M., Aubin, J.B., Clemens, F.H.L.R.: Interpolation in time series: an introductive overview of existing methods, their performance criteria and uncertainty assessment. Water **9**(10), 796 (2017). https://doi.org/10.3390/w9100796
57. Lim, B., Zohren, S.: Time-series forecasting with deep learning: a survey. Philos. Trans. Roy. Soc. A **379**(2194), 20200209 (2021). https://doi.org/10.1098/rsta.2020.0209
58. Lin, J., Keogh, E., Lonardi, S., Chiu, B.: A symbolic representation of time series, with implications for streaming algorithms. In: Proceedings of the 8th CM SIGMOD Workshop on Research Issues in Data Mining and Knowledge Discovery, pp. 2–11. DMKD '03. Association for Computing Machinery (2003). https://doi.org/10.1145/882082.882086
59. Lin, J., Keogh, E., Wei, L., Lonardi, S.: Experiencing SAX: a novel symbolic representation of time series. Data Mining Knowl. Discov. **15**(2), 107–144 (2007). https://doi.org/10.1007/s10618-007-0064-z
60. Linardi, M., Zhu, Y., Palpanas, T., Keogh, E.: Matrix profile goes MAD: variable-length motif and discord discovery in data series. Data Mining Knowl. Discov. **34**(4), 1022–1071 (2020). https://doi.org/10.1007/s10618-020-00685-w
61. Liss, L., Adams, J.N., van der Aalst, W.M.P.: Object-centric alignments. In: Conceptual Modeling: 42nd International Conference, ER 2023, Lisbon, Portugal, November 6–9, 2023, Proceedings. pp. 201–219. Springer, Berlin (2023). https://doi.org/10.1007/978-3-031-47262-6_11
62. Lütkepohl, H.: New Introduction to Multiple Time Series Analysis. Springer, Berlin (2005). https://doi.org/10.1007/978-3-540-27752-1
63. Mangler, J., Grüger, J., Malburg, L., Ehrendorfer, M., Bertrand, Y., Benzin, J.V., Rinderle-Ma, S., Serral Asensio, E., Bergmann, R.: DataStream XES extension: embedding IoT sensor data into extensible event stream logs. Future Internet **15**(3), 109 (2023). https://doi.org/10.3390/fi15030109
64. Mueen, A., Keogh, E., Zhu, Q., Cash, S., Westover, B.: Exact discovery of time series motifs. In: Proceedings of the 2009 SIAM International Conference on Data Mining (SDM), pp. 473–484. Proceedings. Society for Industrial and Applied Mathematics (2009). https://doi.org/10.1137/1.9781611972795.41
65. Nakamura, T., Imamura, M., Mercer, R., Keogh, E.: MERLIN: parameter-free discovery of arbitrary length anomalies in massive time series archives. In: 2020 IEEE International Conference on Data Mining (ICDM), pp. 1190–1195 (2020). https://doi.org/10.1109/ICDM50108.2020.00147
66. Nooijen, E.H.J., van Dongen, B.F., Fahland, D.: Automatic discovery of data-centric and artifact-centric processes. In: La Rosa, M., Soffer, P. (eds.) Business Process Management Workshops, pp. 316–327. Lecture Notes in Business Information Processing. Springer, Berlin (2013). https://doi.org/10.1007/978-3-642-36285-9_36
67. Park, G., van der Aalst, W.M.P.: Realizing digital twin of an organization using action-oriented process mining. In: 2021 3rd International Conference on Process Mining (ICPM), pp. 104–111 (2021). https://doi.org/10.1109/ICPM53251.2021.9576846
68. Podobnik, B., Stanley, H.E.: Detrended cross-correlation analysis: a new method for analyzing two nonstationary time series. Phys. Rev. Lett. **100**(8), 084102 (2008). https://doi.org/10.1103/PhysRevLett.100.084102
69. Ramsay, J.O., Silverman, B.W.: Fitting Differential Equations to Functional Data: Principal Differential Analysis. Springer, Berlin (2005)
70. Rasheed, F., Alhajj, R.: A framework for periodic outlier pattern detection in time-series sequences. IEEE Trans. Cybern. **44**(5), 569–582 (2014). https://doi.org/10.1109/TSMCC.2013.2261984

71. Rozinat, A., van der Aalst, W.M.P.: Conformance checking of processes based on monitoring real behavior. Inform. Syst. **33**(1), 64–95. https://doi.org/10.1016/j.is.2007.07.001
72. Schiekofer, R., Grimm, S., Brandt, M.M., Weyrich, M.: A formal mapping between OPC UA and the Semantic Web. In: 2019 IEEE 17th International Conference on Industrial Informatics (INDIN), vol. 1, pp. 33–40 (2019). https://doi.org/10.1109/INDIN41052.2019.8972102
73. Schmidl, S., Wenig, P., Papenbrock, T.: Anomaly detection in time series: a comprehensive evaluation. Proc. VLDB Endow. **15**(9), 1779–1797 (2022). https://doi.org/10.14778/3538598.3538602
74. Shojaie, A., Fox, E.B.: Granger causality: a review and recent advances. Ann. Rev. Stat. Appl. **9**(1), 289–319 (2022). https://doi.org/10.1146/annurev-statistics-040120-010930
75. Tacke Genannt Unterberg, L., Koren, I., van der Aalst, W.M.P.: Maximizing Reuse and Interoperability in Industry 4.0 with a Minimal Data Exchange Format for Machine Data. In: Modellierung 2024, pp. 103–118. Gesellschaft für Informatik e.V., Bonn (2024). https://doi.org/10.18420/modellierung2024_011
76. Talagala, P.D., Hyndman, R.J., Smith-Miles, K.: Anomaly detection in high-dimensional data. J. Comput. Graph. Stat. **30**(2), 360–374 (2021). https://doi.org/10.1080/10618600.2020.1807997
77. Torrence, C., Compo, G.P.: A practical guide to wavelet analysis. Bull. Am. Meteorol. Soc. **79**(1), 61–78 (1998)
78. Van Kampen, N.G.: Stochastic Processes in Physics and Chemistry, vol. 1. Elsevier, Amsterdam (1992)
79. Welch, G., Bishop, G.: An introduction to the Kalman Filter (1995)
80. Wen, Q., Sun, L., Yang, F., Song, X., Gao, J., Wang, X., Xu, H.: Time series data augmentation for deep learning: a survey. In: Proceedings of the Thirtieth International Joint Conference on Artificial Intelligence, IJCAI 2021, Virtual Event/Montreal, Canada, 19–27 August 2021, pp. 4653–4660 (2021). https://doi.org/10.24963/IJCAI.2021/631
81. Wen, Q., Zhou, T., Zhang, C., Chen, W., Ma, Z., Yan, J., Sun, L.: Transformers in time series: a survey (2023). https://doi.org/10.48550/arXiv.2202.07125
82. van der Werf, J., van Dongen, B., Hurkens, C., Serebrenik, A.: Process discovery using integer linear programming. Fundamenta Informaticae, **94**(3–4), 387–412
83. Zeger, S.L., Qaqish, B.: Markov Regression Models for Time Series: A Quasi-Likelihood Approach, pp. 1019–1031 (1988)
84. Zucchini, W., MacDonald, I.L.: Hidden Markov Models for Time Series: An Introduction Using R. Chapman and Hall/CRC (2009)

A Digital Twin Framework for Bioprocess Development Using IoT Sensor Data

Zahra Dasht Bozorgi, Artem Polyvyanyy, Marcello La Rosa, Ellen Otte, and Abel Armas-Cervantes

1 Introduction

A bioprocess refers to a series of steps or operations designed to harness living organisms (such as bacteria, yeast, or mammalian cells) or their cellular components to produce the desired products or carry out specific biochemical transformations. These processes are commonly employed in various industries, including pharmaceuticals, biotechnology, food and beverage, agriculture, and environmental remediation. In bioprocessing, living organisms are cultivated under controlled conditions, typically in bioreactors or fermentation tanks, where they are provided with nutrients and other necessary conditions for growth and metabolism. The organisms then produce the desired products through metabolic pathways or biochemical reactions. Bioprocesses can involve a wide range of activities, including cell culture, fermentation, purification, and downstream processing.

Typically, bioprocesses are composed of several *unit operations*, which are executed sequentially. The process starts by thawing vials of frozen microorganisms. The cells are then grown in a cell culture where they are fed nutrients to promote growth and target protein production. The cell culture unit operations are known as upstream processing. The goal of upstream processing is to produce as much

Artem Polyvyanyy and Marcello La Rosa authors contributed equally to this work.

Z. D. Bozorgi (✉) · A. Polyvyanyy · M. La Rosa · A. Armas-Cervantes
Computing and Information Systems, The University of Melbourne, Parkville, VIC, Australia
e-mail: zahra.dashtbozorgi@unimelb.edu.au; artem.polyvyanyy@unimelb.edu.au; marcello.larosa@unimelb.edu.au; abel.armas@unimelb.edu.au

E. Otte
Biopharmaceutical Product Development, CSL Limited, Melbourne, VIC, Australia
e-mail: ellen.otte@csl.com.au

G. Fortino, M. Mecella (eds.), *Internet of Things Meets Business Process Management*, Internet of Things, https://doi.org/10.1007/978-3-031-90746-3_12

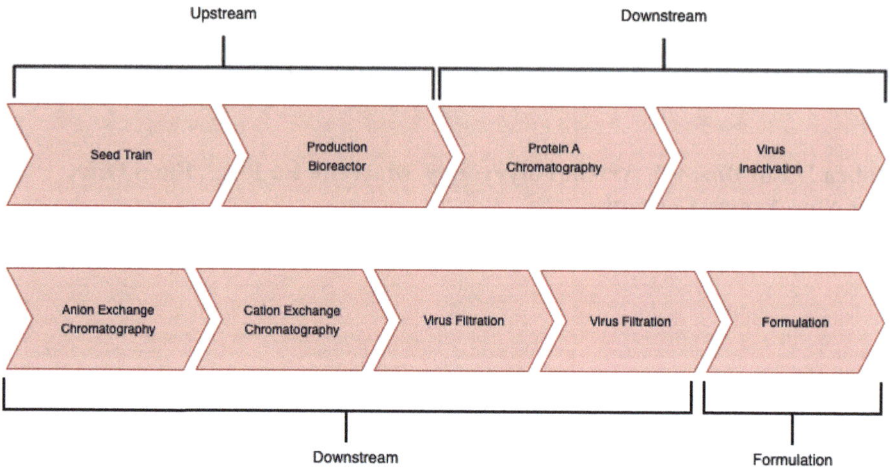

Fig. 1 Unit operation of a typical bioprocess

high-quality product as possible. After the production steps, several purification steps are carried out to remove contaminants and purify the products. These steps are collectively known as downstream processing. The last operation after the downstream steps is known as formulation. In the formulation step, the product of interest is prepared as a substance that can safely be used by customers. Figure 1 shows the upstream, downstream, and formulation unit operations.

In recent years, bioprocess industries have experienced a paradigm shift toward the adoption of digital technologies to streamline and optimize bioprocess development. With the increasing complexity and diversity of bioproducts, there is a pressing need for advanced methodologies to effectively capture, analyze, and optimize the intricate processes involved in bioproduction. Digital twin technology [1] has emerged as a promising approach to address these challenges by creating virtual replicas of physical systems, enabling real-time monitoring, analysis, and optimization. However, the application of digital twins in the realm of bioprocess development remains relatively unexplored and is often confined to a single step in the bioprocess. This results in multiple disconnected digital twins that do not capture the relationship between the various unit operations. This makes optimization and long-term decisions challenging.

This book chapter presents a novel digital twin framework encompassing all unit operations in a typical end-to-end bioprocess. The novelty of our framework lies in the integration of concepts and techniques from business process management and process mining, which are widely used for analysis and improvement of business processes in a variety of industries. Using these concepts, it is possible to represent bioprocesses as a series of human and biological steps, as well as leverage data collected during the execution of the bioprocesses to identify improvement opportunities.

In the proposed digital twin framework, two types of improvements are considered: short-term and long-term. Short-term improvements are done on the fly during process execution. For example, using a controller that learns its strategy from the digital twin's process control component, a short-term strategy is to dynamically adjust the parameters to keep the process in stable condition. Long-term tactical improvements are those that can be derived from the digital twin and are implemented in future instances of the process. For example, using the digital twin, one might determine an optimized feeding strategy, which is then used in future bioprocess executions. Our envisioned digital twin should be equipped to address both types of improvement as will be discussed later in the chapter.

The rest of this chapter is organized as follows. We begin by reviewing the current literature on digital twins in both process mining and bioprocessing domains in Sect. 2. We then present the envisioned framework, its various components, and use cases in Sect. 3. In Sect. 4, we describe the challenges of achieving the envisioned framework. We present an example use case and partial implementation of the proposed digital twin as a proof of concept in Sect. 5 and, finally, conclude the chapter in Sect. 6.

2 Related Work

The related work reviewed in this section is divided into two subsections. Section 2.1 reviews existing works on digital twins of organizational processes. Existing works on digital twins of bioprocesses are reviewed in Sect. 2.2.

2.1 Organizational Digital Twins

Since their conception, digital twins [1] have been adopted in various settings. One such setting is the use of digital twin technology to represent organizational business processes (Fig. 2). While a digital twin is a virtual representation of real-life phenomenon that is indistinguishable from its real-life counterpart, van der Aalst et al. [2] describe digital model and digital shadow as other virtual representations that vary in their level of integration with the real-life phenomena they represent. Figure 3 shows the difference between (1) digital model, (2) digital shadow, and (3) digital twin. A digital model is created using manual and offline collected data from a real process. There is no real-time connection between reality and the digital model. The dotted lines between reality and the digital model represent the loose coupling between them. A digital shadow goes one step further in connection to reality. It is synchronized with reality by having real-time data fed into it (solid arrow from reality to the digital shadow); however, any insights and decisions resulting from the model are fed back to reality offline and manually. The digital twin goes further than digital shadows by having an online feedback loop connecting

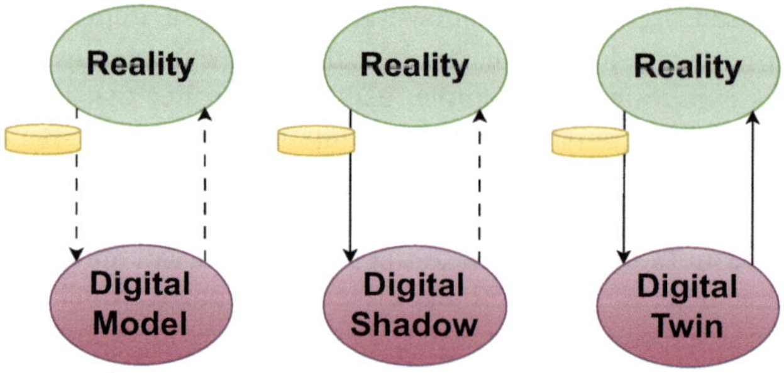

Fig. 2 Differences between a digital model, a digital shadow, and a digital twin (Figure adapted from [2])

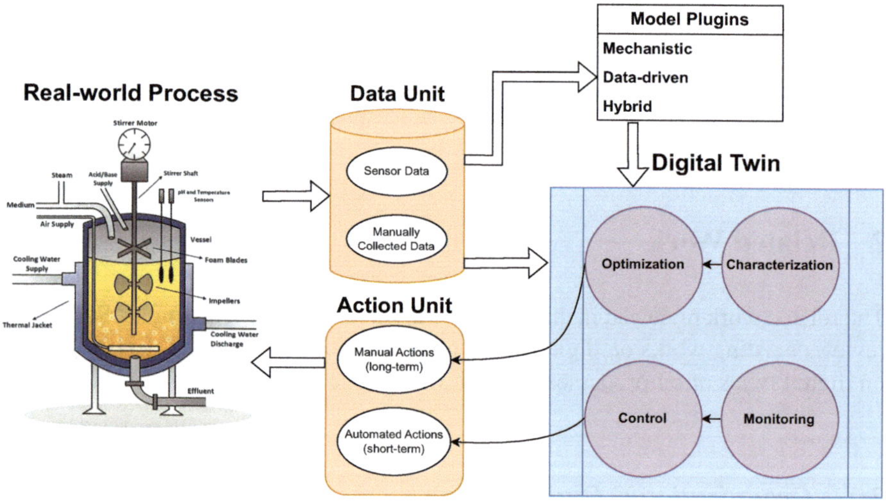

Fig. 3 Proposed framework

the digital twin to reality. Most existing process mining techniques are either digital models or digital shadows, while digital twins currently remain aspirational.

The history of the emergence of the digital twin concept is explored in [3]. The authors explore the role of digital shadows as building blocks for a broader digital twin. Additionally, it builds a connection between digital shadows and mathematical (or physical) models by showing examples of how such models are currently used in building digital shadows. However, such work is mainly concerned with the conceptual link between data management and algorithmic services and is not concerned with the challenges of implementing digital shadows and organizational twins.

Achieving a digital twin of organizations is highly desirable since it provides a virtual environment where different actions and decisions can be tested without sacrificing quality or wasting resources in the real world. Many works using the notion of a process to build a digital replica of a workflow have been proposed in the literature. For example, in the work by van der Aalst et al. [4], the authors discuss the limitations of modelling paradigms such as Turing machines and Markov chains. Such models are limited in their expressive power—e.g., they cannot capture concurrency between activities, a key aspect of organizational business processes—and therefore, are limited in their capability to create a digital twin of an organization. Therefore, the authors propose the use of more sophisticated process modelling languages, such as object-centric Petri nets [5], to represent the workflow and interaction of objects within organizations. The benefits of object-centric Petri nets in digital twinning are further explored in [6].

In another work, Park and van der Aalst [7] present a digital twin of an organization as the digital replica of a production process or the entire organization. Their proposed digital twin includes an interface model representing the business process and its supporting information systems. This representation is in the form of a process model (e.g., a Petri net). The interface model presents the process analyst with the current state of the process and highlights bottlenecks and configurations. The analyst then defines process constraints and suitable actions. The constraints and actions are fed into an action engine that continuously monitors the processes and triggers actions based on the monitoring results. Another work by Park et al. [8] uses digital twins of organizations to assess the effect of information system updates in a process-aware manner. In this work, they define digital twins interface models as object-centric Petri nets. Updates in the information system are modelled as different configurations of the Petri net model.

The connection between process prediction and digital twins is explored in an article by Brockhoff et al. [9]. In this work, the digital twin is defined as a software system that actively represents, controls, and optimizes a cyber-physical system. This digital twin incorporates process mining services such as process discovery and conformance checking components to derive digital shadows and actionable insights. While this work mentions extraction of underlying processes using continuous measurements, they do not discuss the challenges of such extractions and possible solutions to these challenges, which is the focus of this book chapter.

Bano et al. [10] present a digital twin cockpit that is process-aware. They define this cockpit as the user interaction part of the digital twin that can handle processes related to the physical object. Similar to our proposed approach, they incorporate sensor data. They use the sensors to create events and extract event logs from sensors installed in a healthcare setting. However, their work also deals with discrete events (e.g., a nurse entering the patient's room), whereas in our work, we aim to create digital twins of objects involving processes with no discrete events.

Biological processes and their modelling paradigms are mentioned in the work by Becker and Pentland [11]. In this work, they propose to use inspirations from biological models such as regulatory network modelling and incorporate those inspirations in creating digital twins of organizations. However, they do not explore modelling of bioprocesses themselves as we do in this work.

The works mentioned in this section offer valuable insights into how process mining concepts and methodologies can be used in creating digital twins of processes and organizations. However, these works mainly focus on the complexities of various interrelated processes that exist within organizations, while in this book chapter, we seek to use the notion of process and harness the power of process mining techniques to deal with the challenges of twinning biological processes and their interaction to organizational workflows.

2.2 Bioprocess Digital Twins

In recent years, digital twins have gained traction in the field of bioprocess development. As mentioned in Sect. 1, bioprocesses comprised several unit operations each serving a different aim in the development of the bioprocess. Each of these unit operations has its unique associated challenges, such as heterogeneous data types, varying degrees of uncertainty, and complexity of the steps involved. Therefore, the majority of existing works in the bioprocessing field focus on a single unit operation rather than the end-to-end process. In this section, we review existing work on digital twinning in the bioprocess development field grouping them by the unit operation targeted.

2.2.1 Seed Train

After thawing up a small vial of cells, the first step is to increase their number in a unit operation known as the seed train. There are several challenges such as complex cell metabolism, batch-to-batch variation, the uncertainty of cell behavior, and the effects of cultivation conditions. Rodriguez and Frahm [12] outline the necessities of digitizing the seed train step. However, their envisioned digital twin is a mechanistic model fit to observable parameters that predict key performance indicators of the seed train, such as viable cell density. Additionally, their digital twin does not automate decision-making. Decisions such as passaging strategy are still made manually using the prediction model. They also explore the role of data-driven strategies and uncertainty estimation.

2.2.2 Production Bioreactor

This step comes after cell expansion and is one of the more complex unit operations due to its high uncertainty [13]. Consequently, the majority of digital twin literature in the bioprocessing field focuses on this step.

Park et al. [14] propose a digital twin framework that leverages process analytical technologies (PATs). In particular, they explore new data collection technologies such as soft sensors and their impact on achieving real-time synchronization between the real process and the digital twin. Their envisioned framework considers the digital twin to be a mechanistic or data-driven model or a hybrid of the two. However, this view is limiting because such models are developed for single unit operations. To the best of our knowledge, no mechanistic or data-driven model presents the process end-to-end. Therefore, using this framework, we would need to have multiple unconnected digital twins for each unit operation, making optimization of the entire process challenging. Nevertheless, this framework is widely adopted in the bioprocess literature with many works proposing to create digital twins of the bioreactor step [15–18].

2.2.3 Downstream Purification

The downstream steps occur after the bioreactor step to separate the product of interest from other byproducts and remove impurities. Khuat et al. [19] identify typical problems addressed in the downstream process step that benefits from digitization and twinning:

- Monitoring and prediction problems in the capture chromatography step
- Monitoring and prediction problems in the polishing chromatography step
- Control of a chromatography process
- Scale-up and prediction problems in filtration
- Optimization of purification sequences

A rich body of literature proposes to tackle the problems stated above. For instance, Tiwari et al. [20] propose digital twins as a means to design a control strategy for the chromatography step. Similar works propose data-driven approaches for prediction problems in chromatography [21], scale-up problems in filtration [22], and various optimization problems in the downstream [23, 24]. In the majority of these works, building a real-time connection between the models and the real process is not clear. Moreover, the models are developed for various purposes using heterogeneous data sources. Therefore, there is no clear link between the various steps of the downstream.

2.2.4 Generic Digital Twin Frameworks

It is clear that while having unit operation digital twins is very useful for solving various tasks, it is not sufficient for optimizing the entire process. Another line of work centers on a more generic idea of digital twinning. For instance, Taylor et al. [25] suggest holistic process modelling as the main engine for digital twins. They suggest improvements such as combining small and large-scale data for model building, improving uncertainty intervals, and establishing extrapolation procedures for noncontrollable parameters. However, they do not specify a modelling paradigm for multiunit operation digital twins.

In [26, 27], the authors investigate digital twins for design of experiments. Their goal is to reduce the number of experiments by identifying critical process parameters via in silico modelling. Another work by Sokolov et al. [28] argues for the use of hybrid modelling (models consisting of both data-driven and mechanistic components) as the enabler of digital twins. This opinion is supported in other works as well [29, 30]. None of the works mentioned above, however, propose any methodology for connecting the various model components and representing the end-to-end process as a unified entity. We aim to address this gap via the framework proposed in the next section.

3 Framework

In the realm of process industries that utilize biologics, the digital twin concept has emerged as a transformative methodology, aiming to revolutionize processes to enhance efficiency, quality assurance, and adaptability. In this section, we clarify the context and scope of the digital twin and propose a conceptual framework and its various use cases. The digital twin concept involves the creation of a dynamic and interactive virtual model that emulates the behavior of its physical counterpart in real time. By combining data from sensors and manual inputs, the digital twin becomes an accurate and responsive representation of the bioprocess. This virtual replica serves as a powerful tool for various tasks. We describe a comprehensive end-to-end digital twin that contains upstream, downstream, and formulation unit operations to provide an overall picture of the process to end users. There are many tasks that can utilize a digital twin. We group all these tasks into four categories, namely, process characterization, process optimization, process monitoring, and process control. Therefore, the envisioned digital twin consists of four key components each dedicated to one category.

We define the digital twin as a process model that acts as the digital replica of the physical process. This digital replica is close enough to reality that it can be used for process characterization, control, monitoring, and optimization. The envisioned digital twin is connected to the physical process in real time in a bidirectional way. It receives data from the physical process in an online manner and provides feedback either directly or indirectly to the physical process via automated or manual actions. This information transfer occurs via the data and action units as described in Sect. 3.2. Knowledge about specific tasks and unit operations can be incorporated via model plug-ins as described in Sect. 3.3

3.1 The Digital Twin

3.1.1 Process Characterization Component

Understanding the inherent characteristics of each unit operation is the primary focus of the process characterization component. It involves understanding and quantifying the behavior of a bioprocess under various conditions to ensure consistent and reproducible production of bioproducts. This is usually done by defining critical process parameters (CPPs) and critical quality attributes (CQAs). CQAs are the attributes of the final product that determine its quality, safety, and efficacy. CPPs are process parameters that have a significant impact on CQAs. The effect of CPPs on CQAs is commonly done via a procedure called the design of experiments (DoE). This involves systematically running experiments each with varying parameters. This process can be expensive, so the digital twin facilitates DoE by providing a detailed, real-time simulation that allows for in-depth analysis. Engineers and operators can explore the digital twin to gain insights into the underlying dynamics of the process, identifying the most likely CPPs, and specific ranges for each CPP. This involves utilizing models to characterize raw material behavior, product formation kinetics, and scalability. For this component, it is beneficial to have data coming from various process conditions to capture the underlying dynamics of the physical system.

3.1.2 Process Optimization Component

When the dynamics of the process are sufficiently understood in the characterization component, this knowledge can be used to improve the process. The process optimization component identifies opportunities for process improvement and efficiency enhancement. This involves leveraging the digital twin for scenario analysis, what-if simulations, and optimization strategies to improve overall process performance. Below are some example scenarios for process optimization:

- Scenario 1: Raw Material Variability. Simulation of variations in raw material quality provides insights into critical control points, allowing for the development of strategies to mitigate the effects of variability.
- Scenario 2: Equipment Downtime. Simulating scenarios with equipment failures allows the assessment of the impact on overall production timelines, aiding in the design of robust manufacturing schedules and maintenance plans.
- Scenario 3: Process Parameter Optimization. The digital twin supports the optimization of process parameters to maximize product yield and quality. Scenarios involving changes in feeding strategy, temperature, pressure, or flow rates can be simulated to identify optimal operating conditions.

Using process mining simulation engines, it is possible to simulate the above scenarios and confirm hypotheses in silico before running costly wet lab experiments.

3.1.3 Process Monitoring Component

Continuously monitoring and assessing the ongoing manufacturing process is the role of the process monitoring component. This includes employing sensor data and real-time analytics to detect deviations, anomalies, or potential issues during the execution of the physical process. In this component, the digital twin serves as a live dashboard, offering a comprehensive overview of the ongoing bioprocess. By continuously comparing the digital twin's behavior with real-time sensor data, operators can quickly detect anomalies, deviations, or inefficiencies, allowing for proactive intervention before issues escalate. A wide range of process mining techniques can be applied in this setting. For example, conformance checking can be used to determine if the ongoing bioprocesses are progressing according to guidelines. Predictive process monitoring can also be used to predict the conditions of the bioprocess in the near future. Having access to prediction allows for proactive intervention in the process, which is done in the control component described next.

3.1.4 Process Control Component

Process control in the context of bioprocessing refers to the systematic management and regulation of various parameters, conditions, and variables within a production process to ensure that it operates within desired specifications. It involves monitoring taking the results of the monitoring component and making adjustments in real time to maintain optimal performance. The real-time synchronization between the digital twin and the physical process enables operators to make informed decisions promptly. With the ability to predict potential issues and simulate different control strategies, the digital twin empowers operators to optimize process parameters, improve efficiency, and minimize downtime. Below the main aspects of the control components are described:

- **Real-Time Monitoring:** The process control component closely collaborates with the process monitoring component. The control component receives the results of the monitoring component and uses prescriptive process monitoring methodologies to determine actions based on the monitoring results.
- **Feedback Control Mechanisms:** The digital twin facilitates the implementation of feedback control mechanisms. By comparing real-time data with the expected or optimal conditions, the digital twin can automatically trigger adjustments to operating parameters. The digital twin also serves as a testing platform for various control strategies, before they are applied to the physical bioprocess.
- **Optimization Algorithms:** The optimization component discovers optimal parameters and conditions offline. This knowledge can be used by sophisticated algorithms embedded in the control component to optimize process parameters based on real-time data. This adaptive optimization ensures that the manufacturing process operates efficiently under varying conditions.
- **Integration with Control Systems:** The control component needs to seamlessly integrate with existing process control systems. This integration enhances the capabilities of traditional control systems by providing a more detailed and adaptive understanding of the manufacturing environment.

By continuously learning from data and adapting to changing conditions, the digital twin contributes to a culture of continuous improvement. Insights gained from the digital twin can inform adjustments to standard operating procedures, leading to enhanced efficiency and product quality.

3.2 *Data and Action Units*

The data and action units are the links between the physical and virtual systems. The digital twin is built using data collected from the real process and in turn, it provides input to the real process. All communication from the physical process to the digital twin goes through the data unit and from the digital twin to the physical system through the action unit.

3.2.1 The Data Unit

Traditionally, data from bioprocesses has been collected manually by scientists. Typically, a daily sample is taken and analyzed offline. However, in recent years, more efficient and frequent techniques for sampling have been introduced via sensors. These sensors monitor media conditions such as pH, temperature, and dissolved oxygen, as well as cellular processes and feed media addition. The main benefit of using sensors is that they collect process information in or near real time, making it possible to sync the digital twin with the physical process constantly.

Another category of sensors, known as soft sensors, may be used to collect information from the bioprocess. One popular soft sensing technique that is successfully used in bioprocess industries is Raman spectroscopy [19, 31]. Raman can be used to measure a variety of process parameters and variables, such as cell profiles, nutrients, and metabolites.

The data unit collates information from sensors and offline measurements and converts them to event logs. The event log can then be used to discover the process that serves as the digital twin. As more information arrives from the sensors, the event log is expanded, and the digital twin is rebuilt to keep it synced with the physical process.

3.2.2 The Action Unit

The action unit handles the information flow from the digital twin to the physical process. This information can be in the form of new experimental designs resulting from the characterization component of the digital twin, or the control strategy derived from the process control component. Some information is fed directly to the physical process, such as the parameter values coming from the control component. But other information such as the experimental design influences the physical process indirectly via human actors. The information communicated to the physical process directly must be transferred in real time to ensure interventions are executed in a timely manner while also adhering to strict guidelines defined in the design of experiments stage.

3.3 Model Plug-Ins

Here, we describe the relationship between prior knowledge about specific unit operations and the proposed digital twin. The digital twin should utilize various modelling techniques to represent the behavior of the microorganisms and their interaction with the cell culture environment and materials. Three main paradigms of modelling are typically used in the literature: (1) data-driven modelling, (2) mechanistic modelling, and (3) hybrid modelling.

In the upstream steps, machine learning algorithms analyze historical data on raw materials, fermentation, and cell culture. These data-driven models predict variables such as cell growth, nutrient consumption, and metabolite production based on historical patterns. Mechanistic models delve into the biological intricacies of fermentation and cell culture, capturing the underlying kinetics and dynamics of microbial growth, substrate utilization, and product formation. The hybrid modelling approach blends data-driven and mechanistic components. For instance, a machine learning model can be used to empirically estimate the parameter values of mechanistic equations based on data.

In the downstream phase, data-driven models come into play to predict product yield, purity, and quality attributes. These models integrate historical data on equipment performance, process conditions, and product characteristics. Complementing this, mechanistic models simulate the physical and chemical processes involved in downstream operations, considering mass transfer, chromatography column dynamics, and filtration kinetics.

These models, regardless of which paradigm is used, typically describe one unit operation and are only concerned about optimizing one step. However, such models do not provide information about how each unit operation affects the subsequent ones, and global optimization across all steps is still under-explored. The end-to-end digital twin serves as an overall platform where the entire process can be studied with various modelling techniques. For instance, such models can be used to generate data under various conditions. This generated data can be the input to the digital twin to see the impact of each model on the process as a whole.

4 Challenges and Guidelines

Each unit operation in the process gives rise to specific challenges. There are also challenges associated with the end-to-end process. In this section, we provide an overview of some of these challenges and propose directions for future research to tackle them.

4.1 Building Process Models from Real-Valued Data

Traditionally, BPM and process mining techniques have been developed to manage organizational workflows. Workflow models typically deal with discrete steps carried out by actors and involving one or many objects. As a result, most process mining techniques are equipped with ways of dealing with categorical sequences, and not continuous numeric values such as multidimensional time series. Bioprocess data, whether collected manually or via sensors, has a unique structure that workflow models alone are not sufficient to handle. Similar to a typical organization's workflow, a bioprocess involves activities carried out by actors and involves objects. However, these actions are fixed to comply with strict regulatory and safety requirements. Therefore, in an event log from a bioprocess manufacturing facility, there are few process variants, and any suggested change to the workflow by the digital twin needs to go through a comprehensive and lengthy regulatory approval. In such a setting, conformance checking—one of the main operations in process mining—may be used to detect any deviations from the regulatory requirements.

One major challenge in bioprocess development is that such processes involve living microorganisms. These organisms' behavior can be unpredictable and difficult to control. Also, determining the best conditions for their performance is nontrivial. The behavior of these living cells is captured by data collected manually or from sensors installed in the equipment. Typical sensor measurements consist of process parameters that are fixed throughout a specific unit operation and process variables that are subject to change. These data are in the form of multidimensional time series taken at fixed time intervals depending on the sensor type. One major challenge when representing the bioprocess digitally using a BPM notion of process is that the sensor data needs to be converted into event log format since process models are formed using discrete activities, actors, or objects. This is a significant challenge because if not done correctly, important information may be lost in the process of conversion to event logs. One possible solution is to use unsupervised clustering algorithms to determine latent activities in the measurements, as done in [32].

4.2 Generating Control Signals

As mentioned in Sect. 3.1, the process control component of the digital twin is used to test control strategies before it is applied to the physical process. One concern is that the control strategy heavily relies on the quality of the model on which the digital twin is built. Low-quality data and models can lead to sub-optimal control decisions. So, a major challenge is to ensure that the model is of sufficient quality to be used in control decisions. To address this challenge, a rigorous quality assurance procedure must be implemented that ensures that the output of the digital twin meets the standards and requirements.

The other challenge of generating control signals relates to the previous challenge regarding the digital twin being discrete. Typically, control signals specify fine-grained numeric values for controlled process parameters. However, a discretized digital twin can only recommend a range of values. One way to overcome this challenge is to combine it with a numerical model that can determine the exact value of the controlled parameter after the process model has determined the range.

4.3 Shifting from Batch Processing to Continuous Processing

Currently, the majority of bioprocesses are batch processes. This means a batch of material goes through the unit operations serially. Each batch has its unique identifier. In the context of process mining, each batch can correspond to a case. However, there has been a gradual shift toward continuous manufacturing due to its benefits of improving product quality and processing times [33]. For such continuous processes, one challenge is to identify the notion of case identifier as the material continually enters the processing reactor and its waste and unwanted material are continuously removed. This problem can be addressed via segmentation techniques in the robotic process mining literature.

4.4 Determining Simulation Fidelity

Another main challenge in building a digital twin of bioprocesses is to determine the fidelity of the models built into the digital twin. Simply, we need to determine how close the digital twin should be to the physical process. This depends on what the digital twin is being used for and on the unit operation being modelled. Some tasks require higher fidelity, while for other tasks, low-fidelity models might suffice.

To address this challenge, the first step is to review industry standards for each unit operation and for the overall end-to-end process. Also, error bounds must be established for each task, and the digital twin's performance should be assessed against these bounds.

4.5 Incorporating Models of Specific Unit Operations

We might wish to use the knowledge gained from other modelling techniques that describe the bioprocess via the model plug-ins component of the framework. There is extensive literature on using data-driven, mechanistic, and hybrid techniques to model specific aspects of the process. While we envision the end-to-end bioprocess as a process model, some specific tasks will benefit from using the models that have been developed in the literature. So a key challenge is to identify opportunities and frameworks for blending such existing models with the end-to-end process model.

4.6 Degree of Hybridization

High-quality data availability is a major challenge in modelling bioprocesses. Due to the tight regulatory controls placed in bioprocess industries, the data coming from such facilities do not contain many variations. This makes creating what-if scenarios a challenging task. Users might wish to use the digital twin to analyze what will happen if something goes wrong. For example, how would the developed control strategy perform if there is a power outage? Or how should the actors proceed if one of the critical process parameters goes out of the established bounds? If the digital twin solely relies on data, it might not perform optimally unless a similar situation has occurred in the past. In such situations, it might be beneficial to incorporate process knowledge from mechanistic models. One way to achieve this is to generate data from such mechanistic models and discover the digital twin from the generated data. This is also not optimal since existing mechanistic models are too simple to model all aspects of the process. Therefore, the most obvious solution is to use a combination of real data and the generated data from mechanistic models. Nevertheless, the challenge remains, to what degree should these data be combined? Which process behaviors should be coming from real data and which from the generated data?

4.7 Scale-Up

Much of the existing models and knowledge about bioprocesses come from lab experiments. Before a product is manufactured at a large scale, smaller lab and pilot-scale experiments are conducted, and much of the existing data comes from such experiments. It is reasonable to assume that the envisioned digital twin will also be developed using small-scale data since manufacturing scale data is scarce. So another challenge that needs to be addressed is the scalability of the digital twin. Is a digital twin developed by data coming from a 5 L bioreactor going to accurately model a 5000 L reactor? Such questions should be answered using a pilot-scale experiment for the digital twin. Once the safety of the digital twin on all other aspects is determined, it should be trialed in a large manufacturing facility before it can be deployed for regular use. Also, the mechanics of the larger environment should be incorporated into the process model to account for variations that might occur due to the larger equipment and material used in manufacturing.

4.8 Unit Operation-Specific Challenges

Each unit operation comes with its own specific challenges. Depending on the level of detail included in the digital twin, we might need to model individual unit operations. In the upstream phase, the microorganisms used in the bioprocess are grown in a culture and are encouraged to make a target product by tightly controlling the culture environment. The upstream is the step where the process heavily relies on the behavior of the living microbes to achieve the desired outcome. So this step has a high level of uncertainty and a complicated underlying behavior, which is difficult to capture by a model. On the other hand, the downstream steps rely less on living organisms and more on efficient use of equipment and understanding of fluid dynamics to successfully carry out the filtrations required. When designing the digital twin for the upstream steps, it is important to include uncertainty quantification techniques and investigate how the level of uncertainty can impact the downstream steps. For the downstream steps, the digital twin benefits from an object-centric approach where all the equipment and materials are central to the design of the process model and their impact is well established.

5 Case Study with a Pharmaceutical Company

In this section, we demonstrate the application of our envisioned digital twin framework via a case study with a large pharmaceutical company in Australia. This case study focuses on the monitoring component of the digital twin. In the first instance as a proof of concept, we focus on a single unit operation rather than the entire end-to-end process. However, the unit operation we have chosen is the production bioreactor, a very complex step in the bioprocess. In this case study, we show how monitoring of bioprocess parameters and quality attributes is possible via existing process mining solutions.

5.1 Soft Sensors and Bioprocess Monitoring

During the bioreactor phase, monitoring various process parameters and quality attributes is essential. This is because monitoring these parameters provides insights into cell activity and how the bioprocess unfolds. Typically, nutrients, metabolites, protein concentrations, and information about cell growth are monitored in real time during bioreactor runs. Monitoring these parameters allows scientists to make crucial decisions on how to intervene in the process according to indications of cell behavior. In particular, predictive monitoring is especially beneficial in this scenario because it allows for preemptive measures to be taken and for proactive interventions to be possible. For example, consider the metabolite lactate. It is a natural byproduct of cell activity when it is producing a target protein. Small amounts of lactate in the cell culture are normal. However, excessive levels of lactate acidify the cell culture medium, leading to low overall performance [34]. So, lactate levels are constantly monitored during bioreactor runs, and if they start to go high, actions are taken to reduce them. Predictive monitoring of lactate concentrations allows such actions to be taken before lactate reaches undesired levels.

Traditionally, monitoring has been carried out by manually collecting samples from the bioreactor and analyzing the samples offline. However, process analytical technology (PAT) tools are becoming increasingly common to measure monitored parameters during various stages of the bioprocess, including the bioreactor. One subset of such tools is spectroscopy. Various spectroscopy techniques have emerged in the bioprocessing domain, allowing for precise and real-time monitoring of process parameters. A comprehensive review of spectroscopy use cases in the bioprocess industries is presented in [19].

Raman spectroscopy is a PAT technique used in bioprocess development for various purposes. During the bioreactor phase, it is used for real-time monitoring of critical process parameters such as glucose, lactate, glutamine, glutamate, ammonia, and viable cell density (VCD) [35]. To achieve this, Raman probes are inserted into the bioreactor. These probes collect spectral measurements at specific time intervals (every 15 minutes in our experiments). Spectral measurements need to be correlated to analytical measurements via statistical or machine learning models. These models are calibrated on manually collected measurements or data coming from other sensors. The calibration step is typically done offline and then the model is deployed for real-time monitoring. Once the calibrated model is applied to spectral measurements, the concentrations of the aforementioned parameters can be estimated in real time. These estimates form a multivariate time series as they are concentration values collected at a fixed time interval.

5.2 *Predictive Process Monitoring*

The calibrated model provides current concentration values from spectral data. But it does not provide predictive capabilities. So, for predictive monitoring of critical process parameters, predictive techniques can be used to estimate future concentration values. Since the concentration values are time series, one natural way of predicting future values is time series forecasting. We can find various time series forecasting techniques applied in the bioprocessing domain, particularly for forecasting critical process parameters during the bioreactor phase [36, 37].

There are a few limitations in applying time series forecasting techniques for predictive monitoring. First, these techniques do not incorporate the notion of a case. The input type assumed by time series forecasting methods does not account for the fact that measurements might come from independent experiments. For example, if we have glucose concentrations $< a_1, a_2, \dots, a_t >$ for bioreactor A, and $< b_1, b_2, \dots, b_t >$ for bioreactor B, and we wish to predict a_{t+1} and b_{t+1}, we cannot pool both sequences together as input to the forecast model. We need to train separate models for each bioreactor. This means that each model needs to be trained using data from the initial stages of the bioreactor and then applied to get predictions for the remainder of the bioreactor run. This is not optimal because we cannot use patterns and knowledge gained from previous bioreactor runs in the prediction task.

Second, the accuracy of the forecasts typically decreases as the prediction horizon increases. For instance, suppose that we train the model on glucose concentration values of the first 2 days. We can get highly accurate forecasts for day 3. But the forecasts may be highly unreliable for day13. This means that to get accurate results, models need to be updated at least every few days for each bioreactor run. This can be costly and inefficient.

In this case study, we propose using the notion of a process as it is defined in the business process management and process mining communities for the predictive monitoring task. Given data collected using Raman spectroscopy, we aim to predict glucose and lactate concentrations in the next time step. To achieve this, we follow the pipeline outlined in Fig. 4.

5.2.1 Data Pre-processing

The raw Raman spectra do not include the substance concentrations. Several pre-processing steps need to be carried out to obtain concentration values. The first step is clipping. Clipping involves removing data points that are outliers or contain artifacts or other noise. This step helps to clean the spectrum in improve subsequent analyses. The second step is smoothing. Smoothing techniques, such as moving average, Savitzky-Golay, or Gaussian filtering, are applied to reduce high-frequency noise while preserving the spectral features. Smoothing improves signal-to-noise ratio (SNR) and enhances the visibility of peaks corresponding to molecular vibrations (which helps to distinguish concentrations of different substances in the cell culture). The third step, namely, baseline subtraction, is performed to remove the background signal caused by fluorescence, Rayleigh scattering, or other nonresonant processes. There are various methods for doing this such as polynomial fitting, asymmetric least squares, or adaptive iteratively reweighted penalized least squares (airPLS). Baseline subtraction reveals the true spectral features and facilitates accurate peak identification and quantification. Next, in the fourth step, normalization is applied to adjust for intensity variations between spectra caused by factors such as sample concentration, instrument settings, or laser power fluctuations. Common normalization techniques include area normalization, maximum normalization, vector normalization, or probabilistic quotient normalization. But for Raman spectra, in particular, vector normalization is preferred. Normalization ensures that spectral comparisons are not biased by differences in overall intensity and improves the consistency of analyses across samples. It should be noted that normalization should be done after baseline correction [38]. The last and fifth step is dimensionality reduction. In this step, techniques, such as principal

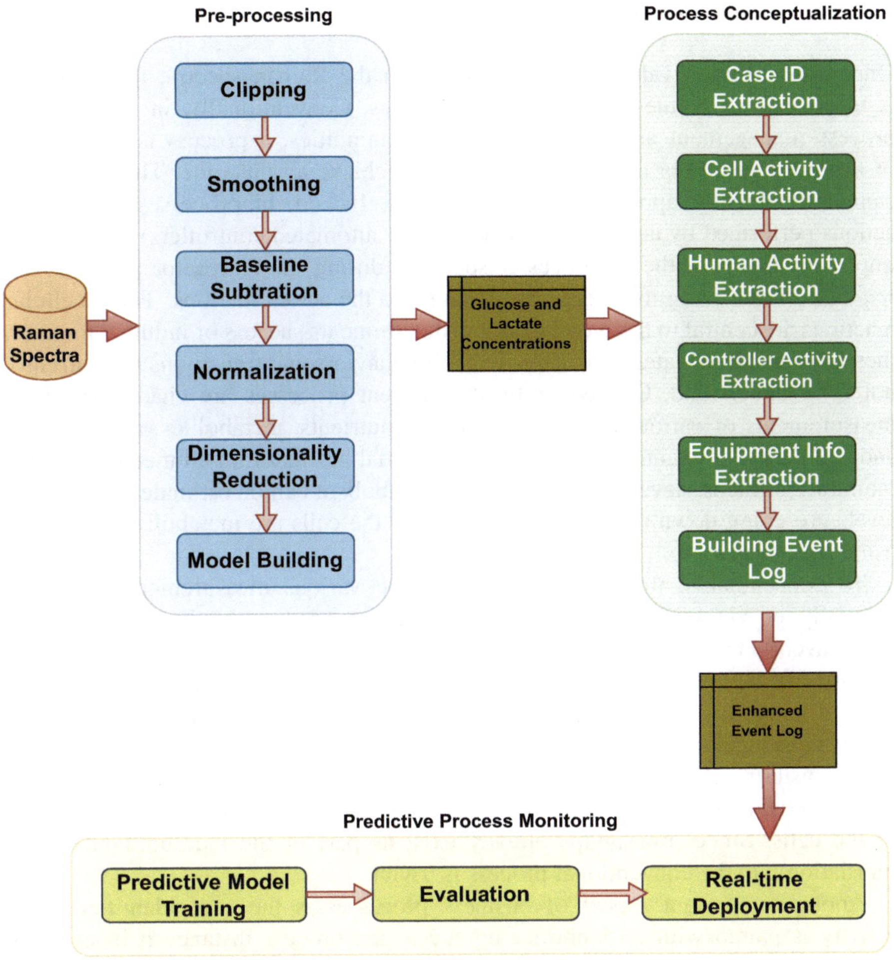

Fig. 4 Predictive bioprocess monitoring pipeline: The activities shown in blue are the pre-processing steps described in Sect. 5.2.1, the activities shown in green are the process conceptualization steps described in Sect. 5.2.2, and the activities in yellow are the process monitoring steps of Sect. 5.2.3

component analysis (PCA) or partial least squares (PLS), are employed to reduce the number of variables while retaining most of the spectral information. PCA identifies principal components that capture the majority of variance in the data, enabling visualization and interpretation of spectral differences. PLS regression combines dimensionality reduction with regression analysis to model relationships between spectral data and sample properties, facilitating quantitative analysis and predictive modelling. PLS is often used to calibrate a model that maps spectral features to concentration values [35].

5.2.2 Process Conceptualization

Once concentration values are estimated from the Raman spectra, the next step is to conceptualize these readings as a process. Conventionally, in the business process management and process mining communities, a process is a sequence of activities involving actions and objects to achieve an outcome. This definition cannot be directly applied to the bioprocess. Indeed, bioprocesses do involve actions performed by actors (e.g., scientists or automated controllers), but another important aspect of the bioprocess, especially during the bioreactor phase, is the *reaction* of the living microorganisms (cells) to the actors' actions. These cellular reactions are central to the bioprocess and determine its success or failure. However, they are considered latent processes as their behaviors and transitions are currently not fully understood. Consequently, these latent processes are characterized by measurements of various substances, such as nutrients, metabolites, amino acids, and cell data. For example, glucose is an essential nutrient for cell metabolism. By monitoring glucose levels, the rate of cell metabolism can be estimated. If glucose levels are going down rapidly, it indicates that the cells are metabolizing fast and using a lot of glucose.

To conceptualize the latent bioprocess, the various measurements collected manually or via sensors (e.g., Raman sensors) need to be incorporated into the conventional definition of a process. The conventional business process has activity labels, whereas we have time series measurements. Therefore, to create activity labels for the bioprocess, we use clustering. The benefit of clustering is that it is unsupervised, so no prior knowledge is required to construct the labels. Also, clustering captures the natural shape of the data, so the discovered activities correspond directly to the collected measurements and, by extension, to the behavior of the cells. Since timestamps already exist as part of the measurements, their translation into the conventional process is trivial.

Another important aspect of business processes is the case identifier. Each activity is paired with an identifier unique to the process instance it belongs to. In this case study, each bioreactor run is an independent instance of the bioprocess. Each run is characterized by the vessel used, the experiment it belongs to, and the project that contains that experiment. Therefore, in this setting, the case ID is a tuple of the following form: (project_ID, experiment_ID, vessel_ID).

Using the abovementioned methods, we can construct event logs consisting of case IDs, activity labels, timestamps, and optional case attributes. Additional information can be added to enhance the event log. For example, information about the initial conditions of the bioreactor, or instrument settings, can be added to the event log as case attributes. In this case study, we use Raman probe settings as case attributes.

5.2.3 Predictive Model Training

Once the bioprocess event log is created, the next step is training a predictive model to predict the next activity (cluster) given a prefix. Within the area of process mining, predictive process monitoring is experiencing a surge in activity, with numerous new techniques being proposed every year. These techniques use a variety of machine learning and deep learning methods for predicting various components of the process, with next activity prediction being one of the main focuses.

Predicting the next activity in this setting is not sufficient. This is because the predictive monitoring method only predicts the next cluster, but not the concentration value. In many cases, external actors of the bioprocess make decisions based on concentration values. Moreover, many control strategies are implemented with concentration values as input. Therefore, the predicted cluster needs to be converted to concentration values. To achieve this, we consider the predicted concentration to be the average value of the given substance in the predicted cluster:

$$Y_{s|c} := \frac{1}{n} \sum_{i=1}^{n} s_i \tag{1}$$

where $Y_{s|c}$ is the predicted concentration of substance s given cluster c, s_i are concentration values of substance s in cluster c, and n is the number of data points in cluster c.

5.3 *Experimental Setup*

We apply the above methodology to a real-life dataset from a pharmaceutical company. The dataset consists of 35 bioreactor runs for the production of monoclonal antibodies. The bioreactor runs comprise different projects and different experiments within the same project. For each bioreactor, Raman spectra are available with a sampling frequency of 15 minutes. Moreover, we use daily manually collected samples to calibrate the model mapping spectra to concentration values.

We performed the pre-processing steps using the following methods. Clipping is carried out to include spectra only within the range of 600–2000. Smoothing is done using a one-dimensional Gaussian filter and baseline correction with the airPLS method. After that, vector normalization is performed. To obtain concentration values, a PLS model is trained using spectral data as input and the daily concentration values of glucose and lactate as output.

Once glucose and lactate values are obtained at 15 minute intervals, we discover latent cellular activities via clustering. We use two common clustering methods to investigate the effect of clustering performance on the overall performance. The methods used are Kmeans and DBSCAN. We select the value of k for Kmeans through various strategies and compare the results. We use the elbow method and the Bayesian information criterion (BIC). In addition, we use the same number of clusters automatically discovered by DBSCAN.

We use a neural network to perform predictive monitoring. We use an embedding layer to represent the traces as a fixed-length vector of size 50. This vector is the input to an LSTM layer consisting of 100 units. It is then followed by a dense layer consisting of n units with n being the number of identified clusters. The softmax activation function is used to get probabilities for each cluster.

We use leave-one-out cross-validation (LOO-CV) to evaluate model performance. This means that for each training iteration, one of the bioreactors is reserved as the validation set, and the rest of the data are used in training. We report the average performance across all bioreactors. The clustering results are evaluated visually, as the input is two-dimensional (glucose and lactate).

We compare our results with time series forecasting since predictive monitoring is done via this approach in the bioprocessing literature [36]. For each bioreactor, a separate forecasting model is trained on the first 2days of glucose and lactate concentrations, and forecasting is done on the subsequent days until the end of the bioreactor run.

5.4 Results

To evaluate the performance of the PLS model, we calculate the mean squared error (MSE) and R2 score for the validation set and compare the results with a random regressor. Table 1 shows that the PLS model significantly outperforms the random classifier, and hence it can be used to obtain concentration values from the Raman spectra.

Figure 5 shows the results of the elbow method and the Bayesian information criterion for selecting the number of k for Kmeans. It can be seen that using the elbow method, the optimal number of clusters is 3, whereas the BIC shows that performance improvement diminishes after 14 clusters. 3 and 14 were picked as possible values for the number of clusters in Kmeans. Also, using DBSCAN, the number of clusters was automatically selected as 8. For comparison purposes, we used 8 clusters for Kmeans as well. Figure 6 shows the clustering results using DBSCAN and Kmeans using 3, 8, and 14 clusters.

It can be seen visually that DBSCAN best captures the shape of the data and the clusters are well-separated. But in Kmeans, the cluster boundaries seem arbitrary. Nevertheless, we proceed to build event logs using both methods and perform

Table 1 Performance of the PLS model in terms of MSE and R2 score

	MSE	R2 score
PLS	23.53	0.48
Random Regressor	48.1	−0.13

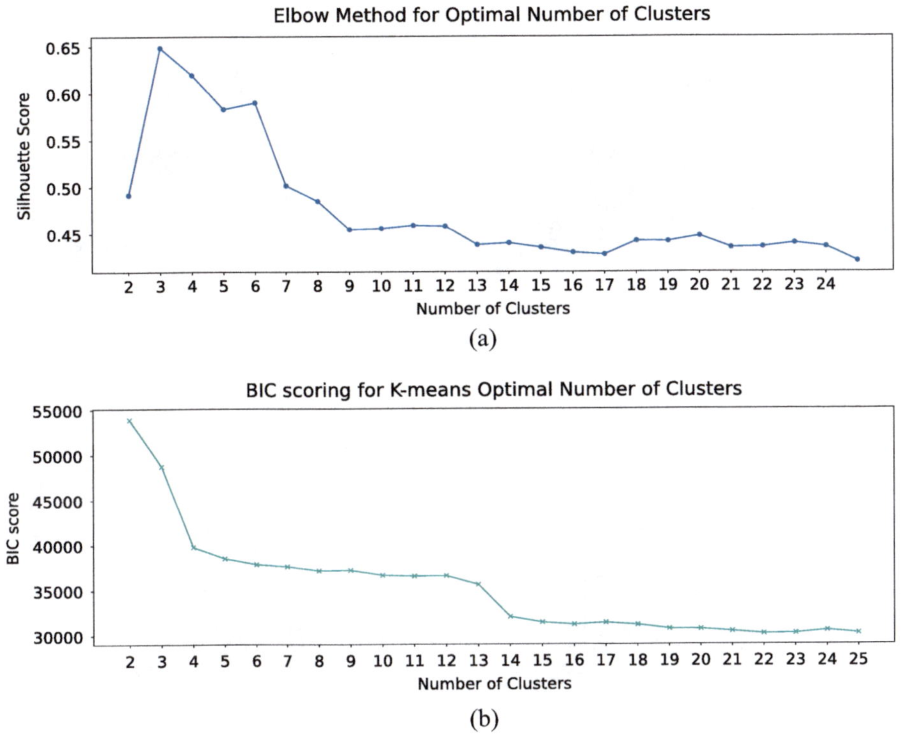

Fig. 5 (**a**) The elbow method, showing the silhouette score for different numbers of clusters (higher is better). (**b**) The BIC score for different numbers of clusters (lower is better)

predictive monitoring. Table 2 shows the results of both predictive monitoring using our proposed approach (with various clustering configurations) and time series forecasting. The metrics used are mean absolute error (MAE), means squared error (MSE), and root mean squared error (RMSE).

5.5 Discussion

It can be seen from the results in Table 2 that overall the proposed approach outperforms the time series baseline for both glucose and lactate. It empirically confirms the hypothesis that using the notion of a process as part of prediction can improve results. For glucose, the best results in terms of MSE and RMSE are obtained using Kmeans with 3 clusters, even though clustering performance is below DBSCAN. This is because the LSTM performance is significantly higher when using 3 clusters. As we have limited data available (35 bioreactors), more complex traces are harder to predict. In the case of glucose prediction, even though

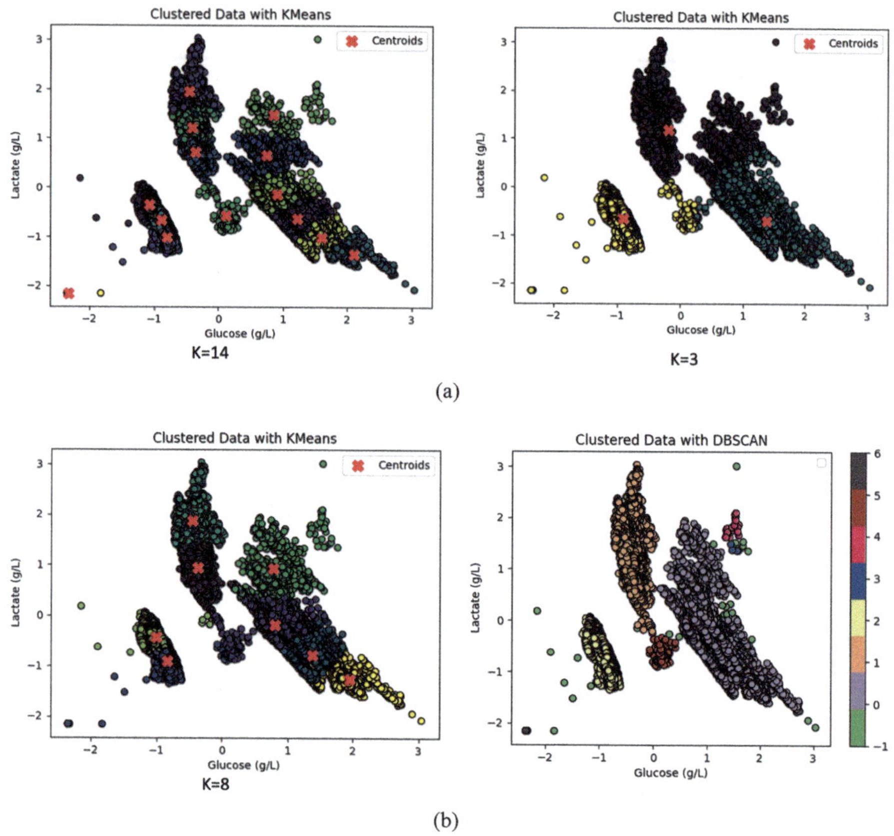

Fig. 6 (**a**) Clustering results using Kmeans with 3 and 14 clusters. (**b**) Clustering results using DBSCAN and Kmeans with 8 clusters

Table 2 Comparison of results using the proposed approach and time series forecasting

	Number of clusters	LSTM accuracy	Glucose MAE	Glucose MSE	Glucose RMSE	Lactate MAE	Lactate MSE	Lactate RMSE
PPM with DBSCAN	8	0.78	4.33	29.92	5.46	**4.20**	**31.65**	**5.62**
PPM with Kmeans	8	0.86	3.04	16.71	4.08	6.62	45.93	6.77
PPM with Kmeans	3	**0.95**	3.05	**16.15**	**4.02**	5.84	51.03	7.14
PPM with Kmeans	14	0.79	**2.8**	16.38	4.04	5.0	44.49	6.67
Time series baseline	–	–	2.94	31.64	5.63	10.63	379.59	19.48

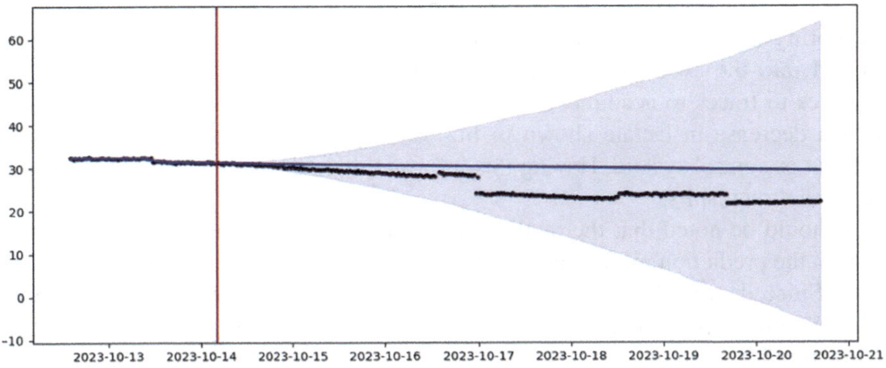

(a)

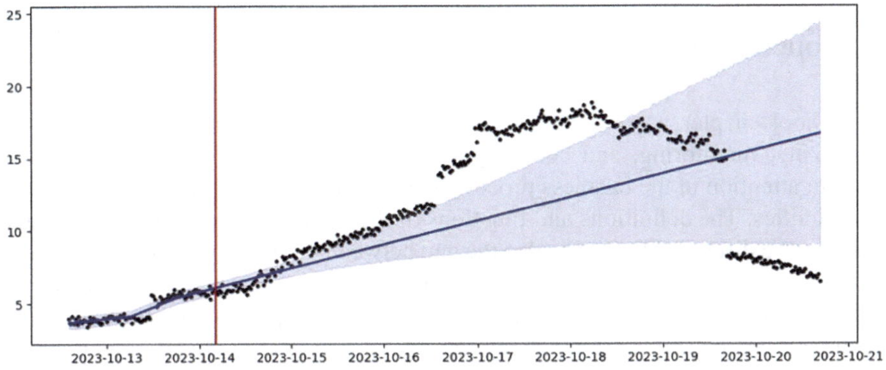

(b)

Fig. 7 (**a**) Example of glucose levels over time. (**b**) Example of lactate levels over time

clustering quality is not optimal, it is sufficient to achieve superior performance to other methods because the correct cluster is identified the majority of the time. Also, it should be noted that glucose levels are relatively stable during the bioreactor run as shown in Fig. 7a.

The best results for lactate in terms of MAE, MSE, and RMSE are obtained using our approach with DBSCAN. It can be seen that across all clustering configurations, the performance is significantly higher than the baseline. This can be explained by the fact that lactate levels fluctuate during the bioreactor run. So the time series model trained on the first 2 days of the run does not account for this non-stationarity, and so, the performance deteriorates rapidly and significantly as time progresses. This more complex pattern of lactate levels requires higher clustering quality, meaning that the clusters should more faithfully capture the natural shape of the data. This is why the best performance is observed using DBSCAN.

Overall, it can be seen that for both glucose and lactate, the performance improves over the baseline across all metrics regardless of the choice of clustering method. This demonstrates that incorporating the notion of process in the predictive monitoring task is crucial. Moreover, other aspects of the process could be incorporated into the event log to improve results. For example, we can add human activities to traces to account for sudden fluctuations in concentration levels. The sudden decrease in lactate shown in Fig. 7b may be due to an intervention by a scientist or control system. Having this information in the trace is likely to improve the prediction results.

It should be noted that the quality of the predictive monitoring approach relies both on the prediction model and the choice of clustering method for more complex tasks. Since data in this domain is expensive to acquire, event logs can often be small. So one important area for future improvement is to design new methods that can work with small data.

6 Conclusion

In this book chapter, we presented a novel digital twin framework for understanding, optimizing, monitoring, and controlling bioprocesses. Digital twins have seen a surge in attention in the business process management and bioprocess development communities. The definitions and functions of the digital twin are varied across disciplines. We discussed how to bridge the gap between these two related communities and present a framework to develop digital twins of bioprocesses using concepts and methods in the business process management and process mining fields. We discussed the challenges of achieving the digital twin and provided guidelines for those challenges. It should be noted that although the proposed framework is geared toward bioprocesses, it applies to other continuous processes that are characterized by regular measurements. Indeed, any process with similar data types can benefit from this framework, which could include other forms of manufacturing.

Through a case study conducted in collaboration with a prominent pharmaceutical company, we have demonstrated the practical applicability and tangible benefits of our proposed framework. The results of the case study showed that through an approach based on the proposed framework, we can achieve high prediction performance in the process monitoring task compared to a time series forecasting baseline.

Looking ahead, further research into the application of the proposed framework is warranted. In the future, we plan to address the challenges mentioned in the book chapter and offer concrete solutions to them.

Acknowledgments This research was supported under the Australian Research Council's Industrial Transformation Research Program (ITRP) funding scheme (project number IH210100051). The ARC Digital Bioprocess Development Hub is a collaboration between the University of Melbourne, the University of Technology Sydney, RMIT University, CSL Innovation Pty Ltd, Cytiva (Global Life Science Solutions Australia Pty Ltd), and Patheon Biologics Australia Pty Ltd.

References

1. Grieves, M.: Digital Twin: Manufacturing Excellence through Virtual Factory Replication. A Whitepaper by Dr. Michael Grieves (2014)
2. van der Aalst, W.M.P., Hinz, O., Weinhardt, C.: Resilient digital twins: organizations need to prepare for the unexpected. Bus. Inform. Syst. Eng. **63**(6), 615–619 (2021). https://doi.org/10.1007/s12599-021-00721-z. Accessed 26 Mar 2024
3. Liebenberg, M., Jarke, M.: Information systems engineering with Digital Shadows: concept and use cases in the Internet of Production. Inform. Syst. **114**, 102182 (2023). https://doi.org/10.1016/j.is.2023.102182. Accessed 14 Mar 2024
4. van der Aalst, W.M.P.: Concurrency and objects matter! Disentangling the fabric of real operational processes to create digital twins. In: Cerone, A., Ölveczky, P.C. (eds.) Theoretical Aspects of Computing—ICTAC 2021. Lecture Notes in Computer Science, pp. 3–17. Springer, Cham (2021). https://doi.org/10.1007/978-3-030-85315-0_1
5. Aalst, W.M.P., Berti, A.: Discovering object-centric petri nets. Fundamenta Informaticae**175**(1–4), 1–40 (2020). https://doi.org/10.3233/FI-2020-1946
6. van der Aalst, W.M.P.: Twin transitions powered by event data—using object-centric process mining to make processes digital and sustainable. In: Joint Proceedings of the Workshop on Algorithms & Theories for the Analysis of Event Data and the International Workshop on Petri Nets for Twin Transition Co-located with the 44th International Conference on Application and Theory of Petri Nets and Concurrency (Petri Nets 2023). CEUR Workshop Proceedings, vol. 3424. CEUR-WS.org, Caparica, Portugal (2023). https://ceur-ws.org/Vol-3424/invited2.pdf
7. Park, G., Van Der Aalst, W.M.P.: Realizing a digital twin of an organization using action-oriented process mining. In: 2021 3rd International Conference on Process Mining (ICPM), pp. 104–111 (2021). https://doi.org/10.1109/ICPM53251.2021.9576846
8. Park, G., Comuzzi, M., van der Aalst, W.M.P.: Analyzing process-aware information system updates using digital twins of organizations. In: Guizzardi, R., Ralyté, J., Franch, X. (eds.) Research Challenges in Information Science. Lecture Notes in Business Information Processing, pp. 159–176. Springer, Cham (2022). https://doi.org/10.1007/978-3-031-05760-1_10
9. Brockhoff, T., Heithoff, M., Koren, I., Michael, J., Pfeiffer, J., Rumpe, B., Uysal, M.S., Van Der Aalst, W.M.P., Wortmann, A.: Process prediction with digital twins. In: 2021 ACM/IEEE International Conference on Model Driven Engineering Languages and Systems Companion (MODELS-C), pp. 182–187 (2021). https://doi.org/10.1109/MODELS-C53483.2021.00032
10. Bano, D., Michael, J.,Rumpe, B., Varga, S., Weske, M.: Process-aware digital twin cockpit synthesis from event logs. J. Comput. Lang. **70**, 101121(2022). https://doi.org/10.1016/j.cola.2022.101121. Accessed 31 July 2023
11. Becker, M.C., Pentland, B.T.: Digital twin of an organization: are you serious? In: Marrella, A., Weber, B. (eds.) Business Process Management Workshops. Lecture Notes in Business Information Processing, pp. 243–254. Springer, Cham (2022). https://doi.org/10.1007/978-3-030-94343-1_19
12. Hernández Rodríguez, T., Frahm, B.: Digital seed train twins and statistical methods. In: Herwig, C., Pörtner, R., Möller, J. (eds.) Digital Twins: Tools and Concepts for Smart Biomanufacturing. Advances in Biochemical Engineering/Biotechnology, pp. 97–131. Springer, Cham (2021). https://doi.org/10.1007/10_2020_137. Accessed 28 July 2023
13. Pham, T.D., Manapragada, C., Sun, Y., Bassett, R., Aickelin, U.: A scoping review of supervised learning modelling and data-driven optimisation in monoclonal antibody process development. Digit. Chem. Eng. **7**, 100080 (2023). https://doi.org/10.1016/j.dche.2022.100080
14. Park, S.-Y., Park, C.-H., Choi, D.-H., Hong, J.K., Lee, D.-Y.: Bioprocess digital twins of mammalian cell culture for advanced biomanufacturing. Curr. Opin. Chem. Eng. **33**, 100702 (2021). https://doi.org/10.1016/j.coche.2021.100702. Accessed 28 July 2023
15. Zhao, B., Li, X., Sun, W., Qian, J., Liu, J., Gao, M., Guan, X., Ma, Z.,Li, J.: BioDT: An integrated digital-twin-based framework for intelligent biomanufacturing. Processes**11**(4), 1213 (2023). https://doi.org/10.3390/pr11041213. Accessed 28 July 2023

16. Whitford, B.: Bioprocess intensification: aspirations and achievements. BioTechniques **69**(2), 84–87 (2020). https://doi.org/10.2144/btn-2020-0072. Accessed 31 July 2023
17. Appl, C.,Baganz, F., Hass, V.C.: Development of a digital twin for enzymatic hydrolysis processes. Processes **9**(10), 1734 (2021). https://doi.org/10.3390/pr9101734. Accessed 28 July 2023
18. Gomis-Fons, J., Schwarz, H., Zhang, L., Andersson, N., Nilsson, B., Castan, A.,Solbrand, A., Stevenson, J., Chotteau, V.:Model-based design and control of a small-scale integrated continuous end-to-end span style="font-variant:small-caps;" mAb/span platform. Biotechnol. Progress **36**(4), 2995 (2020). https://doi.org/10.1002/btpr.2995. Accessed 14 Mar 2024
19. Khuat, T.T., Bassett, R., Otte, E., Grevis-James, A., Gabrys, B.: Applications of machine learning in antibody discovery, process development, manufacturing and formulation: current trends, challenges, and opportunities. Comput. Chem. Eng. **182**, 108585 (2024). https://doi.org/10.1016/j.compchemeng.2024.108585. Accessed 14 Mar 2024
20. Tiwari, A., Masampally, V.S., Agarwal, A., Rathore, A.S.: Digital twin of a continuous chromatography process for mAb purification: design and model-based control. Biotechnol. Bioeng. **120**(3), 748–766 (2023). https://doi.org/10.1002/bit.28307. Accessed 14 Mar 2024
21. Feidl, F., Garbellini, S., Luna, M.F., Vogg, S., Souquet, J., Broly, H., Morbidelli, M., Butté, A.: Combining mechanistic modeling and raman spectroscopy for monitoring antibody chromatographic purification. Processes **7**(10), 683 (2019). https://doi.org/10.3390/pr7100683. Accessed 8 Apr 2024
22. Agarwal, H., Rathore, A.S., Hadpe, S.R., Alva, S.J.: Artificial neural network (ANN)-based prediction of depth filter loading capacity for filter sizing. Biotechnol. Progress **32**(6), 1436–1443 (2016). https://doi.org/10.1002/btpr.2329. Accessed 8 Apr 2024
23. Pirrung, S.M., Van Der Wielen, L.A.M., Van Beckhoven, R.F.W.C., Van De Sandt, E.J.A.X., Eppink, M.H.M., Ottens, M.: Optimization of biopharmaceutical downstream processes supported by mechanistic models and artificial neural networks. Biotechnol. Progress **33**(3), 696–707 (2017). https://doi.org/10.1002/btpr.2435. Accessed 8 Apr 2024
24. Liu, S.,Papageorgiou, L.G.: Optimal antibody purification strategies using data-driven models. Engineering **5**(6), 1077–1092 (2019). https://doi.org/10.1016/j.eng.2019.10.011. Accessed 8 Apr 2024
25. Taylor, C., Pretzner, B., Zahel, T., Herwig, C.: Architectural and technological improvements to integrated bioprocess models towards real-time applications. Bioengineering **9**(10), 534 (2022). https://doi.org/10.3390/bioengineering9100534. Accessed 28 July 2023
26. Bayer, B., Dalmau Diaz, R., Melcher, M., Striedner, G., Duerkop, M.: Digital twin application for model-based DoE to rapidly identify ideal process conditions for space-time yield optimization. Processes **9**(7), 1109 (2021). https://doi.org/10.3390/pr9071109. Accessed 28 July 2023
27. Kuchemüller, K.B., Pörtner, R., Möller, J.: Digital twins and their role in model-assisted design of experiments. In: Herwig, C., Pörtner, R., Möller, J. (eds.) Digital Twins: Applications to the Design and Optimization of Bioprocesses. Advances in Biochemical Engineering/Biotechnology, pp. 29–61. Springer, Cham (2021). https://doi.org/10.1007/10_2020_136. Accessed 28 July 2023
28. Sokolov, M., Stosch, M., Narayanan, H., Feidl, F., Butté, A.: Hybrid modeling—a key enabler towards realizing digital twins in biopharma? Curr. Opin. Chem. Eng. **34**, 100715 (2021). https://doi.org/10.1016/j.coche.2021.100715. Accessed 28 July 2023
29. Mahanty, B.: Hybrid modeling in bioprocess dynamics: structural variabilities, implementation strategies, and practical challenges. Biotechnol. Bioeng. **120**(8), 2072–2091 (2023). https://doi.org/10.1002/bit.28503. Accessed 31 July 2023
30. Narayanan, H., Luna, M., Sokolov, M., Butté, A., Morbidelli, M.: Hybrid models based on machine learning and an increasing degree of process knowledge: application to cell culture processes. Ind. Eng. Chem. Res. **61**(25), 8658–8672 (2022). https://doi.org/10.1021/acs.iecr.1c04507. Accessed 28 Mar 2024

31. Gerzon, G., Sheng, Y., Kirkitadze, M.: Process analytical technologies–advances in bioprocess integration and future perspectives. J. Pharm. Biomed. Anal. **207**, 114379 (2022). https://doi.org/10.1016/j.jpba.2021.114379. Accessed 28 July 2023
32. Su, Z., Yu, T., Lipovetzky, N., Mohammadi, A., Oetomo, D., Polyvyanyy, A., Sardiña, S., Tan, Y., Van Beest, N.: Data-driven goal recognition in transhumeral prostheses using process mining techniques. In: 2023 5th International Conference on Process Mining (ICPM), pp. 25–32. IEEE, Rome, Italy (2023). https://doi.org/10.1109/ICPM60904.2023.10271945. Accessed 15 Apr 2024
33. Smiatek, J., Jung, A., Bluhmki, E.: Towards a digital bioprocess replica: computational approaches in biopharmaceutical development and manufacturing. Trends Biotechnol. **38**(10), 1141–1153 (2020). https://doi.org/10.1016/j.tibtech.2020.05.008. Accessed 28 July 2023
34. Bioprocess engineering: basic concepts. J. Controlled Release **22**(3), 293 (1992). https://doi.org/10.1016/0168-3659(92)90106-2. Accessed 16 Apr 2024
35. Tulsyan, A., Khodabandehlou, H., Wang, T., Schorner, G., Coufal, M., Undey, C.: Spectroscopic models for real-time monitoring of cell culture processes using spatiotemporal just-in-time Gaussian processes. AIChE J. **67**(5), 17210 (2021). https://doi.org/10.1002/aic.17210. Accessed 28 Mar 2024
36. Gan, J., Parulekar, S.J., Cinar, A.: Development of a recursive time series model for fed-batch mammalian cell culture. Comput. Chem. Eng. **109**, 289–298 (2018)
37. Alinaghi, M., Surowiec, I., Scholze, S., McCready, C., Zehe, C., Johansson, E., Trygg, J., Cloarec, O.: Hierarchical time-series analysis of dynamic bioprocess systems. Biotechnol. J. **17**(12), 2200237 (2022)
38. Afseth, N.K., Segtnan, V.H., Wold, J.P.: Raman spectra of biological samples: a study of preprocessing methods. Appl. Spectr. **60**(12), 1358–1367 (2006)

Index

G. Fortino, M. Mecella (eds.), *Internet of Things Meets Business Process Management*, Internet of Things, https://doi.org/10.1007/978-3-031-90746-3

R

S

U

GPSR Compliance

The European Union's (EU) General Product Safety Regulation (GPSR) is a set of rules that requires consumer products to be safe and our obligations to ensure this.

If you have any concerns about our products, you can contact us on ProductSafety@springernature.com

In case Publisher is established outside the EU, the EU authorized representative is:

Springer Nature Customer Service Center GmbH
Europaplatz 3
69115 Heidelberg, Germany

Batch number: 10755329

Printed by Printforce, the Netherlands